Gas Chromatography in
Food Analysis

or before

Gas Chromatography in Food Analysis

G. J. DICKES

M. Chem. A., C. Chem., F.R.I.C.,
F.I.F.S.T., F.R.S.H.

Assistant Scientific Adviser,
Scientific Adviser's Department,
Avon County Council

and

P. V. NICHOLAS

Principal Analyst,
Scientific Adviser's Department,
Avon County Council

Butterworths

LONDON - BOSTON

Sydney - Wellington - Durban - Toronto

THE BUTTERWORTH GROUP

United Kingdom	**Butterworth & Co (Publishers) Ltd** **London:** 88 Kingsway, WC2B 6AB
Australia	**Butterworths Pty Ltd** **Sydney:** 586 Pacific Highway, Chatswood, NSW 2067 Also at Melbourne, Brisbane, Adelaide and Perth
Canada	**Butterworth & Co (Canada) Ltd** **Toronto:** 2265 Midland Avenue, Scarborough, Ontario, M1P 4S1
New Zealand	**Butterworths of New Zealand Ltd** **Wellington:** T & W Young Building, 77–85 Customhouse Quay, 1, CPO Box 472
South Africa	**Butterworth & Co (South Africa) (Pty) Ltd** **Durban:** 152–154 Gale Street
USA	**Butterworth (Publishers) Inc** **Boston:** 19 Cummings Park, Woburn, Mass. 01801

First published 1976
Reprinted 1978

ISBN 0 408 70781 X

© Butterworth & Co (Publishers) Ltd 1976

Library of Congress Cataloging in Publication Data

Dickes, G. J.
Gas chromatography in food analysis.
Includes index.
1. Food—Analysis. 2. Gas chromatography.
I. Nicholas, Peter Victor, joint author. II. Title.
TX545.D45 664'.07 76-20824
ISBN 0-408-70781-X

*Printed in Great Britain by William Clowes & Sons Limited
London, Beccles and Colchester*

Preface

Many books have been written about the technique of gas chromatography (GC) and also about chemical and physical methods in food analysis, but few attempts have been made to bring the two subjects together. In attempting to do this in a concise fashion, this book includes most aspects of GC as applied to food analysis, together with reference to allied techniques and to sample extraction and preparation. It is our aim to provide useful information for all chromatographers engaged in food analysis, and in so doing we have anticipated a working knowledge of GC.

We have found one of the most serious criticisms by analysts engaged in GC work to be that the literature contains a welter of code numbers and names of the essential working materials. There are large numbers of similar methods per analysis disguised under these codings. We have attempted to simplify this aspect and to guide the novice chromatographer along the right lines.

There is not enough space in this book to describe methods in depth and it will be necessary to consult the original literature in order to obtain full working details.

This book is divided into five Parts.

The first Part, Chapters 1–7, summarises the practical aspects which contribute to the GC apparatus and technique. This Part goes beyond those aspects of GC which are only pertinent in food analysis because it was felt that many food analysts have been forced to take up GC methods without being able to take time out to critically appraise exactly what they are doing. For this reason, emphasis is placed on certain aspects of GC where technology is continually developing, e.g. stationary phases, detectors and derivatisation.

The second Part, Chapter 8, gives a GC consumer guide to those methods which, in our opinion, are the most effective in food analysis. These recommended methods are taken from Chapters 9 to 21, which cite most of the methods found in the literature. This chapter is intended to help analysts who do not have the time to evaluate methods and, where possible, these methods take into account accuracy, speed and simplicity.

The third Part comprises Chapters 9 to 16 and deals with those aspects of food quality assessment which involve GC methods. The chapters are titled according to the type of food.

The fourth and fifth Parts comprise Chapters 17 to 21 and deal with the GC determination of food additives and contaminants, the chapters being titled and divided according to the various kinds of these substances, irrespective of the foods in which they may be found.

In order to find out what methods are available for a complete analysis of a food, the chromatographer may therefore refer to up to three chapters, viz. one for methods pertaining to composition, another for methods relevant to additives and possibly a third dealing with contaminants. As an example, GC methods applicable for the analysis of lard are found in Chapters 10, 17 and 20: Chapter 10 cites methods dealing with basic composition; Chapter 17 cites methods for the determination of antioxidants; Chapter 20 cites methods for the detection and determination of organochlorine pesticide residues.

In general, those methods which are described in comparatively more detail than others are those which have more to commend them; secondary methods merit a few lines only.

In many instances it has been difficult to divorce extraction and clean-up procedures from the actual GC analyses. It is vitally important that these isolation procedures are both practical and efficient, the final analytical technique, GC included, being completely dependent on them. It has only been possible to mention extraction and clean-up techniques without giving full accounts of methods, and we believe that they constitute a subject of their own, best covered in a separate publication.

After the Preface there is a list of abbreviations which are used throughout and the three Appendixes at the end of the book comprise equivalent stationary phases, equivalent support materials and nomenclatures of the fatty acids.

Any publication written in this style relies on the permission of experts, authors and editors of scientific journals to reproduce their tables and figures. Our grateful thanks go to all those who have helped us in this manner and, in singling out the following, we have done so because we have drawn considerably on their resources:

The American Chemical Society, Washington (*Anal. Chem.* and *J. Agric. Food Chem.*)
The American Oil Chemists' Society, Illinois (*J. Amer. Oil Chem. Soc.*)
The Association of Public Analysts, London (*J. Ass. Publ. Anal.*)
The Association of Official Analytical Chemists, Washington (*J. Ass. Off. Anal. Chem.*)
The Chemical Society, London (*Analyst*)
Elsevier Publishing Company, Amsterdam (*J. Chromatog.*)
Preston Technical Abstracts Company, Illinois (*J. Chromat. Sci.*)

We are greatly appreciative for the technical advice, constructive criticism and proof reading carried out by I. Dembrey, the former Additional Public Analyst for Bristol and Gloucestershire, and O. G. Tucknott, Higher Scientific Officer, University of Bristol, Long Ashton Research Station.

Finally, we thank our wives, to whom we owe an enormous debt of gratitude for their work in the preparation of the manuscript.

It is our sincere hope that any errors found in this book will be few and insignificant, but we should be grateful for these to be pointed out, so that they can be corrected in later editions.

G. J. Dickes and P. V. Nicholas

Abbreviations

ACD 15M	2-chloro-4-isopropylamino-6-hydroxy-methylamino-1,3,5-triazine
AFID	alkali flame ionisation detector/detection
AFPD	alkali flame photometric detector/detection
AI	argon ionisation
AOAC	Association of Official Analytical Chemists, Washington, DC, USA
BDS	butanediol succinate polyester
BHA	butylated hydroxyanisole
BHC	1,2,3,4,5,6-hexachlorocyclohexane
BHT	butylated hydroxytoluene
BSA	bis- (trimethylsilyl)acetamide
BSTFA	bis- (trimethylsilyl)trifluoroacetamide
CCD	Coulson conductivity detector/detection
CDAA	*N,N*-diallyl-chloroacetamide
CDEC	2-chloroallyl *N,N*-diethyldithiocarbamate
CI–MS	chemical ionisation mass spectrum/spectra
2,4-D	2,4-dichlorophenoxyacetic acid
2,4-DB	4-(2,4-dichlorophenoxy)butyric acid
D-D	1,3-dichloro-1-propene plus 1,2-dichloropropane
DDE	1,1-dichloro-2,2-di-(4-chlorophenyl)ethylene
DDT	1,1,1-trichloro-2,2-di-(4-chlorophenyl)ethane
DDTS	4-dodecyldiethylene triamine succinamide
DEGS	diethylene glycol succinate polyester
DEN	*N*-nitroso-diethylamine
DFPD	dual-flame photometric detector/detection
DLTDP	dilauryl 3,3'-thiodipropionate
DMCS	dimethyldichlorosilane/silanised
DMN	*N*-nitroso-dimethylamine
DNOC	2-methyl-4,6-dinitrophenol
DNP	di-nonyl phthalate

EC	electron capture
EGA	ethylene glycol adipate polyester
EGS	ethylene glycol succinate polyester
EI–MS	electron impact mass spectrum/spectra
EPN	ethyl 4-nitrophenyl phenylphosphonothionate
EPTC	S-ethyldipropylthiocarbamate
ETFA	ethyl trifluoroacetate
EXD	di-(ethoxythiocarbonyl)disulphide
FAO	Food and Agriculture Organisation of the United Nations
FDA	Food and Drug Administration of the USA
FFAP	free fatty acid phase
FID	flame ionisation detector/detection
FPD	flame photometric detector/detection
GC	gas chromatograph/chromatography/chromatographic
GLC	gas liquid chromatograph/chromatography/chromatographic
GSC	gas solid chromatograph/chromatography/chromatographic
HCB	hexachlorobenzene
HETP	height equivalent to a theoretical plate
HMDS	hexamethyldisilazane/silazanised
HPLC	high-performance liquid chromatography
IR	infra-red
LTTNC	length–temperature time normalisation chromatography
MCD	microcoulometric detector/detection
MCPA	4-chloro-2-methylphenoxyacetic acid
MCPB	4-(4-chloro-2-methylphenoxy)butyric acid
MS	mass spectrometer/spectrometry/spectrum/spectra
NDGA	nordihydroguaiaretic acid
NMR	nuclear magnetic resonance
NPGS	neopentyl glycol succinate polyester
OC	organochlorine
ON	organonitrogen
OP	organophosphorus
PC	paper chromatography/chromatographic
PCB	polychlorinated biphenyl
PCNB	pentachloronitrobenzene
PDEAS	phenyldiethanolamine succinate polyester
PEG	polyethylene glycol
PEGA	polyethylene glycol adipate
PEGS	polyethylene glycol succinate
Petrol	normally the petroleum fraction boiling between 30 and 60°C

PG	propyl gallate
PIP	*N*-nitroso-piperidine
PLOT	porous layer open tubular
PPG	polypropylene glycol
PPGA	polypropylene glycol adipate
PTFA	peroxytrifluoroacetic acid
PVP	polyvinylpyrrolidone
PYR	*N*-nitroso-pyrrolidine
SCOT	support-coated open tubular
2,4,5-T	2,4,5-trichlorophenoxyacetic acid
2,3,6-TBA	2,3,6-trichlorobenzoic acid
TBHQ	mono-tert.-butylhydroquinone
TC	thermal conductivity
TCA	trichloroacetyl, sodium trichloroacetate
TCEP	1,2,3-tris(2-cyanoethoxy)propane
TCNB	1,2,4,5-tetrachloro-3-nitrobenzene
TDE	1,1-dichloro-2,2-di-(4-chlorophenyl)ethane
TDPA	3,3'-thiodipropionic acid
TEPP	bis-*O,O*-diethylphosphoric anhydride
TFA	trifluoroacetyl
TFAA	trifluoroacetic anhydride .
THBP	2,4,5-trihydroxybutyrophenone
THEED	tetrahydroxy ethyl ethylenediamine
TLC	thin layer chromatography/chromatographic
TMCS	trimethylchlorosilane/silanised
TMDS	tetramethyldisilazane/silazanised
TMS	trimethylsilyl
TMSDEA	trimethylsilyldiethylamine
2,4,5-TP	2-(2,4,5-trichlorophenoxy)propionic acid
TSIM	*N*-trimethylsilylimidazole
USP	United States Pharmacopoeia
UV	ultra-violet
WHO	World Health Organization

Contents

PART 4

Food Additives

PART 5

Food Contaminants

Part 1

Practical Aspects of Gas Chromatography

1
The Apparatus

A. Introduction

Since its appearance in 1952, gas chromatography (GC) has probably become the most widely used modern analytical technique and is part of the equipment used by virtually all industrial, academic and government laboratories engaged in the analysis of food. It offers rapid qualitative and quantitative analysis of complex mixtures with precision and sensitivity, yet the equipment is relatively cheap and inexpensive to operate.

GC is a separation technique, probably the best available for the separation of organic compounds. The separation of a mixture containing many volatile components may be achieved by introducing it as a single 'plug' into a continuously moving carrier gas flow, which passes through a column of material whose properties may be chosen to bring about this separation. In time, the individual components are separated and emerge from the column for evaluation. For the purpose of analysis, the separated components are detected and electronically displayed on to a recorder in the form of peaks of generally Gaussian form. The time of emergence of each component, referred to as its elution or retention time, is characteristic of that component, and the area under the peak is proportional to its quantity. These parameters yield qualitative and quantitative data, respectively.

Compounds may have to be derivatised, and GC of the products, particularly when hetero-atoms such as halogen, phosphorus or nitrogen are incorporated into the derivatives, will usually show increased sensitivity or specificity when compared with the parent compounds, with certain detectors. Either derivatisation is made prior to GC, when the products are taken to the chromatograph for analysis, or, as in reaction chromatography, the derivatives are formed and chromatographed within the GC unit.

Qualitative assessment depends on some measure of the position of peaks, whether this be retention time, retention volume or variations of these parameters. Quantitative interpretation of results depends, primarily, on the size, shape and separation of peaks and also the accuracy of measurement of peak areas, all these factors being used with a direct comparison between sample compound and standard material.

Confirmation of identities of the separated components is often carried out by additional analytical techniques. This usually requires fraction collection of the GC eluants, in concentration at either the milligram or the microgram level. It is beyond the scope of this book to describe, in any detail, the analytical techniques involved, but thermal fragmentation, mass spectrometry (MS), infra-red (IR) spectrometry and chemical micro-techniques are worthy of note in this connection.

A schematic diagram of a basic chromatograph is illustrated in *Figure 1.1.*

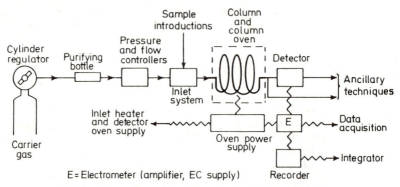

Figure 1.1. A schematic diagram of a basic chromatograph

Sample insertion into a gas chromatograph takes place at the inlet system. It is possible to quantitatively reproduce the sample introduction into the GC for gaseous, liquid or solid samples. Under ideal conditions the sample should vaporise easily without decomposition or chemical reaction, and should be swept into the column as a homogeneous 'plug' of minimum volume to avoid peak broadening.

Gaseous samples are usually flushed into the column by means of a sampling loop system. These are extremely variable in design and often specific to a particular problem. Extensive literature is available on this topic, and as the number of applications in food analysis appears to be minimal, only a few examples are given in this book.

Liquids are introduced into the system by injection with a microsyringe through a self-sealing septum. The liquid injection area, referred to as the injection port, is often equipped with a separate heater to raise its temperature above that of the oven and facilitate rapid vaporisation of the sample. Large-volume gas-tight syringes can also be obtained for the direct introduction of gas samples without resorting to the sampling loops mentioned previously and are used in the technique known as head space analysis.

Solid samples are normally examined in GC by dissolving them in a suitable solvent and injecting on to the instrument, as for liquids. There are techniques for direct solid sampling which invariably use a source of heat to vaporise the sample, and a few examples are given later (Chapter 4, p. 63).

The sample, mixed with carrier gas, enters the column, which is often referred to as the 'heart' of the chromatograph. The column is sited in the chromatographic oven, the temperature of which is usually capable of selection from room temperature to 400°C by means of an oven controller. The column consists of a container, usually metal or glass tubing, in which granular material is packed. The material may be an adsorbent such as alumina, charcoal or silica gel, when the technique is known as gas–solid adsorption chromatography or gas–solid chromatography (GSC). The column material may also be an inert solid such as a diatomaceous earth (Celite) with a liquid coating. In this case the inert solid or support plays no active part in the chromatography, but acts as a base for the effective distribution of the liquid on its surface. The liquid is referred to as the stationary liquid phase, and this technique is known as gas–liquid partition chromatography or gas–liquid chromatography (GLC). In general, liquid phases have a maximum and minimum usable temperature, which is an important consideration in their selection, as this limits the operating temperature range of the column.

After the sample is separated by the column material, the components enter a detector, the function of which is to locate, with the necessary sensitivity, the emergence of these components. There are many possible detection systems, several of which are beyond the scope of this book, but, in addition to the commercial detectors, a few are discussed which may have possible applications in food analysis.

Detectors are often heated independently of the oven unit. The detector temperature is usually just above the operating oven temperature to avoid undesirable condensation of the components. The detector is usually sited as close to the column exit as possible, to avoid producing unnecessary dead volume which would bring about peak broadening.

The gases helium, hydrogen, argon and nitrogen are commonly used as carrier gases, and are selected according to the detector being used and the particular application being carried out. For most purposes the carrier gases mentioned are considered inert and so will not react with the column or the sample. It is essential that the gases should be dry and free from impurities to avoid malfunction of the equipment, such as increased noise level and detector contamination. The use of a gas-purifying bottle containing a suitable molecular sieve is usually sufficient to remove water vapour and residual organic impurities, but it should be reconditioned periodically by heating to approximately 400°C for a few hours with an inert gas passing through it. Careful attention should be paid to the cleanliness of gas lines to avoid contamination, and scrubbing with solvents and flaming with a burner for metal lines is recommended before use.

The carrier gas must be regulated to provide constant pressure and constant mass flow, the former being usually achieved by a two-stage diaphragm regulator valve at the cylinder, and the latter by the use of a mass flow controller, which is usually a precision needle valve coupled to a constant differential pressure device. The properties of devices for setting or regulating carrier gas flow were discussed by Oster[1], and a theoretical comparison was made of several units that provide flow constancy.

B. Instrumentation

GC instrumentation has inevitably become the province of the manufacturers and there is ample commercial literature available describing their products in great detail. Many instruments have been based on a modular design, enabling the analyst to select and construct a chromatographic system according to his own particular requirements. This measure of flexibility can be a useful asset in an analytical laboratory engaged on various types of analysis, although the authors would always advocate in the interests of efficiency 'one instrument—one job', given the necessary financial assistance and a steady flow of samples.

McNair and Chandler[2] gave a well-presented account of GC equipment. They included a table listing most of the major suppliers, their models and the salient features of each model. Its purpose was to provide potential customers with a source of unbiased information as to the features and specifications available. The sensitivity, linear range and selectivity of various detectors were given, as found by personal experience rather than by quoting values from manufacturers' literature. Obviously, the choice of detectors was dependent on their suitability for the customers' applications. They considered oven size to be an important factor, adequate size being required for the use of larger-diameter columns, long columns, multiple columns or analyses requiring the use of valves. Also, it was considered necessary to be able to install or to exchange columns readily. Linear temperature programming was described as probably the most useful single accessory for a GC unit, although it required considerable additional equipment. Reviews of this nature can be invaluable as an aid to the selection of the correct equipment when the choice seems infinite.

Column temperature control in GC analysis is an important factor for obtaining chromatographic separation. There have been many variations on the type of control used, in an attempt to establish the optimum conditions for any particular analysis, ranging from isothermal operation to linear and non-linear temperature programming. Isothermal procedures are preferred, especially for quantitative work, since the usual problems associated with temperature programming, such as equilibration, baseline shift, retention variations and poor precision, are minimised. Unfortunately, the analysis of organic mixtures of wide boiling range generally necessitates some sort of temperature programming and therefore this technique is still widely used.

The 'time—temperature cycle technique' was claimed to be far superior to that of commercial temperature programmers by Hancock and Cataldi[3]. The technique produced a modified stair-step time—temperature profile, as opposed to the ramp function generated by the linear programmer.

GC analysis at subambient temperatures is not required for most analyses but it can be useful for trace analyses of volatile compounds such as those found in flavour analysis. It was in this field of study that Merritt and Walsh[4] reported some studies which had been made with a subambient technique.

This technique was called cryogenic temperature programming, and involved the temperature programming of samples commencing at subambient temperatures of approximately $-80°C$. It required a method of cooling that was able to provide a reasonably precise temperature control and was capable of a quick response to temperature changes. A possible cryogenic modification suitable for any GC equipped with a programmer was described by Dammeyer[5].

Cryogenic temperature programmed chromatographs are now available commercially, often employing an excess of cooling material located in the oven, such as solid carbon dioxide. This procedure was developed by Robertson, Issenberg and Merritt[6] for the modification of a Hewlett-Packard instrument.

6

Careful location of the cooling material was required to avoid undesirable temperature gradients.

The range of programmed cryogenic temperature GC was widened by Merritt *et al.*[7] to include temperatures commencing at − 196°C, using liquid nitrogen to cool the column chamber. They showed the necessity for employing very low starting temperatures for complete resolution of mixtures containing many volatile components. Cryogenic separations were shown to be reproducible, and *Figure 1.2* illustrates two chromatograms of aliquots of irradiated beef obtained

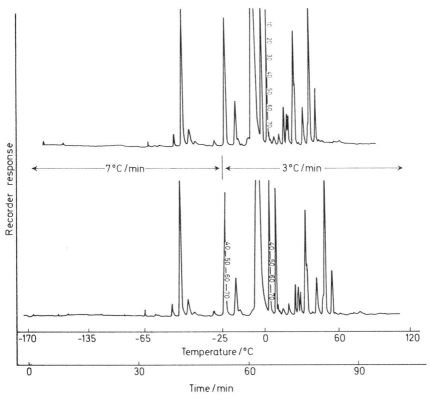

Figure 1.2. Duplicate programmed cryogenic temperature gas chromatograms of irradiated beef centre cut fraction aliquots. Column, 5% TRIS on 60–80 mesh firebrick: Aerograph, F & M apparatus. (After Merritt et al.[7])

by these workers. The temperature-rise rate for the upper chromatogram was slightly faster than that for the lower one. The use of subambient and wide-range temperature programming has helped to provide the type of separation necessary to get definitive mass spectra of components eluted from the chromatograph.

An alternative to column temperature programming, in order to reduce analysis time and to improve resolution, is the application of pressure programming, usually referred to as flow programming. The effects are similar to temperature programming, with the added advantages of instant reset following a particular analysis, and with no exposure of the column stationary phase to high

7

temperatures, thus allowing the use of a wider choice of phases with less chance of column bleed. The use of high gas flow rates necessitates a high column inlet pressure, and so particular attention must be paid to the design of the instrument. Commercial instruments are available, and Scott[8] has given details of a flow programmer, which could be readily made, at little cost, by laboratories with engineering facilities.

The separation of thermally labile samples is a particular advantage of flow programming, and *Figure 1.3* illustrates a comparison of temperature programming with flow programming for a sample of lemon grass oil. Thermal degrada-

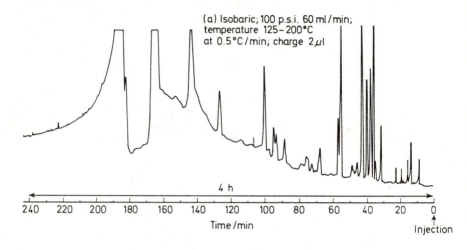

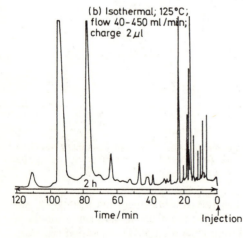

Figure 1.3. Separation of lemon grass oil by (a) temperature programming, (b) flow programming (after Scott[8])

tion has accounted for the changing baseline shown in the temperature programmed example. One disadvantage of flow programming is often a reduction in the column efficiency towards the end of the chromatogram.

As a technique, flow programming may be used alone, or in combination with temperature programming, for separating mixtures with a wide range of boiling points. This would entail the use of exceedingly high inlet pressures if flow programming was used alone. Zlatkis *et al.*[9] adapted the technique of both temperature and flow programming to open tubular columns, and illustrated the advantages of this combination, namely increased resolution with a shorter analysis time. This point was further exemplified by MacLeod[10], who used the technique in the analysis of complex essences on capillary columns and so shortened the analysis time without adversely affecting the resolution.

Preparative GC provides a means of isolating small amounts of pure components for further study, such as their examination by spectrophotometric methods and other techniques to establish their identities. The outlet of the chromatographic column enters a cold-trap or other suitable trapping device to enable a portion of the eluate to be recovered. In general, large-diameter columns are recommended for the collection of trace components for identification purposes[11], but if only small quantities of a major sample constituent are required, small-diameter columns are usually satisfactory.

There are many applications in food analysis where sufficient material can be collected for identification using the 'long-narrow' preparative columns, e.g. 20 ft × 0.375 in, which are accommodated in an analytical oven, being referred to as 'analytical prep.' columns.

In laboratories carrying out many repetitive analyses the use of fully automatic GCs has obvious advantages. Furthermore, it would appear to offer a logical alternative to extending the working day or to purchasing additional equipment. Automation allows a degree of flexibility, enables the output from individual chromatographs to be increased and improves the accuracy of the results owing to the exclusion of manual operational errors. To be fully effective, an automatic GC must embody many features, but often in this field the term 'automation' is used in a restricted manner, and may exclude desirable automated facilities such as the injection procedure.

Many commercial automatic GCs are currently available and were reviewed by Stockwell and Sawyer[12] with particular reference to requirements in the analyses of foodstuffs for fatty acid composition, and tinctures and essences for dutiable spirit. The conclusions on the applicability of commercial units to these particular problems were summarised, and the design of instruments for specific purposes was proposed, since each application needed a slightly different set of operating parameters.

Another paper published by these same authors[13] described the development of a new approach in the automation of GC. A complete system to monitor samples was evolved, using a fully automatic liquid injector, handling discrete samples, admixing with internal standards, transporting to a mixing vessel, injecting and allowing the compounds to elute before repeating the cycle in a continuous manner. The injection system was a modification of one designed by Boer[14] for an automatic preparative unit marketed by Pye Unicam (UK).

In addition to the GC, other associated equipment is required which can be obtained from the manufacturers or their suppliers, and is too well known to any gas chromatographer to warrant any lengthy discussion. Chart recorders, equipped with such facilities as event pen markers for marking the point of injection of the sample, are available, but for analysts wishing to avail themselves of this extra, Hoffmann and Evans[15] described an inexpensive and readily made article suitable for use with any recorder.

Integrators or, better still, computer integrators, often described as aids to GC, are very much dependent on the type of problem in hand, and the manufacturers' advice should be sought together with a laboratory evaluation before embarking on their usage.

C. Injection systems

The injection of liquids, or solids in solvents, with a microsyringe is carried out by inserting the needle through a self-sealing silicone rubber septum, situated in the injection port of the GC. In order to rapidly vaporise the sample, one of two systems is used.

In the first system the sample is injected into a heated stainless steel block, where it is vaporised and swept on to the column. This is not recommended for thermolabile compounds, unless a glass liner, called an insert, is placed into the block. The insert also traps any non-volatile material, and this insert can then be subsequently rejected.

In the second system the column extends from the oven up to the septum, so that the sample is placed directly on to the column. This arrangement, called on-column injection, is preferred for thermolabile compounds.

A good injection technique is essential for reproducible sample introduction into the GC column. A frequently used method has been one based on a smooth rhythm that allows the injection of the sample to take place as quickly as possible. However, the technique for sample injection is gained by experience and a reproducibility accuracy of about 2% should be achieved with practice.

Anomalies have arisen in injection technique, and both de Faubert Maunder, Egan and Roburn[16] and Mendoza[17] reported a significant difference in the electron-capture detector response to compounds injected at fast and slow rates, the latter giving a greater response.

For the injection of solid samples the solvent should be selected with care. It should have a low boiling point to be eluted quickly, but not too low, which would make the manipulation of the samples difficult, particularly with respect to the solvent's behaviour in the syringe. A solvent which gives a low response with the detector being used often enables a lower limit of detection to be obtained for the solute, e.g. carbon disulphide is a good choice of solvent for flame ionisation work, as it is virtually insensitive to this detector, and, similarly, aliphatic solvents give the least response in electron-capture work.

The injection system can be a source of problems, particularly in analysis at high sensitivity using temperature programming. Deposited volatile material in the injection system may bleed through the GC column and give rise to spurious peaks on the chromatograms, known as 'ghost peaks'. This is the result of the cooling period, which causes condensation and accumulation of this material at the front of the column, which is then analysed at the next programme. In contrast, under the conditions of isothermal operation the volatile material contributes continuously to the level of background noise and so is not generally apparent as peaks, but may result in an unstable or drifting baseline.

The injection port septa can provide a source of volatile material. Preheating the septa to high temperatures for a period of time before use to overcome the problem has been suggested by several workers. Kolloff[18] heated the septa to

300°C for 90 h in a stream of nitrogen, and Tamsma *et al.*[19] heated them to 250°C overnight under vacuum in a rotary evaporator. Grob and Grob[20] used a vaporising attachment, with built-in septum flushing to eliminate septum bleeding, and applied it to capillary GC.

Callery[21] reported on the effect of time and temperature on the bleed rate of volatile material from several types of septa and concluded that the maximum bleed occurred when the septa were first installed. The problem of septum bleeding was encountered by Smith and Sorrells[22] in temperature programming TMS derivatives of amino acids at the submicrogram level, and, following an investigation into the bleed characteristics of several commercial septa, it was concluded that none were satisfactory for use with conventional port equipment.

The increase in the bleed rate of volatile material from septa together with the shortening of their useful life have also arisen as the result of progressively higher inlet temperatures associated with the advent of the high-temperature stationary phases, e.g. Dexsil 300. The reduction in septum life can be attributed to the increasing hardness of the silicone rubber and to the overtightening of the septum on installation. The latter fault can be remedied by hand tightening at room temperature.

Tucknott and Williams[23] had little success with the methods of preheating and found that the life of the septum was shortened by the process. They prevented septum bleeding by a modification to the injection port, which required no septum pretreatment, and consisted of brass cooling fins placed between the septum securing nut and the injector port. Consequently, 'ghost peaks' were eliminated by keeping the septum cool.

Kishimoto and Kinoshita[24] modified the injection system to eliminate 'ghost peaks' by using a portion of the carrier gas to 'wash' over the septum, passing it through an activated charcoal tube to remove impurities, and then combining it with the main stream.

A solution to the problems of septum bleed and septum life is the use of a water-cooled inlet system for high-temperature GC operation[25].

Anomalous peaks in chromatograms may arise from the elution of material that has been adsorbed previously by the GC system, e.g. portions of certain solutes. These peaks are also referred to as 'ghost peaks', and often exhibit the retention characteristics of the solutes. Even in isothermal operation 'ghost peaks' arising from adsorption phenomena can be a problem in certain assays and can give rise to a major source of error in quantitative work. Geddes and Gilmour[26] reported that such a problem had arisen in the determination of C_2-C_5 acids and that the adsorption sites responsible had been located in the glass wool column plugs and in the charred deposits formed at the injection area. Periodic cleaning of the injection area and the use of Teflon tape plugs had helped to eliminate the problem.

Apart from 'ghost peaks', adsorption effects taking place at active sites in the injection area brought about by the deposition or degradation of components may lead to extreme peak tailing and reduce the apparent performance of the column.

Thompson, Walker and Moseman[27] carried out a study on the performance of various conventional GC columns under extreme injection loading of concentrated, cleaned and uncleaned extracts. They indicated that the daily changing of glass inserts in the injection port, used for off-column injection, together with the glass wool plug in the column inlet, would maintain and restore the overall performance of the columns.

It would, therefore, seem that the operation of high-temperature, high-sensitivity GCs can be adversely affected by the shortcomings of the injection area and by the unsatisfactory condition of the septum. Baseline instability, anomalous peaks, sample adsorption and leakage are examples of problems encountered with septa. Leakage terminates the life of the septa and is indicated by variations in retention time, resolution and peak amplitude. Leakage is usually detected by the application of a drop of soap solution to the injection port, but, although effective, it may itself introduce contamination. An alternative method was suggested by MacDonell[28], who used a small rubber bulb containing a drop of volatile solvent held firmly against the injection port. When the bulb was quickly squeezed, a high pressure was produced on the septum, giving a response on the recorder, so confirming a leakage.

References

1. Oster, H., *Chromatographia,* **1,** 144 (1968)
2. McNair, H. M. and Chandler, C. D., *J. Chromat. Sci.,* **11,** 454 (1973)
3. Hancock, H. A. and Cataldi, A., *J. Gas Chromat.,* **5,** 406 (1967)
4. Merritt, C. and Walsh, J. T., *Anal. Chem.,* **35,** 110 (1963)
5. Dammeyer, W., *Anal. Chem.,* **39,** 1339 (1967)
6. Robertson, D. H., Issenberg, P. and Merritt, C., *Facts and Methods,* **5,** No. 1 (1964)
7. Merritt, C., Walsh, J. T., Forss, D. A., Angelini, P. and Swift, S. M., *Anal. Chem.,* **36,** 1502 (1964)
8. Scott, R. P. W., in *Gas Chromatography 1964* (A. Goldup, Ed.), Institute of Petroleum, London, 25 (1965)
9. Zlatkis, A., Fenimore, D. C., Ettre, L. S. and Purcell, J. E., *J. Gas Chromat.,* **3,** 75 (1965)
10. MacLeod, W. D., *J. Agric. Food Chem.,* **16,** 884 (1968)
11. Anon., *Facts and Methods,* **6,** No. 5 (1965)
12. Stockwell, P. B. and Sawyer, R., *Lab. Pract.,* **19,** 279 (1970)
13. Stockwell, P. B. and Sawyer, R., *Anal. Chem.,* **42,** 1136 (1970)
14. Boer, H., *J. Sci. Instrum.,* **41,** 365 (1964)
15. Hoffmann, R. L. and Evans, C. D., *J. Gas Chromat.,* **6,** 63 (1968)
16. de Faubert Maunder, M. J., Egan, H. and Roburn, J., *Analyst,* **89,** 157 (1964)
17. Mendoza, C. E., *J. Chromat. Sci.,* **9,** 753 (1971)
18. Kolloff, R. H., *Anal. Chem.,* **34,** 1840 (1962)
19. Tamsma, A., Kurtz, F. E., Rainey, N. and Pallansch, M. J., *J. Gas Chromat.,* **5,** 271 (1967)
20. Grob, K. and Grob, G., *Chromatographia,* **5,** 3 (1972)
21. Callery, I. M., *J. Chromat. Sci.,* **8,** 408 (1970)
22. Smith, E. D. and Sorrells, K. E., *J. Chromat. Sci.,* **9,** 15 (1971)
23. Tucknott, O. G. and Williams, A. A., *Anal. Chem.,* **41,** 2086 (1969)
24. Kishimoto, K. and Kinoshita, K., *Bunseki Kagaku,* **18,** 588 (1969)
25. van Rensburg, J. F. J., Mouton, P. L. and Pretorius, V., *J. Chromat. Sci.,* **10,** 580 (1972)
26. Geddes, D. A. M. and Gilmour, M. N., *J. Chromat. Sci.,* **8,** 394 (1970)
27. Thompson, J. F., Walker, A. C. and Moseman, R. F., *J. Ass. Off. Anal. Chem.,* **52,** 1251 (1969)
28. MacDonell, H. L., *J. Gas Chromat.,* **6,** 411 (1968)

2
Columns

A. Introduction

GC columns are usually glass or metal, which is generally stainless steel or copper, although aluminium and other metals have been used. Copper columns are notorious for catalytically breaking down many sample constituents, e.g. pesticides and steroids. Aluminium columns may behave similarly. These particular metal columns should be avoided if at all possible. Glass columns are preferred, as generally there are few problems of adsorption, although stainless steel ones are suitable if they are thoroughly cleaned, and deactivated by silanisation before filling. Crossley[1] made columns from Teflon tubing and filled them with standard packing materials; he found that Teflon columns were more suitable than glass ones for the analysis of thermally labile compounds.

Columns are generally classified into two types, viz. packed and open tubular (capillary). Open tubular columns can provide very high resolution, but they can be expensive. Low dead volume connections and usually an inlet splitter are required for open tubular columns, since they are only able to cope with small amounts of sample. It is usually essential in GC requiring a high degree of resolution to use open tubular columns or other columns of high efficiency.

Standard analytical packed columns may be filled with an active solid support or an inert solid support coated with a liquid phase; they are generally 2–6 mm inside diameter and usually 1 to 3 m in length. The particle size of the column material is usually expressed in British Standard mesh size, and 60/80, 80/100 and 100/120 mesh are generally used for analytical work. The loading by weight of liquid phase applied to the solid support is generally between 3 and 20%, except for the separation of less volatile samples and for fast analyses, where lower loadings may be used.

It is generally accepted that a GC column gives its best performance with the smallest possible sample size, although Albrecht and Verzele[2] showed that there was an optimum sample size for any column, which was not necessarily the minimum. A column of 2 mm diameter functioned most efficiently with a sample size of 0.2 μl.

The maximum amount of sample which can be separated effectively by a column (sample capacity) is proportional to the amount of stationary phase present and to approximately the square of the column diameter. Van den Heuvel and Kuron[3] studied the relationship between column performance, diameter and sample size. They concluded that columns of 4 mm internal diameter were the most efficient with sample quantities of less than 30 μg. Column overloading is indicated by poor performance, which is shown by loss of resolution, tailing and shift in retention times. These effects can be caused by major components, e.g. the solvent, when trace components are being separated.

Although prepacked and pretested columns are available commercially, many workers, including the authors, prefer to use their own prepared units, tailored according to their particular needs. The preparation of columns requires extreme care to produce satisfactory results, although the methods employed are usually straightforward.

Several methods are used for coating the support with the liquid phase, some more laborious than others. A technique in frequent use, often referred to as the batch coating method, involves the soaking of the support with a solution of the liquid phase and evaporating the solvent by agitating the slurry. The solvent must be volatile and be readily evaporated by means of the heat from an IR lamp, for example. Agitation should be gentle to avoid any break-up of the support material which might produce active sites which would give rise to tailing of the sample components on the chromatogram. A low-speed rotary evaporator may also be used for removing the solvent under vacuum, usually operated at room temperature.

An alternative technique employs filtration through a sintered glass funnel. The weighed support is placed in the funnel and the stationary phase dissolved in a solvent is added. The solvent is removed by aspiration and the amount of stationary phase deposited on the support is deduced by reweighing the support after drying it with an IR lamp. Additional applications can be made to the support if the concentration of stationary phase is found to be insufficient.

In addition to these methods, Averill[4] and Lysyj and Newton[5] described a means of coating the support already in the column. The solution of stationary phase was forced through the column, previously filled with the support, and excess material, together with the solvent, were removed by an inert gas. This method was found to be particularly useful for the coating of microbeads, which are normally difficult to pack after preparation by conventional means, owing to their stickiness.

Careful packing of the column material is essential to ensure a uniform packing density and particle size throughout the length of the column. It is usually desirable to sieve coated supports to remove any fines and oversize material that may have arisen in its preparation. Packing of the column may be achieved by plugging one end with a porous plug, e.g. quartz wool or sintered stainless steel, attaching a funnel to the other end, pouring the material into the funnel and tapping and/or vibrating it into the column. Excessive use of vibration should be avoided, since this may cause disintegration of the column material. The application of vacuum at the porous plug end may help to assist the packing of awkwardly shaped columns. When the consolidation of the packing material is complete, the funnel is removed and another plug is placed in the open end of the fully packed column.

The packing of a column may also be carried out by applying a pressure of inert gas to a reservoir containing the packing material, which has been attached to one end of the column. There are several commercially available packing devices, based upon this method, which are capable of packing columns with a good reproducibility. This principle was used in the apparatus described by Villalobos[6], which was designed to produce closely matched columns. The apparatus permitted the tightness of the column bed packing to be adjusted while the packing operation proceeded. It allowed the change in the flow resistance of the column to be monitored at the same time as it was vibrated, its packing density being increased accordingly, by noting the difference in upstream and downstream pressures of a fixed restrictor, placed in series with the column. The packing density, flow rate, retention times and height equivalent to a theoretical plate (HETP) of replica columns packed with this apparatus, using a single batch of column material, were reproducible to better than 1%. It was suggested that this apparatus could be used to investigate the effect of packing density on HETP, with a view to establishing the correct packing density for optimum efficiency.

All new columns require a period of conditioning before being used for GC work. This is essential to ensure the complete removal of any solvent, water and volatile contaminants that may have been retained during the preparation of the column material. Conditioning is often carried out in the chromatographic oven with the column *in situ,* but disconnected at the detection end to avoid detector contamination. It is usually heated for at least 24 h at approximately $20-30°C$ above its proposed operating temperature, with a stream of carrier gas passing through it. Many silicone-coated supports appear to benefit from an initial period of several hours' 'baking' at a temperature higher than the conditioning temperature and in the absence of carrier gas. A steady recorder baseline at the required sensitivity of the GC is an indication of adequate column conditioning. If longer conditioning periods are warranted, a conventional laboratory oven could be utilised for the purpose, or a thermoregulated sand-bath used as described by Mussini *et al.*[7]

The lifetime of a column is extremely variable and depends on the packing material, its operational temperature and the cleanliness of the sample extracts. The replacement of a column, or attempts to rejuvenate it, should be carried out when there are indications of deterioration in column efficiency or changes in relative retention data. It is possible to regenerate columns by various means. The repacking of the first few inches at the column inlet invariably removes the non-volatile sample constituents, which accumulate and create contamination problems by producing active adsorption sites for volatile components. The use of a small replaceable column before the analytical one is a possible alternative. Active adsorption sites resulting from excessive liquid phase bleeding or decomposition can sometimes be reduced by injecting small quantities of silylating reagent, but this is only applicable to non-polar columns. For this purpose, the oven temperature should be approximately $100°C$ and the carrier gas flow reduced to a low rate.

Braddock and Marec[8] overcame the problem of adsorption on free substrate sites by successive injections of a benzene solution of silicone fluid into their SE-30 column. The column was then conditioned in the usual manner, followed by repeated washings by injecting chloroform, until a steady baseline was achieved.

The gradual bleeding of liquid phase from a column obviously takes place in the direction of the gas flow, resulting in an almost naked support at the column inlet. Christophe[9, 10] reported a significant improvement in the resolution of temperature-programmed GC by reversing the gas flow in the column after a period of time. In some instruments this can be easily achieved by removing the column and reorientating it so that the column inlet coupling is attached to the detector base and the column detector coupling to the inlet system.

B. Solid stationary phases

GSC utilises a column packed with an active solid adsorbent, more aptly called the solid stationary phase, and it has become an established technique for the analysis of very volatile samples, such as gases. The introduction of new and modified adsorbents has led to their use in the analysis of high-boiling and polar samples. However, the applications in food analysis seem, at present, to be few, although it is likely that this field will be extended in the future.

The main advantages of solid stationary phases are the absence of column bleed, and therefore the potential of high operational temperatures, and their high selectivity. The solid phases that have been used include carbonaceous material, alumina, silica and molecular sieves, e.g. zeolites and porous polymers.

Sometimes it has been found advantageous to coat these adsorbents with a liquid phase, and the technique has been referred to as gas–solid–liquid chromatography (GSLC) or gas adsorption layer chromatography. This was first introduced to eliminate the strong peak tailing often observed when compounds of low volatility were chromatographed on conventional adsorbents. Many liquid phases have been employed for this purpose, e.g. squalane, silicone oil and Carbowax, in quantities of 1–2% w/w. This combination often results in a stationary phase with new properties, resembling neither the adsorbent nor the liquid phase in its application to the separation of solutes.

1. Carbonaceous material

Active carbon has been used as a solid phase in the analysis of gases and low-boiling hydrocarbons, the most active material providing the highest column efficiency. To broaden its applications, several attempts were made to reduce the activity by various means, including liquid loading, but in this respect carbon black is a better phase. However, graphited carbon black, produced by the crystallisation of carbon black to a finely dispersed graphite when it is heated to a high temperature in an inert gas, has a large specific surface area and is a useful material for GSC. Unfortunately, tailing peaks are often produced, although Di Corcia, Fritz and Bruner[11] showed that this problem could be reduced by the addition of small amounts of polar stationary liquid phases. The material was then found to be suitable for separating alcohols, amines and acids. Graphited carbon black has also been used by Frycka[12], who deposited it on to Chromosorb W for the separation of polynuclear aromatic hydrocarbons.

2. Alumina

Alumina can be dehydrated by heating at 200–1000°C, becoming extremely active, but such highly active alumina causes severe tailing. Thus it is only suitable as a stationary phase when it has been partially deactivated by such means as hydration. The extent of the hydration, and, hence, the activity, governs the retention values, the selectivity and the column efficiency. One method of maintaining the activity of the adsorbent at a constant level has been to moisten the carrier gas by the addition of steam.

Several methods have been described for the modification of alumina; these include treatment with sodium hydroxide and metal salts, and the deactivation with either an alcohol, pyridine or acid anhydride[13]. Vernon[14] used a sodium chloride-impregnated, sodium hydroxide-modified alumina for the separation of polycyclic aromatic hydrocarbons, the peaks being symmetrical even for hydrocarbons with boiling points as high as 350°C.

3. Silica

The chromatographic performance of silica gels depends on their pore structure and on the extent of their hydration. In GSC work peak tailing often results, and various modifications to the silica gel surface have been proposed to extend the applications of this material[15].

Spherosil and Porasil are based on porous silica beads and have applications both as adsorbent in GSC and as support in GLC. They are very widely employed as adsorbents in GSC because of their spectrum of activity and their porosity. The beads are solid and incompressible, despite their considerable porosity. They are chemically quite inert, heat and wear resistant, and do not swell in liquids.

Guillemin, Le Page and de Vries[16] examined the relationship between the physical and chromatographic properties of Spherosil, and gave a scheme for assisting the user to make the correct choice of Spherosil for a particular analysis. They suggested that any Spherosil should be initially activated at 150°C for 2 h to limit peak asymmetry.

4. Molecular sieves

Molecular sieves, such as zeolites, have found limited applications in GSC, usually being confined to gas analysis. Being highly hygroscopic, they are mainly used for drying the carrier gas or as a precolumn for removing water from samples. They are activated by heating to approximately 400°C in a stream of inert gas. McKinney and Jordan[17] evaluated coated molecular sieves for GC, and found that they provided a useful column packing material for the analysis of many types of aqueous samples. A coating of 10% Carbowax 400 on a molecular sieve, grade 569 (pore diameter considerably smaller than 5A type sieve), was found to be extremely effective for the analysis of an aqueous solution of alcohols, as no water peak was indicated owing to its removal by the sieve, symmetrical peaks being obtained.

5. Porous polymers

Organic polymers, based on polystyrene, were first reported by Hollis[18] and Hollis and Hayes[19], and are now manufactured under various commercial names such as Porapak, Chromosorbs 101 to 107, Synachrom, Polypak, etc. They are mechanically strong and so can be packed by vibration in the normal manner. The thermal stability of these materials is good and most of them can be operated up to temperatures of 250°C, with little or no bleeding. They have excellent separation properties, especially for polar compounds, and have found extensive applications as column packings in GC. Their individual properties vary according to their chemical nature, pore structure, surface area and particle size.

Retention data are often provided in the manufacturers' leaflets, and there are numerous papers in the literature [20—23], which tabulate these in the form of retention volumes and Kováts indices. Dave[24] evaluated all the porous polymers commercially available at the time, and presented retention data and chromatograms for many classes of compounds as an aid to analysts in the selection of an appropriate packing for specific separations.

The lack of reproducibility of the retention data for these materials has been the subject of some concern. Gough and Simpson[25] showed that the retention behaviour of compounds on Porapak might be dependent on the conditioning of the polymer. This was substantiated by Dressler, Vespalec and Janak[26], who also showed that the data depended on the individual batch of materials used, and that the retention differences were too large for the data to be used for the qualitative identification of compounds on other batches. Other abnormalities in the behaviour of these materials, particularly peak broadening of certain types of compounds, were reported by Dressler, Guha and Janak[27] and Ackman[28].

In an attempt to improve separations on porous polymers, many methods for their deactivation have been used, including coating them with various liquid phases and, in effect, using them as support material, although with the solid adsorbent still fulfilling its part in contributing to the separation of the components. The addition of 2% Carbowax 20M is useful for this purpose, and is effective in shortening extremely long retention times. Phosphoric acid treatment of the polymers is frequently used to reduce the tailing of acids[29], and polyamine treatment has been used to improve peak shapes of amines[19].

Besides porous polymer beads, polymeric foam has been used with some success in GSC[30, 31]. This appears to be a promising material for the analysis of both polar and non-polar compounds with relatively low vapour pressures at low column temperatures.

C. Liquid stationary phases

The versatility of GLC is undoubtedly due to the availability of numerous liquid phases, covering a complete range of polarity. The greatest disadvantage of liquid phases is their volatility, and another is that they sometimes lack thermal stability. Fortunately, there are several liquid phases with a sufficiently low vapour pressure and also with adequate thermal stability, even when operated at moderately high column temperatures. Column temperatures in excess of 300°C are often required in GLC, and although the number of high-temperature phases available is increasing, they are still minimal. High column temperature produces higher liquid phase vapour pressures and decomposition, which lead to high column bleed rates, with subsequent depletion of the stationary phase. This is of major concern when maximum detector sensitivity is required, or when the column is coupled via a separator to a mass spectrometer. In addition, contamination of the detectors, collected fractions and other connected equipment may result.

The choice of the correct liquid phase is one of the most important decisions in GLC, and should be made with consideration given to its volatility, stability, viscosity, solubility and selectivity. Selection of a suitable material is often on a trial and error basis, with experience and prejudice playing a considerable part. The literature provides a wealth of information for the chromatographer, but to attempt to use all the phases recommended would be costly and could probably prove to be unnecessary.

Many different chemical compounds have been evaluated as liquid partition phases, and a few, such as silicones, polyether glycols and certain polyesters, have become well established in different types of analysis. However, many workers have drawn attention to the problems created by the use of too many different liquid phases in GC and to the fact that it would be an advantage to reduce their number. Gas chromatographers continue to use new liquid phases in their work, where these are often unnecessary.

The most important factors governing the selection of a liquid phase are its polarity and its operational temperature range. With regard to temperature limits for a particular stationary phase, the recommended maximum operating temperature may be found in tables or from the supplier. The values should be considered only as a guide, as the detector sensitivity, the mode of preparation and column construction are all factors which will affect the maximum practical operating temperature. For example, the maximum operating temperature for a liquid phase in an open tubular column would be lower than that obtainable in a packed column. The lowest operating temperature for a given phase is restricted by its melting point or its viscosity. The latter will be increased by lowering the temperature and thereby reduce the column efficiency. This is most noticeable with the high-temperature non-polar phase, i.e. Apiezon L, which ought not be used below 75°C, and SE-30, which ought not be used below 125°C.

From a practical point of view, the greater the polarity of the liquid phase, the greater the retention of a polar solute relative to that of a non-polar solute with a similar boiling point. Like is dissolved by like, is the general rule. Thus, if the sample components are similar in chemical structure but have different volatilities, a non-polar liquid phase is generally better. Conversely, if the components have different functional groups, but are of similar boiling points, then a polar phase is generally more suitable.

Stationary liquid phases are generally divided into three major groups, viz. non-polar, semi-polar and polar. Non-polar, such as methyl silicones, Apiezon greases and squalane, tend to fractionate solutes by boiling point; semi-polar, such as long-chain aliphatic esters, dissolve both polar and non-polar solutes equally; polar, such as polyethylene glycol and polyesters, retain polar solutes only.

19

Although the polarity is such an important factor in the selection of a liquid phase, it is not possible to calculate it by any simple measurement of dipole moment. However, the relative polarities of liquid phases have been determined in a number of ways in an attempt to characterise them. Rohrschneider[32] has suggested an empirical method which has proved to be a useful guide, but has been inadequate in many cases. The polarity, P, of the stationary phase under investigation is given by the expression:

$$P = 100 \left(\frac{q_x - q_2}{q_1 - q_2} \right) \tag{2.1}$$

where q_1 is the ratio of the specific retention volumes of butadiene: butane on a column coated with $\beta\beta'$-oxydipropionitrile; q_2 is the ratio of the specific retention volumes of butadiene:butane on a column coated with squalane; and q_x is the ratio of the specific retention volumes of butadiene:butane on a column coated with the phase under investigation.

All the retention volumes are determined at the same temperature. A better standard pair of solutes, benzene: cyclohexane, makes the method more generally applicable and allows the polarity to be measured at higher temperatures. At a given temperature, one liquid phase could be more polar than another, and the reverse could be true at another temperature[33].

Later Rohrschneider[34] developed a system based on the idea that the polarity of a column was dependent on the substance being analysed, in addition to the type of stationary phase. This system required the use of retention indices, I (see Chapter 6, p. 88).

When the retention index of a substance is determined on a non-polar stationary phase, e.g. squalane, and on a relatively more polar stationary phase, the difference,

$$\Delta I = I_{\text{polar}} - I_{\text{non-polar}} \tag{2.2}$$

is proportional to the column polarity. Rohrschneider suggested that this expression ignored the influence of the substance being analysed on the polarity of the stationary phase, and that it would be desirable to determine the retention index difference, ΔI, with several different types of compounds. For this, he selected benzene, ethanol, methyl ethyl ketone, nitromethane and pyridine. The ΔI term was then redefined as:

$$\Delta I = ax + by + cz + du + es \tag{2.3}$$

where x is the polarity of a column when benzene is analysed and is equal to $\Delta I/100$. Similarly, y is $\Delta I/100$ for ethanol, z is $\Delta I/100$ for methyl ethyl ketone, etc. x, y, z, u and s are known as the Rohrschneider constants for a specific column. These constants can be determined from the retention indices of benzene, ethanol, methyl ethyl ketone, etc., on the column, in association with the retention indices of benzene, ethanol, methyl ethyl ketone, etc., on a squalane column under identical conditions. Rohrschneider constants may be found for many stationary phases in the literature and also in some suppliers' lists. The other terms, *a, b, c, d* and *e*, are also constants and characterise the compound being chromatographed. By definition, for benzene, *a* = 100 and *b,c,d* and *e* = 0; for ethanol, *b* = 100 and *a,c,d* and *e* = 0; etc. The *a,b,c,d* and *e* terms, once obtained, are valid for the compounds, regardless of the column being used. The full mathematical treatment, and the necessary equations for the determination of these values, were given by Rohrschneider. Supina and Rose[35] gave a table of Rohrschneider constants for several phases, and described their use in the selection of suitable stationary phases for several separations.

McReynolds[36] recommended the use of ten standard substances to provide Δ*I* values for stationary phase characterisation. Over 200 liquid phases were characterised by this method and the Δ*I* values tabulated, all the necessary data being obtained at 120°C. The average polarity of each liquid phase was determined from the sum of the Δ*I* values of benzene, butanol, 2-pentanone, nitropropane and pyridine, and the results arranged and tabulated according to increasing polarity, i.e. increasing Δ*I* values. The work was presented to show the similarities of many liquid phases now in use.

Bearing in mind the important factors of polarity and temperature limits, and taking into account the work of both Rohrschneider and McReynolds, the task of deciding upon the most suitable stationary phase for a particular analysis has been made very much easier for the gas chromatographer.

Table 2.1 gives a limited number of stationary liquid phases, which the authors believe can achieve most of the separations required for routine analysis provided suitable selections are made. ΣΔ*I* values indicate the polarity

Table 2.1 A selection of stationary liquid phases for general use

Liquid phase	*Suggested operational temperature/°C*	*Phase solvent*	*Polarity (ΣΔI values)*
Squalane	20–100	toluene	0
Apiezon 'L'	50–250	dichloromethane or toluene	143
Methyl silicone, e.g. OV-101	0–350	chloroform	229
Dexsil 300 GC	50–500	chloroform	474
Silicone OV-17	0–375	chloroform	884
Silicone OV-210	0–300	acetone	1520
Silicone OV-225	0–275	acetone or chloroform	1813
Carbowax 20M	60–225	chloroform	2308
Carbowax 600	20–120	chloroform	2646
PDEAS	0–200	chloroform	2710
DEGS	20–200	acetone	3543
TCEP	0–175	chloroform	4518

and were determined from McReynold's data by the summation of Δ*I* (benzene, butanol, 2-pentanone, nitropropane and pyridine). Appendix 1 lists some of the many available liquid phases, giving their various equivalent trade names and equating their polarity with the aid of McReynolds data.

Recently a greater interest has been shown in the development of more stable and selective polymeric materials, designed especially for use as liquid partition phases in GLC. Beeson and Pecsar[37] gave an account of a new high-temperature material, poly-M-phenoxylene, which is commerically available.

Several special polyamide-type polymers were prepared and evaluated as liquid phases by Matthews *et al.*[38] These polymers had low melting points, good thermal stability, a high degree of selectivity and the ability to coat siliceous supports or metal capillary tubing with a uniform thin film.

Liquid crystals, or, more correctly, mesomorphic compounds, have also been used by several workers as stationary phases in GLC[39, 40]. They are particularly useful for separating isomers, the different geometric structures of which exhibit different solubilities in the crystal lattice.

Richmond[39] used three liquid crystals, namely 4,4'-azoxydianisole, 4,4'-azoxy-diphenetole and 4,4'-biphenylene bis[p-(heptyloxy)benzoate] coated on 60/80 mesh Chromosorb W, for the separation of positional isomers of 25 di-substituted benzenes. Sample components were not found to change the behaviour of the stationary phases towards subsequent samples of the same or different chemical type, but small sample volumes were used to avoid over-loading the columns with the solvent. Efficient temperature control was necessary, since a variation of a few degrees at or near a transition point could cause loss of separation.

Any chemical bonding of stationary phases to solid supports gives a packing material which is more thermally stable than the conventional stationary phase-coated material. In the first instance, this was achieved by producing poly-siloxanes, chemically bonded to the silicone surfaces of chromatographic supports. Celite was modified by treatment with hexadecyltrichlorosilane[41] and with octadecyltrichlorosilane[42] followed by hydrolysis with concentrated hydrochloric acid to give packing materials which were equivalent to Celite coated with silicone elastomers. Both these methods utilised monomers of low volatility, but Hastings, Aue and Augl[43] used highly volatile silicone monomers, such as dimethyldichlorosilane, to achieve a similar end-product. As the pro-cedures often involved apparatus not readily available in all laboratories, Hastings, Aue and Larsen[44] introduced an *in situ* method, which polymerised silicones on the adsorbent particles in the chromatographic column. The column was filled with the monomeric-coated support and then water-saturated nitrogen was blown through at room temperature until polymerisation was complete. Most polymerisations were complete within 20 h, as indicated by the absence of hydrochloric acid in the effluent gas. The GLC performance of most of the phases produced by this method were similar to those of well-conditioned commercial silicones, coated on highly deactivated supports, very little tailing being observed. Although the benefits from this process were small in comparison with conventional columns, it should be possible to produce silicone liquid phases not available commerically, which might be worthy of consideration.

New stationary phases, chemically bonded to a support, are now com-mercially available, e.g. Durapak. These are the products of the esterification of an alcohol with a porous silica surface, and in the case of Durapak, Porasil is used as the base material. Since the orientation of the organic molecules on the silica is like bristles, there is no pooling as with conventional liquid phases, and so mass transfer takes place more quickly. These packings were evaluated by Little *et al.*[45], and the elution behaviour of various classes of polar and non-polar solutes was presented, to enable the user to select an appropriate packing for a particular separation. The thermal stability of the Durapak was found to be much greater than the equivalent in conventionally coated packings, and it also displayed high efficiency.

Another type of chemically bonded stationary phase was described by Neff *et al.*[46], based on the findings of Davison and Moore[47]. They used polyester-acetals cross-linked and bonded to silanol groups of an acidic siliceous support, and this had the properties of an immobile liquid, thus avoiding the bleeding characteristics of most liquid stationary phases. The packing obtained was relatively thermally stable, capable of being used up to 290°C in dual-column operation, being ideally suited to subambient use. It possessed polar characteristics and could separate many classes of compound. The packing gave excellent separation, with good efficiency, for non-polar and polar solutes, and was capable of separating both gases and higher-boiling compounds with equal effectiveness.

D. Support materials

The function of the support in GLC is to hold the stationary liquid phase in a finely dispersed form, thereby providing a high interfacial area in the column. The support should have a large surface area per unit volume, it should be chemically inert and thermally stable, and it should not adsorb the sample components. In addition, it should possess sufficient mechanical strength to withstand the procedure of coating and packing into the column, without disintegration.

Unfortunately, the ideal support material has yet to be found, although there are many available which will give satisfactory results for most analyses. Most of the supports commonly used frequently interact with highly polar substances, a fact which is often responsible for the peak tailing during elution, the variation in retention times and the low column efficiencies sometimes experienced in GLC analyses. The most commonly used supports consist of porous granules, but non-porous materials, such as glass microbeads, are also used.

1. Diatomaceous earths

Diatomaceous earth, otherwise diatomite, which consists of amorphous silica with minor impurities, is the raw material of the most frequently used supports, marketed under such names as Celite, Chromosorb, Diatomite, Anakrom, Embacel and Celaton.

The two basic types of support made from diatomite by heat treatment are referred to as white and pink supports. The white support is made by calcination of the diatomite with a flux of sodium carbonate at temperatures above 900°C, and the pink one is made from pressed bricks of diatomite, fired at a temperature above 900°C after the addition of a small amount of clay. The products are then crushed and graded for size. The best-known white material is Chromosorb W and the best-known pink one is Chromosorb P, although many similar supports are known under other designations.

The pink supports are generally dense materials of good mechanical strength and are efficient, as far as column performance is concerned, with the exception that they exhibit undesirable tailing when polar samples are analysed. The white supports, in contrast, have poor mechanical strength, lower specific surface area, fairly non-adsorptive surfaces and a poorer column efficiency.

Another support, Chromosorb G, is regarded by the manufacturers as being intermediate between white and pink, having the less adsorptive surface of the former and the mechanical strength of the latter. This is the most inactive support of its type, but it suffers from the disadvantage that good efficiency is difficult to obtain. Chromosorb G has a density of about two and a half times that of Chromosorb W, and so a 5% w/w loading of liquid phase on the former would be equivalent to approximately a 12% w/w loading on the latter.

The structural differences between various supports, i.e. density, porosity and specific surface area, restrict the maximum amount of liquid phase that can be applied to each support without producing deterioration of the column performance.

Appendix 2 gives a list of some of the more readily available supports and their equivalents. Whichever support is chosen, interactions between the support and solutes frequently occur, unless steps have been taken to minimise the 'active sites' present on their surfaces. This becomes more apparent as the polarity of the solute is increased, as the amount of stationary phase is decreased and as the sample size is increased. The adsorptions on active sites are attributed to hydrogen-bonding effects, e.g. between the surface silanol groups and the solute, and these sites would be partly deactivated by coating the support with the liquid phase. In order to suppress this activity, several methods have been developed, although few are completely effective for the separation of highly polar solutes. Unfortunately, the suppression of adsorption activity by most methods often results in a loss in column efficiency, i.e. greater HETP.

Acid washing is the most commonly used treatment of diatomites, and can be carried out by heating a slurry of the support in concentrated hydrochloric acid for approximately 30 min. It is washed with water, then methanol-rinsed and dried. Alkali washing, with or without acid treatment, is also beneficial sometimes, and is achieved by refluxing the support with a 10% solution of sodium hydroxide in methanol, washing with aqueous methanol and then drying. Acid washing by refluxing the support with glacial acetic acid has also been suggested by Bombaugh[48], this treated support being used for improved separation of methanol and ethanol.

The most widely used method for the deactivation of the support is the treatment called silanisation, using such reagents as dimethyldichlorosilane (DMCS), trimethylchlorosilane (TMCS) and hexamethyldisilazane (HMDS). This converts the surface silanol groups to silyl ethers, and effectively eliminates the hydrogen-bonding potential of the support. Care should be taken in using the reagents, as they are very volatile, toxic and inflammable, HMDS being the least toxic and volatile.

Before treatment the support should be dried, since the reagents are decomposed by water. DMCS treatment may be performed by making a slurry of the support in a 5% solution of DMCS in toluene under vacuum, filtering, then washing with toluene and methanol, and finally drying. HMDS treatment may be carried out by refluxing the support with a 5% solution of HMDS in hexane for several hours, filtering, washing with hexane and methanol, and drying. Alternatively, the supports may be treated with the reagents in the vapour phase, by exposing the support to approximately 5% by weight of reagent in a sealed glass container, at room temperature for DMCS and TMCS, and at 100°C for HMDS. Finally, they are washed with methanol and dried. Both acid-washed and silanised materials are commercially available.

A critical appraisal of these methods of deactivation was given by Ottenstein[49]. An alcohol was used as a test sample, and treatments such as acid washing, alkali washing and a combination of the two were ineffective in reducing tailing. In contrast, silanisation was very effective in reducing tailing, DMCS being more effective than HMDS, which, in turn, was more effective than TMCS. However, the real benefit of acid washing was observed when the material was silanised, since acid-washed silanised supports were more effective in reducing tailing than non-acid-washed silanised supports. The best deactivation was achieved with acid washing and treatment with DMCS.

The process of silanisation changes the character of the support surface from hydrophilic to hydrophobic, making it particularly effective for the retention of non-polar and moderately polar phases. However, this treatment is not recommended for use with very polar phases, when other means of deactivation might be found more suitable.

The subcoating of the support for the suppression of surface activity, before coating with the stationary liquid phase, has been used as an alternative by many workers. These types of materials are often referred to as 'tail reducers' or support deactivators. The use of 1% diglycerol as a tail reducer prior to coating the support with squalane was suggested by Evans[50] to reduce the peak asymmetry of alcohols.

Impregnation of the support with methanolic solutions of alkalis before coating with the liquid phase has often been used in order to reduce tailing of amines. Amines are a problem to separate, and it appears that both the support and the liquid phase must be more basic than the solute to prevent tailing.

Polymeric organic materials have been used for support subcoating, and Van den Heuvel, Gardiner and Horning[51] found that the use of an addition of polyvinylpyrrolidone (PVP) in methanol was effective prior to coating with a polar liquid phase on an acid-washed diatomite support, particularly for columns with a low percentage of liquid phase. It was essential to dry the support thoroughly by heating at 200°C for 4 h before coating it with PVP. This coating technique has been used successfully for preparing polar columns for steroid analysis, and is similar to the use of Epikote 1001 resin for addition to the column materials used for pesticide residue analysis.

The selection of a suitable support, and the method of treatment for improving a given support, depend on the type of analysis. A loss in column efficiency may have to be tolerated in order to carry out certain difficult analyses.

2. Fluorocarbons

Fluorine-containing polymers are considered to be the most inert of the support materials but they have a low specific surface area. They were first introduced as GC supports for the analysis of highly polar samples.

There are two main types: polytetrafluoroethylene, e.g. Teflon-6 and Fluoropak-80; and a polymer of chlorotrifluoroethylene, namely Kel-F. The chlorofluorocarbon supports are not so inactive as the Teflon supports, but are harder. Both types are not so readily 'wetted' by liquid phases as the conventional diatomites, and need cooling below their transition points of 19°C before handling since they are rather fragile.

Kirkland[52] suggested that the coating of Teflon supports should be carried out by a special technique to avoid agglomeration of the particles. The support was cooled to 0°C before mixing with the solution of the stationary phase, and the evaporation of the solvent was carried out by passing a stream of nitrogen over the surface without applying any heat. The coated support was then cooled to 0°C before packing the columns.

The column efficiencies attainable with Teflon-type supports have been the subject of some controversy, but it now appears to be generally accepted that the efficiencies are lower in comparison with conventional supports. They are, however, relatively inert and have been considered useful in the analysis of aqueous samples[53]. Unlike the more frequently used supports, Teflon shows little adsorption of polar solutes, such as amines, and appears to be the nearest to the ideal inert non-adsorbent support. However, its apparent inertness to polar solutes would imply a complete inertness to non-polar solutes, and yet both Evans and Smith[54] and Jequier and Robin[55] observed substantial adsorption of hydrocarbons on Teflon coated with PEG 400 and also with squalane. This led to the conclusion that even amines were probably adsorbed also, but did not exhibit the tailing usually associated with adsorption. Conder[56] studied the adsorption of non-polar and polar solutes on Teflon, and he also concluded that the absence of tailing should not be taken as evidence of an inert support.

Teflon's mechanical properties are quite poor, and two methods for improving them were suggested by Brazhnikov, Moseva and Sakodynskii[57]. One method involved a thermal treatment, the other the addition of a soluble fluorinated copolymer to join together the particles of polytetrafluoroethylene.

3. Other porous materials

A great variety of porous materials — such as unglazed tile[58]; microporous polyethylene[59]; a commercial household detergent, 'Tide'[60]; and naturally occurring sterrasters, obtained from marine sponges[61] — have been found to be useful supports. The sterrasters, in particular, were found to give columns of high efficiency, similar to those of glass beads, and showed little tendency to produce tailing peaks with relatively polar solutes, such as the higher alcohols.

Terephthalic acid was found to have the ability to hold a large amount of liquid phase[62]. The surface area and pore volume were found to be comparable with those of diatomaceous earth, and the support gave good baseline stability at operating temperatures of less than 200°C.

Uncoated terephthalic acid strongly retained solutes and gave considerable tailing. From observations, it was deduced that the carbonyl groups on the surface of the terephthalic acid were the main tailing-producing sites for hydrogen-bonding solutes, these being eliminated by esterification with the hydroxyl groups of stationary phases. It was therefore necessary, at all times with this support, to use stationary phases having hydroxyl groups, which would appear to restrict its use. However, it was found that the addition of only a small amount of polyethylene glycol effectively deactivated the adsorption sites, so that it could be used successfully for other stationary liquids, even non-polar Apiezon.

4. Non-porous materials

The use of glass beads as a solid support is well known, having been first used by Callear and Cvetanovic[63] in an effort to reduce the tailing associated with the porous supports. The glass beads originally used possessed a very low surface area, and it was difficult to get the liquid phase uniformly distributed over the surface, with the result that the column performance was often poor. Particular attention given to the proper conditioning did, however, improve the distribution and, hence, the column performance.

Although glass beads were originally considered non-active, this is now known to be untrue. When the quantity of stationary liquid phase is diminished and the sample size is decreased, interaction between the solute and the bead surface becomes increasingly evident, i.e. polar solutes exhibit peak tailing. Saturation of the active sites with larger amounts of liquid phase to reduce the adsorption is not feasible with glass beads, since 'puddling' at bead contact points and excessive column bleeding occur. Even low loadings on smooth glass beads are frequently too 'tacky' for convenient packing: non-uniform columns and poor efficiency again result. Consequently, these supports are not able to tolerate high stationary phase loadings and the absolute maximum is considered to be 3% w/w. In most instances relatively low loadings of 0.05–0.5% w/w are used, and have been applied to the analysis of high-boiling compounds at low column temperatures.

In attempts to increase the surface area of glass beads and, hence, their loading capacity, surface etching was investigated by several workers[64, 65], resulting in improved column efficiencies and improved packing characteristics. However, the higher surface area of the etched beads exposed large numbers of active sites and so extensive tailing could be expected with polar components.

The glass beads discussed so far are of the soda-lime composition, and Filbert and Hair[66, 67] considered it inevitable that sodium and calcium ions would be present on the surface, possibly functioning as weak Lewis acid sites for adsorption of certain molecules. They compared beads made from soda-lime with beads made from fused silica. Superior results were obtained with the latter, but even better results were obtained when sodium silicate glass beads were used, particularly after leaching them with acid to remove surface ion. Before packing in a chromatographic column, all the beads were silanised to remove surface hydroxyl groups and so were not suitable for use with polar liquid phases.

Alkali-silicate glass beads leached with mineral acids and silanised, as described by Filbert and Hair[67], were used by MacDonell[68] in a study of their performance in the GLC of saturated hydrocarbons and pesticides, and were found to give excellent resolution. It was stated that the use of lightly loaded, textured glass beads permitted GLC separations of steroids and other high molecular weight compounds at lower temperatures than were possible when more porous supports were used.

E. High-efficiency columns

High-efficiency columns, together with specifically designed techniques, are the basis of high-resolution GC. Three types of column have been used: wall-coated capillary, packed capillary and micropacked. To avoid the overloading of both the wall-coated and packed capillary columns, only a small proportion of the injected sample is utilised, which necessitates a device, e.g. a stream-splitter, dividing it in a quantitative manner after the normal injection procedure.

1. Capillary columns

The term 'open tubular' has been associated in the literature with both wall-coated and packed capillary columns, the latter being also known either as porous layer open tubular (PLOT) or support-coated open tubular (SCOT), where the porous layer is support-coated with a stationary liquid phase. To avoid confusion, we have chosen to use the expressions 'wall-coated capillary' where the capillary column is coated with a stationary liquid phase and 'packed capillary' to describe the porous layer open tubular columns, where support material is used in combination with a liquid phase, i.e. PLOT and SCOT.

In these 'open tubular' or 'capillary' columns the stationary phase is situated on the inner wall of the column, so that the carrier gas flow is unobstructed, and there is little flow resistance. Thus columns which are considerably longer than the conventionally packed columns can be used to give improved resolution, and in practice vary from 15 to 250 m in length. The internal diameter of these open tubular columns is usually less than 1 mm; hence the term 'capillary' column. They are also referred to as Golay columns after their designer[69]. In theory, they should provide the best separations, but in practice, difficulties are encountered with the injection of samples and also in their coating with stationary phases.

Wall-coated capillary columns Wall-coated capillary columns have a thin film of stationary phase coated on their inner walls, and are usually 0.25–0.5 mm inside diameter with an outside diameter of a standard 1/16 in. The tubing materials are generally copper, stainless steel or glass. Glass capillaries are drawn from a thick-walled glass capillary tube with the aid of a glass capillary drawing machine as described by Desty, Haresnape and Whyman[70]. They can be coated by either a dynamic method or a static method, and owing to its simplicity the former appears to be the more popular.

The dynamic method was first described by Dijkstra and de Goey[71], in which a solution of the liquid phase, the volume of which should be greater than the total internal volume of the column, is forced through the column with the aid of a dry inert gas. The inside wall of the column is therefore wetted, and the solvent is subsequently removed by the passage of the gas through the column, leaving the liquid phase as an evenly distributed thin film. The coating solution should pass through the column with a constant flow velocity, which is preferably less than 10 cm/s, and the concentration of the solution is normally 10% (w/w).

A variation on this method is the so-called plug method used by Scott and Hazeldean[72], and studied in detail by Kaiser[73], in which a small volume of solution is forced as a plug at constant velocity through the column. This was also the basis of the method described by Teranishi and Mon[74], in which three or four portions of coating solution were pushed through the capillary columns with nitrogen, flushing with nitrogen for several hours between each portion to remove the solvent.

The separation efficiency of the columns produced by these methods depends on the thickness of the stationary phase film, together with its homogeneous spreading over the whole internal capillary surface. The influence of the viscosity of the coating solution and the operational temperature on the thickness of the stationary phase film was studied by Tesarik and Necasova[75]. It was concluded that the film thickness was proportional to the viscosity of the coating solution, and could therefore be influenced by the selection of the solvent, although it was found advantageous to select a solvent with a low viscosity. The operational temperature was also found to be significant, and ideally should be approximately 15°C lower than the boiling point of the solvent. Any higher temperature would cause the complete destruction of the stationary phase film.

It is generally recognised that there is an optimum thickness for the liquid phase, since films which are too thick are unstable and break up after a short operating time, while films which are too thin may not spread uniformly on the capillary wall. Hence, for the dynamic coating methods, the choice of coating velocity may also be critical, and the optimum velocity probably depends on the materials being used. This technique appears to give good results with metal columns and non-polar liquid phases, but it is more a matter of chance with polar phases and glass columns.

The static coating method was first developed by Golay[69], in which the column was completely filled with a solution of the liquid phase and one end was closed. It was then evaporated by slowly drawing the column through a heated chamber, thus leaving a film of liquid phase on the inner wall. The film thickness was predetermined from the concentration of the coating solution and the internal volume of the column. This method has the advantage of guaranteeing a more uniform film, but requires relatively complicated equipment.

An alternative static method, which was considered to be as efficient as Golay's method, was described by Bouche and Verzele[76]. This coating was achieved by filling the column, in its final shape, from a vessel containing a solution of the stationary phase (usually 5–10 mg/ml), with the aid of suction applied at the other end of the column, followed by closing the column at one end and then evaporating the solvent under reduced pressure at a temperature below the boiling point of the solvent. The solvent was evaporated slowly from the column to leave a film of known thickness. The only disadvantage of the method is its length of time but this is more than compensated for because no special equipment is required and the procedure can be used for glass as well as for metal capillary columns. This method was later improved by Boogaerts, Verstappe and Verzele[77] to enable more troublesome phases to be deposited.

In order to get more liquid phase into wall-coated capillary columns, and thus reduce the so-called β-value (the ratio of the volume of the gas to the volume of the liquid in the column), several attempts have been made to increase the interior surface area. Chemical etching or roughening the interior of glass capillaries was carried out by Mohrke and Saffert[78] with an ammonia solution and by Bruner and Cartoni[65] with aqueous sodium hydroxide solution. As a result of these processes, a porous SiO_2 layer was formed on the inside wall.

The wettability of the column walls seems to be the major problem associated with capillary columns, i.e. the ability to coat the inside walls with a uniform layer of the liquid phase. This is particularly difficult with polar phases, and the coating of glass columns is extremely difficult. Even with non-polar liquid phases, the coating of glass columns is not easy and is highly irreproducible.

Apart from etching, the addition of surfactants, such as Alkaterge T, Span 80, Alpet 80 and Igepal, to the liquid phase to facilitate spreading of the liquid was useful up to a temperature of about 170°C, but above this temperature they had a tendency to aid the bleeding of the liquid phase.

The use of the property of long-chain quaternary ammonium compounds to form monomolecular layers on metal and other surfaces was utilised by Metcalfe and Martin[79]. A larger amount of liquid phase could be more evenly coated on the column wall by using these quaternary ammonium salts. These workers used trioctadecylmethyl ammonium bromide added to the liquid phase in a solvent, and applied it to both metal and glass columns. The column performances were considerably improved and the lives of the columns were increased. The addition of 0.1–0.2% quaternary ammonium salt in the solvent was sufficient for the coating, and its effect on the properties of the liquid phase was negligible.

The temperature limit of the quaternary ammonium salt was about 200°C, whereas the limit for benzyltriphenylphosphonium chloride, as used by Malec[80], was in excess of 250°C. The quaternary phosphonium halide gave columns of excellent resolution and extended lifetime, when used with several silicone, Carbowax and polyester phases. Although they are very reactive compounds, no interaction with the samples was noted with concentrations of up to 0.2% w/w in the coating solution. Initial tests with a similar material, tetraphenyl-phosphonium chloride, were also promising.

With glass capillaries, satisfactory results were obtained by Grob[81], using a thin carbon layer coating on the walls, achieved by the pyrolysis of nitrogen saturated with dichloromethane at 600°C, before coating with the liquid phase. This preparation of so-called carbonised columns was used by Zoccolillo, Liberti and Goretti[82], prior to coating glass ones with trimer acid liquid phase, the combination yielding a high degree of resolution in the separation of the components of essential oils. Goretti and Liberti[83] found that the liquid phase, FFAP, was far superior to trimer acid for this particular application.

Cleaning of the capillary tubing is essential to avoid any adsorbed material from hindering the uniform distribution of the liquid phase during coating. New tubing is flushed with solvents, commencing with a non-polar solvent and ending with the solvent which is to be used for the liquid phase, e.g. pentane, chloroform, methanol, acetone and the phase solvent. A similar procedure is followed before the recoating of used capillary tubing after the previously deposited phase has been partially decomposed with nitric acid. An alternative method was described by Lavoue[84], in which the column was cleaned with ammonia and acetone with the aid of ultrasonic vibration. The treatment of stainless steel capillary with phosphoric acid to form an insoluble phosphate layer has been used successfully for the analysis of acidic compounds[85].

Packed capillary columns The early packed capillary columns were PLOT columns, and were prepared from a suspension of a finely divided solid with, or without, a liquid phase. The dynamic coating method was used for preparing the columns lined with an adsorptive layer of such materials as colloidal silica or colloidal boehmite.

The static coating method permitted greater control over the distribution process, and enabled practically any kind of solid support and liquid phase to be used. Halasz and Horvath[86] and Ettre, Purcell and Norem[87] succeeded in getting solids impregnated with liquid to adhere to capillary column walls using this technique. Very fine particles of the support were coated with the liquid phase, forming a stable colloidal suspension, using high-density solvents. The columns produced were considered to be far superior to conventional wall-coated capillary columns, having a greater amount of stationary phase per column length, i.e. lower β-value, and therefore the maximum acceptable sample size was greater. These were SCOT columns.

The application of these columns in trace analysis was the subject of papers by Purcell and Ettre[88] and Ettre, Purcell and Billeb[89]. It was concluded that the permitted sample size with these columns was comparable with that used with standard 0.125 in O.D. packed columns, and that good baseline resolution could be achieved with sample volumes containing as much as 0.15 μl of a single component. The minimum detectable limit with such a column was about 50 times lower than with a well-coated capillary column.

Grant[90] described a method for the preparation of packed capillary columns from Pyrex capillary tubing, using a glass capillary drawing machine. A metal rod was placed along the axis of the tubing to reduce the amount of packing, and the annular space was packed with a finely graded Celite, containing a binding agent. After the tubing was packed, the rod was removed, and the tubing and its packing were drawn over a tungsten wire as illustrated in *Figure 2.1*. The packed volume of the column was considerably higher than that obtained by Halasz and Horvath[86]; hence, the column capacity was greater.

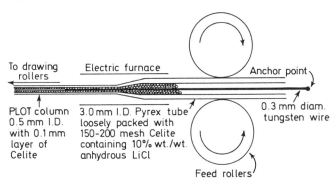

To drawing rollers

Electric furnace

Anchor point

PLOT column 0.5 mm I.D. with 0.1 mm layer of Celite

3.0 mm I.D. Pyrex tube loosely packed with 150-200 mesh Celite containing 10% wt./wt. anhydrous LiCl

0.3 mm diam. tungsten wire

Feed rollers

Figure 2.1. Production of porous layer open tubes (after Grant[90])

Prepared columns were coated with a variety of liquid phases by the dynamic coating method, and gave excellent performance. Although lithium chloride was used as a binding agent, it possesses undesirable adsorption properties, and powdered glass has been used as a successful replacement.

Columns

Packed capillary columns have a considerably higher sample capacity than wall-coated capillary columns, so that it is just possible, although not always convenient, to use them without sample stream-splitters. Conversely, wall-coated columns have a relatively low sample capacity of approximately 10^{-6} g, i.e. sample volumes of the order of 0.01 μl, and so stream-splitters are usually employed. However, the use of a stream-splitter usually brings about the loss of a large portion of the sample, which is not always acceptable; for example, in the analysis of trace constituents, sufficient sample is required to provide adequate sensitivity. Stream-splitters are also unsuitable for the analysis of high-boiling compounds.

In general, packed capillary columns have greater resolution and produce faster elution of compounds than wall-coated capillary columns of the same length. Very long packed capillary columns have reasonably good sample capacity, and permit the best resolution of complex mixtures.

F. Stream-splitters and alternatives

A sample-splitting device for use with capillary columns must be capable of separating 1% of the sample with good reproducibility. This part of the sample must then enter the capillary column as a narrow zone. Many types of splitter have been used but most gave poor reproducibility and were unsatisfactory for quantitative analysis. Two exceptions were the splitters of Dubsky[91] and Bruderreck, Schneider and Halasz[92].

If splitting can be avoided by the use of a suitable direct injection technique, quantitative errors may be reduced, sample losses may be minimised and the sensitivity to trace constituents may be increased.

The loss in resolution in going from 0.01 to 0.03 in internal diameter columns is often more than offset by the possibility of a direct injection of the sample with a small-volume syringe. Direct injection can be used by careful attention to the design of the inlet system and the injector insert to minimise the dead volume.

Other ways have been used to avoid the necessity of using a splitter. Cramers and van Kessel[93] suggested diluting the sample with an excess of non-volatile solvent, e.g. silicone oil. The diluted sample was introduced with a standard syringe into a small packed precolumn to retain the non-volatile solvent, whereas the entire sample evaporated and passed into the column, with no fractionation.

Another splitter injection technique was described by Grob and Grob[94, 95]. The system gave maximum separation efficiency without additional equipment and utilised the column as a trapping system for the sample components. Grob and Grob found that a column trapped the components when held at least 100°C below the boiling point of the most volatile sample component during the injection period, and that on heating to the analysis temperature full separation was still observed. During injection the column temperature was high enough to allow elution of the solvent and by-products and yet low enough to provide efficient trapping of the sample components. An injection split valve was incorporated to allow the injection port to be cleaned of traces of back-diffused sample. The combination of the concentration effect, brought about by this simple technique plus a solvent by-passing technique, avoided any deterioration of the column and was therefore found to broaden the application of capillary columns in trace analysis.

G. Micropacked columns

Micropacked columns are made of small diameter tube with an internal diameter less than 1 mm. They are packed similarly to the conventional packed columns, but smaller particles having a diameter less than one-tenth of the tube diameter are usually used. These columns incorporate both the advantages of ordinary packed columns and, to a large extent, those of capillary columns, and the direct injection of samples presents no problem.

Cramers, Rijks and Bocek[96] discussed the preparation, properties and applications of micropacked columns. The HETP values of the columns were high, being quoted at approximately 3000/m. The columns were up to 15 m in length and the internal diameter was 0.6–0.8 mm. The limiting factor of the pressure drop, when reducing the diameter of ordinary packed columns, was overcome by careful sieving to obtain a particle size distribution smaller than 20 μm. The fines were removed by sieving in vacuum, although the methods of flotation or sedimentation were suggested alternatives. The columns were coiled, then packed by the application of pressure and vibration. This was carried out by the immersion of the column in an ultrasonic bath, ensuring that the plugged end was held above the level of the bath.

H. Adsorption reducers

The elimination or minimising of adsorptive effects in both packed and open tubular columns has been accomplished by the addition of small proportions of tailing reducers to the stationary phase, by the treatment of the solid support or by the saturation of the carrier gas with water vapour, formic acid or some other polar substance.

Many of the tailing reducers are surfactants, which appear to aid the spreading of the liquid phase by reducing the surface tension. They are normally polar compounds, and tend to be used with non-polar phases. Acidic samples require the use of an anionic additive, e.g. terephthalic acid, and basic samples require the use of a cationic additive, e.g. an amine such as Alkaterge T. Some additives are non-ionic, e.g. Span 80 and Igepal CO-880, which can generally be used for liquid phases and samples, both of which may be either polar or non-polar.

Several non-volatile acids and bases have been used as adsorption-reducers, and many of these are applied to the support with the liquid phase, as opposed to the support alone (see also p. 24).

1. Non-ionic additives

Averill[97, 98], showed that the addition of small amounts of surfactants reduced the interactions between the sample and open tubular columns. Igepal CO-880

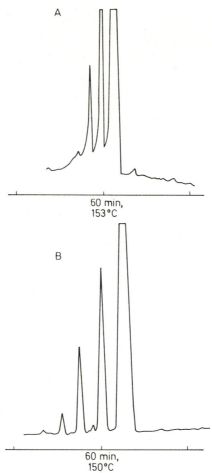

Figure 2.2. Comparison of resolution of isopulegol isomers with 1000 ft, 0.03 in I.D., open tubular columns. Chromatogram A with old SF-96 (50) silicone oil coating. Chromatogram B with same tubing, cleaned and recoated with 10% SF-96 (50), 0.5% Igepal, in ether–acetone solution. (After Mon et al.[99])

was used with open tubular columns coated with Apiezon, and considerably reduced peak tailing. This additive was also used by Mon, Forrey and Teranishi[99], who investigated its effect on the retention times of several components analysed on open tubular columns. Little effect on the retention times was observed, and when combined with a vigorous cleaning procedure for the capillary tubing, an improvement in the resolution was achieved. Packed columns were also found to benefit from the addition of Igepal. *Figure 2.2* compares the resolution before (A) and after cleaning and recoating with 10% SF-96 silicone oil plus 0.5% Igepal (B), during the analysis of isopulegol isomers using a column quoted as being a 1000 ft x 0.03 in internal diameter open tubular column.

2. Acids

The addition of phosphoric acid to the liquid phase has been used by many workers to prevent the peak tailing of free fatty acids by inhibiting the adsorption effects[100-103].

Stainless steel columns are also considered to be extremely adsorptive towards free fatty acids, and phosphoric acid was found to be the most effective material for their deactivation[103]. It was emphasised that silanised surfaces, e.g. supports and column plugs, are incompatible with a system designed for free fatty acid separation.

The treatment of less adsorptive solid supports with phosphoric acid has also found some use. Jowett and Horrocks[104] used glass beads coated with 1% PEGS and 0.4% phosphoric acid for the analysis of lower fatty acids.

Isophthalic acid was also used as an additive to liquid phases by Nickelly[105] and McKinney and Jordan[106]. Nickelly separated fatty acids up to C_{18} on columns packed with glass microbeads coated with Carbowax 20M and isophthalic acid, and McKinney and Jordan used the same additive on their column material for the separation of lower aldehydes and acids in aqueous solution.

Other acidic additives have been used to aid the separation of fatty acids, e.g. stearic acid[107], terephthalic acid[108], dinonylnaphthalenedisulphonic acid[109, 110] and citric acid[111], and, in addition, they were all thought to prevent dimerisation of the fatty acids.

3. Bases

The use of methanolic potassium hydroxide for the treatment of supports used in the analysis of amines has been mentioned elsewhere (p. 25). Its tendency to age the column was overcome by Di Lorenzo and Russo[112] by treating the support with potassium hydroxide after impregnation with the liquid phase; this gave stable columns suitable for quantitative analysis when the addition of the alkali was 20%.

Many amines and imines of low volatility have been used as column packing modifiers for amine analysis, including triethanolamine and polyethylenimine[113, 114].

4. Carrier gas-conveyed reducers

The use of water and other polar liquids in the carrier gas to reduce peak tailing effects has been applied to both packed and open tubular columns with some success. Mon, Forrey and Teranishi[115] used a stainless steel flask containing water between the carrier gas supply and the GC to saturate the carrier gas with water vapour. Changes in the relative retention times of oxygenated compounds and a reduction in the adsorption were apparent.

The addition of formic acid to the carrier gas as recommended by Ackman and his co-workers[116-119], together with the use of a flame ionisation detector, which is insensitive to formic acid, considerably reduces the amount of tailing and eliminates the repeat peak effects in the analysis of volatile free fatty acids. This procedure was successfully used by Geddes and Gilmour[120] to reduce the 'ghosting' during the analysis of C_2 to C_5 acids.

I. Column performance

There are many parameters that affect the overall column efficiency, including HETP, column diameter, column length, temperature, type of carrier gas and its velocity, particle size of the support and sample volume. However, column performance is best evaluated by attention to the efficiency of separation between the components in a particular analysis. It has been expressed, measured and optimised in many ways.

The term 'resolution', R, expresses the extent of the separation of two chromatographic peaks. By definition, resolution is equal to the ratio of the spacing between the peak maxima, Δt, to the mean base width of two neighbouring peaks, w_m, and is deduced from the following equation:

$$R = \frac{\Delta t}{w_m} = \frac{2(t_2 - t_1)}{w_1 + w_2} \tag{2.4}$$

The retention times t_1 and t_2 and the peak widths w_1 and w_2 are measured as shown in *Figure 2.3*. The resolution is a measure of column efficiency for the

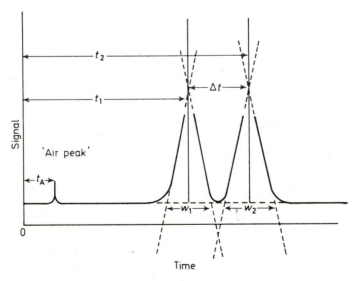

Figure 2.3. Separation of a pair of solutes

separation of two components. The components are considered to be fully resolved when $R = 1.5$, but for practical purposes the value $R = 1$ (98% separation) is considered adequate. Values of R less then unity are considered inefficient.

Grushka[121] considered that chromatographic resolutions greater than unity were wasteful in terms of the number of components that could be resolved within a given time. He suggested that a better measure of column efficiency was the maximum number of solutes that could be resolved in the time span between the inert peak (air peak) and the last eluted component. This was known as the peak capacity of the system for which Giddings[122] had developed a number of mathematical expressions. For optimum conditions of $R = 1$ for all adjacent peaks, the peak capacity was given by:

$$\text{peak capacity} = 1 + \frac{\sqrt{n}}{4} \log_e \frac{t_m}{t_A} \qquad (2.5)$$

where n is the plate number, t_m is the retention time of the last component and t_A is the retention time of 'air' peak. Peak capacity was increased by the judicious choice of the column length, the velocity of the carrier gas, the column temperature and the amount of stationary phase. The use of temperature programming also increased the available peak capacity.

An alternative method for expressing the extent of the separation between two incompletely resolved peaks was suggested by Christophe[123]. The valley to peak ratio (V/P ratio) was proposed to overcome the drawbacks inherent with the calculation of resolution, namely the errors introduced in measuring the width during construction of tangents at the inflection points to the curve, and the failure of the method to take into account any moderate tailing of the peaks, which does, however, affect the separation. The V/P ratio was defined as the ratio of the minimum height between two peaks to the height of one of the peaks, and was expressed by the equation:

$$V/P = \frac{h_i}{h_{0_1}} \qquad (2.6)$$

where h_i = minimum height between two peaks and h_{0_1} = peak height at the maximum of the elution curve of one of the components. When compared with resolution, the V/P ratio was a far more sensitive means of expressing the separation; it was also a better method for measuring the separation where there was appreciable tailing. When the separation was poor, the V/P ratio was easily determined, although it was impossible to determine the resolution. The peak height of either one of the peaks could be used for the calculation, but when the values were being used for establishing the optimum column conditions, it was advantageous to use the peak that gave the most sensitive V/P ratio. When the V/P ratio was used to determine whether a separation was satisfactory or not, the peaks were adjusted to be of similar height, and either value could be used.

Problems associated with resolution are of great importance in GC, and improvements in chromatographic separation are often desired. The efficiency of the system can be improved by careful manipulation of the experimental conditions, but often an increase in resolution means longer analysis time. For example, while an increase in column length will increase the resolution, retention times will also be lengthened. A resolution of $R = 1.5$ allows precise quantitative analysis but needs 3.4 times more time to obtain than a resolution of 1.0, which allows slightly less precision; so the analyst will have to decide whether he really needs the extra resolution at the price of extra analysis time.

Many workers[124-130] have suggested similar processes for achieving the best performance from a column, i.e. the highest resolution in the least time. This technique has been referred to as time normalisation chromatography, and necessitates the optimisation of such factors as column length, column temperature, support particle size, stationary phase loading and mobile phase velocity.

J. Foreflushing and backflushing

Quite frequently, in analysis, only a portion of the sample is of interest. In these circumstances, the use of cutting, flushing and multicolumn techniques may be applicable. These techniques enable a portion of the sample to be separated into individual components, and the remaining part to be either analysed as a single peak, discarded or transferred to another column for further separation. These techniques are particularly applicable to the separation of wide-boiling-range samples, for the determination of trace components and in the analysis of complex mixtures.

Foreflushing and backflushing may be applied to single-column separations, and in both cases some sample components are removed as a group from the column, while the rest are separated and analysed. The carrier gas flow through the column during foreflushing is in the normal direction, and in this case the group of sample components are passed either through the detector, through a vent or into another column, by operation of a valve. In backflushing, the flow direction of the carrier gas in the column is reversed after a predetermined time, so that a group of sample components are swept back towards the inlet and out through a vent, or to a detector. This operation usually takes place after the components of interest have been separated, eluted and analysed, and is particularly useful for reducing the analysis time and preventing column contamination with unwanted higher-boiling material.

The combination of foreflushing and backflushing to transfer one or more selected groups of components from one column to another column is referred to as 'heart cutting' or 'cutting'. This is a well-established technique, and many systems of foreflushing and backflushing have been developed that utilise several separating columns. In the analysis of complex mixtures groups of components which are unresolved on a single column can be selectively transferred to another column of different characteristics, using the technique of 'cutting'. Similarly, in trace analysis this technique may be used to remove a part of the major constituents and prevent column overloading and peak tailing, which would arise from the large amount of sample necessary to produce a significant signal for the trace component. This is particularly useful when the trace component emerges on the tail of a major constituent.

Many of the early systems involved the use of fairly complex multiport taps or valves, situated in the oven. These were expensive, complicated in construction, restricted to fairly low operating temperatures and tended to reduce the separation efficiency because of their inherent dead volume.

Deans[131] described a system which eliminated the use of mechanical valves in the oven, and could be added easily and cheaply to existing chromatographs. The system utilised externally operated pressure controllers, and could be used with capillary columns without any loss of efficiency.

Figure 2.4 illustrates the flow system for cutting, in combination with a backflushing system described previously by Deans[132]. As the on/off taps were located outside the oven, their design was not critical. A capillary flow restrictor

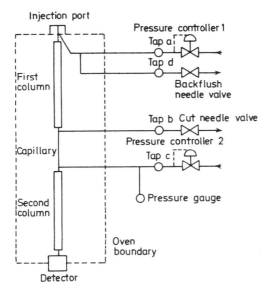

Figure 2.4. Flow system for cutting and back-flushing (after Deans [132])

was connected between the two columns, but its dimensions were also not critical. The new system offered many advantages, e.g. negligible dead volume and the absence of valves or taps in the sample path.

A further application of the system was described in which the capillary connection between the two columns was replaced by a microreactor. This could be used to facilitate the identification of components, e.g. the carbon skeleton of an unknown, unsaturated compound, after hydrogenation, may be deduced from the identification of the saturates eluted on the second column.

Mellor[133] experienced certain limitations with Deans's system, and described a similar system in which only on/off valves and a single pressure-regulated carrier gas supply were used for its operation. The arrangement enabled cutting, backflushing and holding operations (flow in the first column is arrested, while flow down the second column continues, this being used to increase the separation of complex mixtures) to be performed simply, and could easily be fitted to any dual-flame ionisation GC. An interceptive trap, containing silicone-coated Celite and positioned between the two columns, was an optional facility, which allowed any emerging peak or portion of a peak from the first column to be trapped, prior to its examination on the second column. With care this trap could serve as a means of improving peak sharpness. The method proved particularly useful for determining trace impurities that were not completely separated from the tail of a major component.

The use of backflushing with electron-capture GC can increase the duration of the intervals between detector cleaning operations, e.g. in pesticide residue analysis, and it enables likely contamination from long-retained co-extracted materials to be reduced. Unfortunately, the electron-capture detector is extremely sensitive to changes in the carrier gas flow, and most backflushing systems are found to be unsatisfactory. Croll[134] suggested an alternative system, which overcame the limitations of the earlier external methods, and could be very easily constructed and adapted to fit most GCs.

The external switching of gas systems also formed the basis of a method described by Primavesi[135] for the concentration, detection and identification of late peaks. The apparatus enabled a large sample to be used, and the late components, together with the tail of the main component, were retained on the column, backflushed, trapped and rechromatographed. The lowest detectable concentration of a component was restricted only by the amount of sample and available analysis time.

Backflushing was the subject of a review by Eppert[136], which included an examination of the merits of the various 'outside' or tap or valve systems, as opposed to the heated valve systems.

References

1. Crossley, J., *J. Chromat. Sci.*, **8**, 426 (1970)
2. Albrecht, J. and Verzele, M., *Chromatographia*, **4**, 419 (1971)
3. Van den Heuvel, W. J. A. and Kuron, G. W., *J. Chromat. Sci*, **7**, 651 (1969)
4. Averill, W., *J. Gas Chromat.*, **1** (January), 34 (1963)
5. Lysyj, I. and Newton, P. R., *Anal. Chem.*, **36**, 949 (1964)
6. Villalobos, R., *J. Gas Chromat.*, **6**, 367 (1968)
7. Mussini, E., de Nadai, F., Fanelli, R. and Frigerio, A., *Anal. Biochem.*, **47**, 635 (1972)
8. Braddock, L. I. and Marec, N., *J. Gas Chromat.*, **5**, 588 (1967)
9. Christophe, A., *J. Chromat. Sci.*, **8**, 614 (1970)
10. Christophe, A. B., *J. Chromatog.*, **58**, 195 (1971)
11. Di Corcia, A., Fritz, D. and Bruner, F., *Anal. Chem.*, **42**, 1500 (1970)
12. Frycka, J., *J. Chromatog.*, **65**, 432 (1972)
13. Neumann, M. G. and Hertl, W., *J. Chromatog.*, **65**, 467 (1972)
14. Vernon, F., *J. Chromatog.*, **60**, 406 (1971)
15. Rowan, R. and Sorrell, J. B., *Anal. Chem.*, **42**, 1716 (1970)
16. Guillemin, C. L., Le Page, M. and de Vries, A. J., *J. Chromat. Sci.*, **9**, 470 (1971)
17. McKinney, R. W. and Jordan, R. L., *J. Gas Chromat.*, **5**, 501 (1967)
18. Hollis, O. L., *Anal. Chem.*, **38**, 309 (1966)
19. Hollis, O. L. and Hayes, W. V., *J. Gas Chromat.*, **4**, 235 (1966)
20. Burger, J. D., *J. Gas Chromat.*, **6**, 177 (1968)
21. Lindsay Smith, J. R. and Waddington, D. J., *Anal. Chem.*, **40**, 522 (1968)
22. Lindsay Smith, J. R. and Waddington, D. J., *J. Chromatog.*, **36**, 145 (1968)
23. Supina, W. R. and Rose, L. P., *J. Chromat. Sci.*, **7**, 192 (1969)
24. Dave, S. B., *J. Chromat. Sci.*, **7**, 389 (1969)
25. Gough, T. A. and Simpson, C. F., *J. Chromatog.*, **51**, 129 (1970)
26. Dressler, M., Vespalec, R. and Janak, J., *J. Chromatog.*, **59**, 423 (1971)
27. Dressler, M., Guha, O. K. and Janak, J., *J. Chromatog.*, **65**, 261 (1972)
28. Ackman, R. G., *J. Chromat. Sci.*, **10**, 506 (1972)
29. Mahadevan, V. and Stenroos, L., *Anal. Chem.*, **39**, 1652 (1967)
30. Ross, W. D. and Jefferson, R. T., *J. Chromat. Sci.*, **8**, 386 (1970)

31. Schnecko, H. and Bieber, O., *Chromatographia*, **4**, 109 (1971)
32. Rohrschneider, L., *Z. Anal. Chem.*, **170**, 256 (1959)
33. Petsev, N., *J. Chromatog.*, **59**, 21 (1971)
34. Rohrschneider, L., *J. Chromatog.*, **22**, 6 (1966)
35. Supina, W. R. and Rose, L. P., *J. Chromat. Sci.*, **8**, 214 (1970)
36. McReynolds, W. O., *J. Chromat. Sci.*, **8**, 685 (1970)
37. Beeson, J. H. and Pecsar, R. E., *Anal. Chem.*, **41**, 1678 (1969)
38. Mathews, R. G., Schwartz, R. D., Stouffer, J. E. and Pettitt, B. C., *J. Chromat. Sci.*, **8**, 508 (1970)
39. Richmond, A. B., *J. Chromat. Sci.*, **9**, 571 (1971)
40. Porcaro, P. J. and Shubiak, P., *J. Chromat. Sci.*, **9**, 690 (1971)
41. Abel, E. W., Pollard, F. H., Uden, P. C. and Nickless, G., *J. Chromatog.*, **22**, 23 (1966)
42. Aue, W. A. and Hastings, C. R., *J. Chromatog.*, **42**, 319 (1969)
43. Hastings, C. R., Aue, W. A. and Augl, J. M., *J. Chromatog.*, **53**, 487 (1970)
44. Hastings, C. R., Aue, W. A and Larsen, F. N., *J. Chromatog.*, **60**, 329 (1971)
45. Little, J. N., Dark, W. A., Farlinger, P. W. and Bombaugh, K. J., *J. Chromat. Sci.*, **8**, 647 (1970)
46. Neff, W. E., Pryde, E. H., Selke, E. and Cowan, J. C., *J. Chromat. Sci.*, **10**, 512 (1972)
47. Davison, V. L. and Moore, D. J., *J. Gas Chromat.*, **6**, 540 (1968)
48. Bombaugh, K. J., *J. Chromatog.*, **11**, 27 (1963)
49. Ottenstein, D. M., *J. Gas Chromat.*, **6**, 129 (1968)
50. Evans, M. B., *Chromatographia*, **4**, 441 (1971)
51. Van den Heuvel, W. J. A., Gardiner, W. L. and Horning, E. C., *Anal. Chem.*, **35**, 1745 (1963)
52. Kirkland, J. J., *Anal. Chem.*, **35**, 2003 (1963)
53. Landault, C. and Guiochon, G., *J. Chromatog.*, **9**, 133 (1962)
54. Evans, M. B. and Smith, J. F., *J. Chromatog.*, **30**, 325 (1967)
55. Jequier, W. and Robin, J., *Chromatographia*, **1**, 297 (1968)
56. Conder, J. R., *Anal. Chem.*, **43**, 367 (1971)
57. Brazhnikov, V., Moseva, L. and Sakodynski, K., *J. Chromatog.*, **38**, 287 (1968)
58. Lukes, V., Komers, R. and Herout, V., *J. Chromatog.*, **3**, 303 (1960)
59. Baum, E. H., *J. Gas Chromat.*, **1** (November), 13 (1963)
60. Decora, A. W. and Dinneen, G. U., *Anal. Chem.*, **32**, 164 (1960)
61. Webb, J. L., Smith, V. E. and Wells, H. W., *J. Gas Chromat.*, **3**, 384 (1965)
62. Miyake, H., Mitooka, M. and Matsumoto, T., *Anal. Chem.*, **40**, 113 (1968)
63. Callear, A. B. and Cvetanovic, R. J., *Can. J. Chem.*, **33**, 1256 (1955)
64. Ohline, R. W. and Jojola, R., *Anal. Chem.*, **36**, 1681 (1964)
65. Bruner, F. A. and Cartoni, G. P., *Anal. Chem.*, **36**, 1522 (1964)
66. Filbert, A. M. and Hair, M. L., *J. Gas Chromat.*, **6**, 150 (1968)
67. Filbert, A. M. and Hair, M. L., *J. Gas Chromat.*, **6**, 218 (1968)
68. MacDonell, H. L., *Anal. Chem.*, **40**, 221 (1968)
69. Golay, M. J. E., in *Gas Chromatography* (V. J. Coates, H. J. Noebels and I. S. Fagerson, Eds.), Academic Press, New York and London (1958)
70. Desty, D. H., Haresnape, J. N. and Whyman, B. H. F., *Anal. Chem.*, **32**, 302 (1960)
71. Dijkstra, G. and de Goey, J. in *Gas Chromatography 1958* (D. H. Desty, Ed.), Butterworths, London, 56 (1958)
72. Scott, R. P. W. and Hazeldean, G. S. F., in *Gas Chromatography 1960* (R. P. W. Scott, Ed.), Butterworths, Washington DC, 144 (1960)
73. Kaiser, R., in *Gas Phase Chromatography*, Vol. II *Capillary Chromatography*, Butterworths, London, 45 (1963)
74. Teranishi, R. and Mon. T. R., *Anal. Chem.*, **36**, 1490 (1964)
75. Tesarik, K. and Necasova, M., *J. Chromatog.*, **65**, 39 (1972)
76. Bouche, J. and Verzele, M., *J. Gas Chromat.*, **6**, 501 (1968)
77. Boogaerts, T., Verstappe, M. and Verzele, M., *J. Chromat. Sci.*, **10**, 217 (1972)
78. Mohrke, M. and Saffert, W., in *Gas Chromatography 1962* (M. van Swaay, Ed.), Butterworths, London, 216 (1962)
79. Metcalfe, L. D. and Martin, R. J., *Anal. Chem.*, **39**, 1204 (1967)
80. Malec, E. J., *J. Chromat. Sci.*, **9**, 318 (1971)
81. Grob. K., *Helv. Chim. Acta*, **48**, 1362 (1965)
82. Zoccolillo, L., Liberti, A. and Goretti, G. C., *J. Chromatog.*, **43**, 497 (1969)

83. Goretti, G. and Liberti, A., *J. Chromatog.,* **61,** 334 (1971)
84. Lavoue, G., *J. Gas Chromat.,* **6,** 233 (1968)
85. Hrivnak, J., *J. Chromat. Sci.,* 8, 602 (1970)
86. Halasz, I. and Horvath, C., *Anal. Chem.,* **35,** 499 (1963)
87. Ettre, L. S., Purcell, J. E. and Norem, S. D., *J. Gas Chromat.,* **3,** 181 (1965)
88. Purcell, J. E. and Ettre, L. S., *J. Gas Chromat.,* **4,** 23 (1966)
89. Ettre, L. S., Purcell, J. E. and Billeb, K., *J. Chromatog.,* **24,** 335 (1966)
90. Grant, D. W., *J. Gas Chromat.,* **6,** 18 (1968)
91. Dubsky, H., *J. Chromatog.,* **47,** 313 (1970)
92. Bruderreck, H., Schneider, W. and Halasz, I., *J. Gas Chromat.,* **5,** 217 (1967)
93. Cramers, C. A. and van Kessel, M. M., *J. Gas Chromat.,* **6,** 577 (1968)
94. Grob, K. and Grob, G., *J. Chromat. Sci.,* **7,** 584 (1969)
95. Grob, K. and Grob, G., *J. Chromat. Sci.,* **7,** 587 (1969)
96. Cramers, C. A., Rijks, J. and Bocek, P., *J. Chromatog.,* **65,** 29 (1972)
97. Averill, W., in *Gas Chromotography* (N. Brenner, J. E. Callen and M. D. Weiss, Eds.), Academic Press, New York and London, 1 (1962)
98. Averill, W., published at *Pittsburgh Conference on Analytical Chemistry and Applied Spectroscopy,* Pittsburgh, Penn., March 1–5, 1965
99. Mon, T. R., Forrey, R. R. and Teranishi, R., *J. Gas Chromat.,* **5,** 497 (1967)
100. Metcalfe, L. D. *Nature,* **188,** 142 (1960)
101. Metcalfe, L. D., *J. Gas Chromat.,* **1** (January), 7 (1963)
102. Hrivnak, J. and Palo, V., *J. Gas Chromat.,* **5,** 325 (1967)
103. Ottenstein, D. M. and Bartley, D. A., *J. Chromat. Sci.,* **9,** 673 (1971)
104. Jowett, P. and Horrocks, B. J., *Nature,* **192,** 966 (1961)
105. Nickelly, J. G., *Anal. Chem.,* **36,** 2244 (1964)
106. McKinney, R. W. and Jordan, R. L., *J. Gas Chromat.,* **3,** 317 (1965)
107. James, A. T. and Martin, A. J. P., *Biochem, J.,* **50,** 679 (1952)
108. Byars, B. and Jordan, G., *J. Gas Chromat.,* **2,** 304 (1964)
109. Lee, W. K. and Bethea, R. M., *J. Gas Chromat.,* **6,** 582 (1968)
110. Averill, W., *J. Gas Chromat.,* **1** (January), 22 (1963)
111. Kaplanova, B. and Janak, J., *Mikrochim. Acta,* (1–2), 119 (1966)
112. Di Lorenzo, A. and Russo, G., *J. Gas Chromat.,* **6,** 509 (1968)
113. Lindsay Smith, J. R. and Waddington, D. J., *J. Chromatog.,* **42,** 183 (1969)
114. O'Donnell, J. F. and Mann, C. K., *Anal. Chem.,* **36,** 2097 (1964)
115. Mon, T. R., Forrey, R. R. and Teranishi, R., *J. Gas Chromat.,* **4,** 176 (1966)
116. Ackman, R. G. and Burgher, R. D., *Anal. Chem.,* **35,** 647 (1963)
117. Ackman, R. G., Burgher, R. D. and Sipos, J. C., *Nature,* **200,** 777 (1963)
118. Ackman, R. G. and Sipos, J. C., *J. Chromatog.,* **13,** 337 (1964)
119. Ackman, R. G., *J. Chromat. Sci.,* **10,** 560 (1972)
120. Geddes, D. A. M. and Gilmour, M. N., *J. Chromat. Sci.,* **8,** 394 (1970)
121. Grushka, E., *Anal. Chem.,* **42,** 1142 (1970)
122. Giddings, J. C., *Anal. Chem.,* **39,** 1027 (1967)
123. Christophe, A. B., *Chromatographia,* **4,** 455 (1971)
124. Karger, B. L. and Cooke, W. D., *Anal. Chem.,* **36,** 985 (1964)
125. Karger, B. L. and Cooke, W. D., *Anal. Chem.,* **36,** 991 (1964)
126. Guiochon, G., *Anal. Chem.,* **38,** 1020 (1966)
127. Grushka, E., *Anal. Chem.,* **43,** 766 (1971)
128. Grushka, E., *J. Chromat. Sci.,* **9,** 310 (1971)
129. Giddings, J. C., *Anal. Chim. Acta,* **27,** 207 (1962)
130. Grushka, E., Yepes-Baraya, M. and Cooke, W. D., *J. Chromat. Sci.,* **9,** 653 (1971)
131. Deans, D. R., *Chromatographia,* **1,** 18 (1968)
132. Deans, D. R., *J. Chromatog.,* **18,** 477 (1965)
133. Mellor, N., *Analyst,* **96,** 164 (1971)
134. Croll, B. T., *Analyst,* **96,** 810 (1971)
135. Primavesi, G. R., *Analyst,* **95,** 242 (1970)
136. Eppert, G., *J. Gas Chromat.,* **6,** 361 (1968)

3
Detectors

A. Introduction

The literature contains a vast number of references to detection systems of varying complexity and nature for GC, many of which have not gained widespread acceptance, and many of which are not applicable to food analysis. Most of them are considered beyond the scope of this book, and the reader is referred to the excellent review by Krejci and Dressler[1], together with the detailed account of detectors by Gudzinowicz[2]. The more frequently encountered detectors cited in this book are reviewed in detail.

The requirements of a detector are numerous, with the result that no single detector can be used effectively in all circumstances. One important property of the detector is that it should have a predictable response, and ideally this should be linear for all compounds over the whole range of concentrations encountered in GC. Unfortunately, no detector has such an ideal behaviour, although a few do approach this condition. A means of expressing this detector parameter is by the determination of its linear dynamic range. This is determined numerically by dividing the maximum obtainable signal with the detector by the minimum detectable signal, over the range for which the response is linear with respect to sample size. The larger the linear dynamic range, the better the detector.

Other important parameters are sensitivity, stability, response time and the detector signal-to-noise ratio. Sensitivity and stability are self-explanatory. The response time for the detector should be rapid to avoid peak distortions and poor resolution. The response time is dependent on the detector dead volume, the time constant of the associated equipment and the speed of response of the sensing element. A detector must have a high signal-to-noise ratio, so that the signal from the sample is clearly distinguishable from that of the background.

The purpose of the detector is to offer the means of measuring the component concentration and to provide a means of obtaining retention data for the purpose of establishing the identity of the component. For routine qualitative work most of the non-selective detectors are satisfactory, but for the analysis of trace constituents the choice is more limited. Compounds containing specific functional groups may often respond to detectors whose behaviour is selective to these moieties.

Detectors

Common detectors are those of flame ionisation (FID), thermal conductivity (TC) and electron capture (EC), of which the FID is probably the most widely used. The FID responds only to organic compounds, giving little response to water or carbon disulphide. It possesses good stability, and has a linear response over six orders of magnitude, with a detection limit of $10^{-9}-10^{-10}$ g. The TC detector is universal, accommodates large sample sizes, is non-destructive and, in contrast to the FID, is used in the analysis of water and inorganic compounds. The EC detector is selective towards molecules containing the electronegative atoms, nitrogen, oxygen, sulphur and, particularly, halogens. It requires a clean analytical technique to avoid contamination of the radioactive ionisation source, or otherwise it needs frequent cleaning. The detection limit of the EC detector is of the order of $10^{-9}-10^{-12}$ g, whereas that of the TC detector is approximately 10^{-7} g.

Table 3.1 compares important characteristics of some of the detectors used in the GC of food components.

B. Thermal conductivity detector (TC detector)

The TC detector is a widely used, non-destructive detector, often referred to as the katharometer or hot-wire detector. Its response is a function of the difference between the thermal conductivity of pure carrier gas and that of the eluted components in the gas stream.

In the basic detector a heated thin-wire filament or thermistor (semiconductors of fused metal oxides) is positioned in the path of the effluent gas from the column, to act as a detecting element. This forms one arm of a Wheatstone bridge, which is adjusted to balance when pure carrier gas flows. When a sample component is eluted from the column, the thermal conductivity of the gas in the detector cell changes, which causes a change in the temperature and resistance of the detecting element, and, hence, a change in the electrical current flowing in the bridge circuit. Thus the presence of a component is detected when the bridge circuit goes out of balance, and the amount of component is given by the extent of imbalance.

Unfortunately, external temperature fluctuations upset the bridge balance in this simple detector, and give rise to an unstable baseline. This bridge imbalance is reduced by incorporating one other arm of the Wheatstone bridge in a pure carrier gas stream adjacent to the other arm to serve as a reference. Both arms are incorporated in the same metal block detector assembly. Usually, the reference arm is located in the carrier gas stream prior to the sample introduction, while the measuring arm is located at the exit of the column.

The circuitry for incorporating the two detecting elements in a Wheatstone bridge is shown in *Figure 3.1 (a)*[3]. Many thermal conductivity cells use four elements in the detector block to give twice the sensitivity, and to increase the stability. The circuitry for this arrangement is shown in *Figure 3.1 (b)*[3].

Gases of high thermal conductivity, e.g. hydrogen and helium, are preferred to give better sensitivity and linearity of response, since most solutes have low thermal conductivity, and the detector, in effect, measures differential conductivity. Other factors determine the detector sensitivity, e.g. cell geometry, the change in the electrical current in the detecting elements and the detector

Table 3.1 Some detector characteristics

Detector	Type	Sensitivity/gs^{-1}	Linear dynamic range	Approx. upper temp. limit/°C	Responsive molecules or compounds
FID	universal	5×10^{-13}	10^7	400	combustible organic compounds
AFID	selective	5×10^{-13}	10^3	300	combustible phosphorus- and nitrogen-containing compounds
EC	selective	10^{-14}	5×10^2	220 (^{3}H) 350 (^{63}Ni)	electron-absorbing molecules
AI	universal	10^{-13}	10^6	300	molecules, with ionisation potential <11.4 eV
Helium ionisation	universal	10^{-14}	5×10^3	100	molecules, with ionisation potential <19.6 eV
Cross-section	universal	10^{-8}	10^4	300	all compounds
Gas density	universal	10^{-7}	10^3	-	all compounds

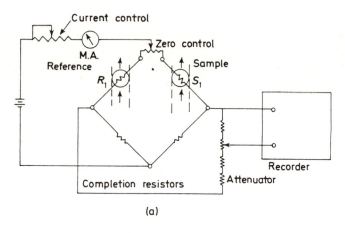

(a)

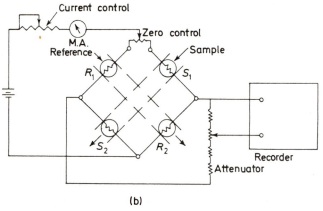

(b)

Figure 3.1. Wheatstone bridge circuit for TC cells (after Lawson and Miller[3]); (a) 2-element; (b) 4-element

block temperature. The latter should be sufficiently high to prevent condensation in the detector.

Sensitivity is increased with hot-wire filaments by increasing the filament current, but the life of the filament is then shortened. Usually, a correspondingly lower baseline stability is also obtained with an increase in filament current. The maximum sensitivity with thermistors is obtained at specific bridge currents, and the manufacturer's recommendations should be carefully observed.

Detector cell geometries can be divided into three main categories, i.e. flow-through, semi-diffusion and diffusion, as shown in *Figure 3.2*[3]. Flow-through detector cells have a rapid response time, but they are sensitive to flow-rate fluctuations. Conversely, diffusion cells have a long response time, which may

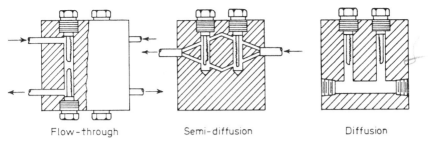

Figure 3.2. Typical detector block geometries (after Lawson and Miller[3])

result in anomalous response, but they show good stability to flow variations. As a compromise, the semi-diffusion type of cell is generally accepted for analytical purposes.

The choice of detecting elements, i.e. thermistors or hot-wire filament, depends upon the particular application. Thermistors are preferred at low temperatures, particularly subambient, where their sensitivity is greatest. Thermistors are also used for the analysis of thermally unstable materials, However, thermistor detectors are usually less stable than hot-wire filaments and they are generally not recommended except for specific applications.

The internal cell volume of TC detectors is generally about 2 ml and, hence, they are unsatisfactory for use with high-resolution columns, e.g. capillary columns. This limitation can be overcome by the reduction of the internal volume, e.g. TC detectors, where the volumes used vary from 100 μl down to 3 μl[4-6]. Commercial examples are available, such as the Gow-Mac microcell[7], which has an internal volume of 80 μl, but the modifications proposed by Farre-Rius and Guiochon[8] were necessary to make it suitable for use with capillary columns.

There have been many publications which have reviewed the design and characteristics of TC detectors, of which the most detailed was that by Lawson and Miller[3].

A TC detector can detect all materials including the permanent gases, but as it is sensitive to changes in temperature and flow rate, it is difficult to obtain accurate quantitative results in applications involving programming. Nevertheless, it has many advantages, viz. simplicity, robustness, a low cost, versatility and sufficient sensitivity for many applications. Its response is linear for most components using hydrogen or helium carrier gas, having a linear dynamic range of approximately 10^4.

C. Gas density balance and mass detector

The gas density balance or meter is a non-destructive detector, which measures a property which may be readily calculated, viz. the function of the density of the carrier gas and the sample vapour passing through it.

The detector is illustrated in *Figure 3.3*[9], and is constructed in such a way that sample components in the column effluent cause density variations in the upper and lower reference arms, producing a flow-rate differential. Two sensing elements are located in the balance arms of the detector, either hot-wire filaments

47

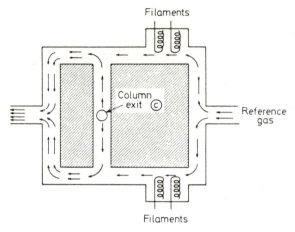

Figure 3.3. Gas density balance detector (after Swingle[9])

or thermistors, which are part of an electrical Wheatstone bridge, the resistance of which increases with increasing flow-rate owing to cooling. Hence, a flow rate differential in the detector balance arms arising from the arrival of a sample component produces a resistance change in the measuring bridge circuit and a chromatographic peak.

The detector was evaluated by Guillemin and Auricourt[10,11] and Guillemin, Auricourt and Vermont[12]. It was shown that the response of the detector was a function of the nature of the carrier gas, of the temperature and of the ratio of the flow rates in the balance arms. Optimum sensitivity was achieved by operating at the highest ratio of flow rates in the reference arm to those in the measuring arm, and the sensitivity was at least equal to, and in some cases superior to, that of TC detectors utilising the same sensing elements. Its linearity was also similar to that of the TC detector.

The balance is robust, stable, reliable and versatile, e.g. capable of operation with any carrier gas. It is non-selective, permitting analysis of all gases or vaporised compounds. The main property of the balance is its predictable response. In the linear range the quantity of solute, Q, is, in effect, proportional to its peak area A, corrected by a factor referred to as the proportionality factor, f, which depends, primarily, on the molecular weights of the solute and the carrier gas:

$$Q = fA \quad \text{and} \quad f = \frac{M_s}{M_s - m_g} \tag{3.1}$$

where M_s = molecular weight of the solute and m_g = molecular weight of the carrier gas.

This property is valid for any gaseous or vaporised solute, irrespective of the carrier gas, and provides the rare opportunity of providing both quantitative analytical information and information regarding molecular weights.

Walsh and Rosie[13] showed that reliable molecular weight measurements could be made with the gas density balance and that the accuracy of such measurements could be optimised by the careful choice of the reference gas for the molecular weight range of a given sample, viz. He, 10–50; N_2, 40–125; and SF_6, 160–300.

A GC detector based on the measurement of weight changes was devised by Bevan and Thorburn[14], and the performance characteristics of this mass detector were reviewed by Bevan, Gough and Thorburn[15]. The detector comprised an automatic recording electro-microbalance, to the arm of which was attached a small cylindrical chamber containing activated charcoal to adsorb the components eluted from the chromatographic column. The charcoal had the capacity to enable the adsorption of components from many runs before regeneration was required. The response of the detector was a function of the weight of the material adsorbed by the charcoal, and was both linear and integral. Hence, quantitative information was obtained by direct measurement without the need for integration or qualitative information. The limit of detection was found to be similar to that of a conventional hot-wire detector. The use of this detector for quantitative analysis was demonstrated by Bevan, Gough and Thorburn[16], who showed that results could be obtained with a precision of 1%.

A comparison of the mass detector and the gas density detector for quantitative analysis was made by Bevan, Gough and Thorburn[17]. They concluded that for all the materials examined the gas density detector gave a response close to the calculated value over a small concentration range and, within this range, could give excellent quantitative results. It was very stable, but had rather a long response time, whereas the mass detector, under the same conditions, gave a significantly better precision and accuracy.

The combination, in series, of the gas density balance and the mass detector was used by Bevan and Thorburn[18] and Bevan, Gough and Thorburn[19] to determine the molecular weights of the constituents of an unknown mixture within a certain molecular range in a single analysis. The appropriate molecular range was determined using the molecular weight of the carrier gas. With nitrogen as a carrier gas, satisfactory molecular weights were obtained over the range 18–100 for multicomponent mixtures. The advantage of the determination of molecular weights by GC is that pure materials are not necessary.

Molecular weight chromatography is a relatively new technique which could prove useful for both qualitative and quantitative analysis. Swingle[9] described the use of two gas density detectors in a dual chromatographic system, utilising two carrier gases of widely different molecular weights, namely SF_6 and CO_2, for molecular weight chromatography. The technique was applicable to most samples, which might normally be analysed using TC detection.

D. Flame ionisation detector (FID)

The FID is one of the most useful detectors in GC, because of its high sensitivity, good stability and wide range of linearity of response. It is, however, destructive and so cannot be used in-line with any ancillary technique, e.g. sample trapping. This particular problem may be overcome by the use of a stream-splitter at the column exit, in which a small proportion is passed to the FID and the bulk diverted to the ancillary technique.

Although there are a number of different designs of FID, the basis of operation is similar, viz. the ionisation of organic molecules by burning them in a flame. The eluant from the column exit is mixed with hydrogen, and the mixture is burnt at the detector jet in an atmosphere of air fed from a gas supply. In addition to the flame jet, the detector contains two electrodes across which a potential is applied. The flame jet is often employed as an electrode, and a metallic loop or cylinder surrounding the flame serves as the other collector electrode. Thus the thermally induced ionisation of an eluted component changes the electrical resistance of the flame, and the resulting increase in the current at the collection electrode is amplified and fed to the recorder. The FID response is directly proportional to the number of carbon atoms in the molecule which are bound directly to hydrogen or other carbon atoms. Consequently, the detector responds to all organic molecules, but is insensitive to water, carbon disulphide, most inorganic compounds and permanent gases.

It is generally considered that, as carbon disulphide has only a low or zero response with the FID, it should be the solvent of choice for all analyses using this particular detector. However, it has been reported that the flame ionisation response to carbon disulphide is dependent on the flow rates of the gases[20] and on the detector geometry[21].

The use of argon or helium as the carrier gas, or the use of oxygen instead of air for combustion, are means of lowering the limit of detection for the FID. Hoffmann and Evans[22] increased the sensitivity of the FID by hyperoxygenation of the chromatographic effluent before its combustion. Hyperoxygenation was accomplished by bleeding pure oxygen into the hydrogen line close to the flame jet, causing it to burn with the intense heat of an oxyhydrogen torch, lowering the carrier gas thermal conductivity and thereby increasing the detector response[23].

E. Alkali flame ionisation detector (AFID)

Several modifications have been made to the basic FID for the specific determination of certain elements. One such modification gave rise to the AFID, also called the 'thermionic detector' or 'sodium detector'. In this detector the ionisation processes in the flame were modified by the presence of an alkali metal, as this was found to bring about an enhancement of the detector response towards certain hetero-atoms, e.g. phosphorus, this enhancement being about 1000 times. However, the mechanism for this enhanced response has not yet been satisfactorily explained, and has been the subject of some divergent opinion.

The AFID can respond in a positive or negative fashion to hetero-atoms, but for both modes of response this is proportional to the amount of hetero-atom introduced into the flame, irrespective of the rest of the organic molecule, e.g. the response to phosphorus is in the positive mode. The other hetero-elements for which enhancement has been observed are halogens, nitrogen and sulphur.

There are many AFIDs of different design, which vary mainly in the manner of the location of the alkali metal salt and in the detector geometry, both of which affect the detector response. The alkali metal salt was originally coated on to a coil, loop, screen or porous metal, and heated either by the flame or independently by electrical means. Greater stability, faster equilibration and a longer life were obtained when a tip of pressed alkali salt containing an orifice was sited on the flame jet such that the flame burnt on the surface of the tip above the orifice through which the gases passed. This design has formed the basis of several commercial detectors, usually with a tip composed of caesium bromide as the alkali salt, and compressed with a diatomaceous earth as a binder. Alternative binders have been used, including alumina.

A three-electrode AFID is produced by the Pye Co. Ltd., in which the alkali metal salt (caesium bromide or rubidium chloride) is compressed with a binder into the shape of a tube. The salt tube is inserted in a metal tube-holder which acts as the third electrode, together with the burner and collector electrodes. The use of this system resulted in an enhancement of stability, sensitivity and select-ivity, compared with the basic two-electrode system. It is suitable for use as both a phosporus or a nitrogen detector[24,25].

Many other AFID designs have been evaluated, and the reader is referred to the excellent review of Brazhnikov, Gurev and Sakodynskii[26], which gives a detailed account of the design, evolution, characteristics and applications of this detector. The review is particularly useful to those analysts wishing to con-struct and experiment with their own detectors.

The response of the AFID detector to a particular hetero-atom depends on a series of parameters, viz. the composition and type of alkali salt, the air, hydrogen and carrier gas flow rates, the distance between the collector electrode and the flame, the polarity of the electrodes, the diameter of the jet and the temperature of the flame and its surroundings. The alkali metal and the hydrogen flow rate have the greatest influence on the response. Most of these parameters are predetermined in commercial detectors, and usually the analyst need only be concerned with the adjustment of flow rates, which depend on the individual design.

The response of an AFID to compounds which contain hetero-atoms can be negative under certain conditions. Lakota and Aue[27] used the negative mode of the AFID for chlorine-containing compounds in the detection of organo-chlorine pesticide residues, and thus eliminated the need of an excessive puri-fication procedure to remove the co-extracted material. In terms of sensitivity, the AFID was inferior to the EC detector, but provided a qualitative indication of chlorine present in trace quantities in a compound. The negative mode of operation was obtained by establishing the necessary flow conditions for hy-drogen, air and carrier gas, together with the selection of an appropriate alkali source.

A systematic study of the various parameters influencing the negative response of the AFID was made by Gerhardt and Aue[28], including the bore of the tip, hydrogen and carrier gas flow rates, and electrode height. In general, a large bead bore, a high hydrogen flow rate and a low carrier gas flow rate favoured negative response. A stronger response in the negative mode than in the positive mode was obtained for both halogens and carbon, but the main potential of the technique lay in the ability to distinguish between several hetero-atoms in organic compounds.

The response of sulphur-containing compounds with the AFID using pot-assium, rubidium or caesium salt tips is also negative. Dressler and Janak[29] observed the dependence of the response to sulphur-containing compounds on the background ionisation current. They found that it was possible to select a range of background ionisation current in which the AFID response to phosphorus, nitrogen and halogens, as well as to carbon, was positive, whereas the response to sulphur was negative.

Aue *et al.*[30] described an AFID for nitrogen-containing compounds, which used a rubidium sulphate bead and a collector electrode consisting of a movable, single platinum loop, to enable the detector to discriminate against other hetero-atoms. The detector has an increased response to nitrogen compounds of two to three orders of magnitude compared with the normal FID. In addition, halides were found to be best determined at low hydrogen flow rates and very low electrode position with respect to the rubidium sulphate bead. Other detectors for nitrogen-containing compounds using rubidium sulphate were described by Hartmann[31] and Craven[32].

F. Flame photometric detector (FPD)

The FPD is a flame emission detector which was described in detail by Juvet and Durbin[33] and by Brody and Chaney[34].

The FPD of Brody and Chaney has extremely high selectivity for phosphorus and high selectivity for sulphur, together with adequate sensitivity. This makes it particularly suitable for analyses in which the sample solution contains large amounts of co-extractives, the chromatograms produced being virtually free from background interference. Thus the application of this detector to the analysis of pesticide residues in foodstuffs has the advantage that there is no need for extensive clean-up prior to chromatography owing to the high specificity. Besides the analysis of pesticide residues, other sulphur determinations are of importance in the food industry. Sulphur-containing compounds are often found in small concentrations as contaminants, and the FPD is the only detector sensitive enough for their direct determination without their prior concentration. The detector can also be made specific to other hetero-atoms, but this can only be attained by sacrificing its high sensitivity. The minimum detectable quantities of sulphur and phosphorus with the Brody and Chaney detector are 200 pg and 40 pg, respectively.

The FPD is, virtually, a combination of a FID and an optical system consisting of a filter and a very sensitive low-noise photomultiplier, to monitor the chemiluminescent emission above the hydrogen-rich, hydrogen–air flame. The optical filter used for the determination of phosphorus has a transparency maximum at 526 nm, while the filter for sulphur determination has a similar maximum at 394 nm. Both filters are narrow-band interference filters.

This detector is noted for its stability, requiring infrequent calibrations and adjustments, this stability being probably due to its insensitivity towards pressure and temperature fluctuations.

Several modifications have been described to overcome the major disadvantage of the self-extinguishing of the flame after sample injection. Fetzer[35] suggested a modification to the igniter, which would allow the flame to stay alight if the igniter button was kept depressed during the passage of the solvent. Moye[36] eliminated the need to relight the flame, even after extremely large injections, by a modification to the burner and housing assembly.

A dual-headed detector based on the dual detector of Bowman and Beroza[37] was introduced by Melpar, Inc. It employed two photomultipliers to monitor the flame, and by using three electrometers the system provided a flame ionisation response which was selective for both phosphorus and sulphur. The response characteristics of this detector were discussed by Grice, Yates and David[38], and since there was mutual interference of phosphorus and sulphur at their respective wavelengths, data were presented for determining the quantities of sulphur and phosphorus that contributed to the response when both were present in the sample.

A rugged and operationally simple FPD based on the Beilstein effect, i.e. a green flame emitted by halogens in the presence of copper, was described by Bowman and Beroza[39]. The detector consisted of a photomultiplier tube and 526 nm interference filter to monitor the light emitted when halogen-containing compounds were burnt in a hydrogen–oxygen flame just above a 30-mesh copper screen. It gave an exponential response for chlorine-, bromine- and iodine-containing compounds and a linear response to fluorine- and non-halogen-containing compounds, and was therefore not selective. However, the stability and sensitivity of the detector was good, and the exponential response could serve to recognise chlorine-, bromine- and iodine-containing compounds. A particular application of this detector is in the analysis of pesticide residues, to verify any that have been indicated by EC detection. Thorough clean-up was required to ensure that the extracts would not deactivate the copper screen in the detector.

This detector was later modified by Bowman, Beroza and Nickless[40] by replacing the copper screen with an indium-coated stainless steel one, and by monitoring the emission at 360 nm. Although noisier, it was more sensitive to halogen-containing compounds than was the copper-sensitised detector, and was not subject to interference by extracts that were given only a partial clean-up.

A combination of the indium-sensitised flame photometric detector for chlorine-containing organic compounds with the standard FPD for phosphorus- and sulphur-containing organic compounds was constructed by Versino and Rossi[41]. The detector was described as a dual-flame photometric detector (DFPD), and, with the aid of three-channel operation, allowed the simultaneous and selective determination of phosphorus-, sulphur- and chlorine-containing organic compounds. Although the sensitivity of this combination was not as good as that of the EC detector and AFID, it had better linearity and specificity, thus allowing more concentrated samples to be used, since co-extracted substances did not cause interference.

G. Emission spectrometric detector

The GC detection of organic compounds containing sulphur, phosphorus and halogens by emission spectrometry in a microwave-powered inert gas plasma was devised by McCormack, Tong and Cooke[42]. The detector is available commercially and is referred to as the microwave plasma detector.

The detector was based on monitoring the intensity of the electron emission spectra of the eluted organic compounds in an argon carrier gas. The spectra were excited in the plasma generated from a 2540 MHz electrodeless discharge, and detected photoelectrically in the UV-visible region. The detector has good sensitivity, rapid response, a small volume and a wide linear dynamic range (approximately four orders of magnitude), and by monitoring at the wavelength of various atomic lines and molecular bands some selectivity could be obtained. Argon was used both as the carrier gas and for sustaining the plasma discharge, since it was found to give a stable discharge at atmospheric pressure when used in conjunction with a small-diameter quartz capillary discharge tube. This avoided the complexity of a vacuum system associated with the use of a low-pressure discharge in helium, although helium was the gas chosen for later applications owing to its having fewer lines in its spectrum than argon. Helium was also found to initiate atomic emission in sulphur, chlorine and bromine, whereas the production of atomic lines for these elements was not observed in an argon plasma. Irrespective of the gas chosen, the discharge had to be initiated by means of a spark coil. The discharge was removed once the self-sustaining plasma had formed.

McCormack et al.[42] showed that the response was dependent on the carrier gas flow and on the column temperature. However, Moye[43] found that a mixture of 85% He and 15% Ar was insensitive to these parameters and that this composition of gas was the optimum for detector selectivity and signal-to-noise ratio for phosphorus detection.

Detection by means of emission spectrometry is based on the same principle as that of FPD, and the factors influencing the selectivity of the latter essentially hold for emission spectrometry. The selectivity is influenced by overlapping of the spectra, by source background, monochromator dispersion and the slit width, e.g. the narrower the width of the slit, the higher the selectivity and the lower the sensitivity. However, the sensitivity of this detector is higher than that of the FPD.

This detector was applied to the determination of pesticide residues in agricultural and food samples by Bache and Lisk[44-47], for both iodine- and phosphorus-containing compounds. Although the pesticide extracts also contained a large quantity of sample material, the chromatograms produced showed only the solvent and pesticide peaks, owing to the high selectivity of the detector towards these compounds.

Bache and Lisk[48] also applied this detector to the determination of organic mercury compounds in fish. It is the only existing selective GC detector for mercury compounds, which is particularly desirable owing to the great need for a specific and sensitive means of detecting these compounds.

H. Electrolytic conductivity detector

The operation of conductometric detectors is based on the measurement of the electrolytic conductivity of water. Organic substances in the column effluent are combusted, the combustion products are dissolved in water and the conductivity of the water is measured. The technique was developed and made specific for halogen-, sulphur- and especially nitrogen-containing compounds by Coulson[49,50], and consequently the detector has become known as the

Coulson conductivity detector (CCD). For this specific mode of operation, the column effluent was combusted under oxidative or reductive conditions, interfering combustion products being selectively removed and the conductivity of the water containing the component of interest being measured.

Compounds were burnt in a stream of oxygen or air over a platinum catalyst, and a response to sulphur- and halogen-containing compounds was produced owing to the absorption of the combustion products, SO_2, SO_3 and HCl, by the water. The other combustion product, CO_2, was not absorbed quickly by water, and the response to carbon was therefore practically nil. The sensitivity was approximately 10^4 times higher for sulphur- and halogen-containing compounds than for compounds of carbon, hydrogen, oxygen and nitrogen. Additional specificity was obtained by either removing the sulphur oxides with calcium oxide on quartz wool at 800°C or removing the hydrogen chloride with silver nitrate on quartz wool at 700°C.

With reducing conditions over a nickel catalyst at 800°C, ammonia was produced from nitrogen-containing compounds, hydrogen sulphide from sulphur-containing compounds, hydrogen chloride from organochlorine compounds and water from oxygen-containing compounds. The detector responded to the nitrogen- and chlorine-containing compounds, but the other compounds were not detectable, since they produced only a weak signal. A scrubber of strontium hydroxide on quartz wool was used to enhance the specificity towards nitrogen-containing compounds, as this removed all the acidic combustion products.

The standard commercial detector is based on Coulson's design and consists, essentially, of the following parts: a combustion furnace containing a quartz reactor tube coupled to the GC effluent, a gas–liquid contactor and phase separator, a measuring cell, a water deionisation system and a pump to recycle the water. The deionisation system is required to take care of the standard quality of the water used for conductivity measurements, as this must remain constant. The quartz reactor tube contained the catalysts and also the scrubber material used for removing the interfering gases.

Patchett[51] introduced several refinements to the Coulson detector, and enabled the limit of detection of nitrogen to be lowered to about 0.1 ng, as opposed to the usual practical limit of 1 ng. The detector was used in the analysis of nitrogen-containing pesticide residues, mostly carbamates.

These refinements were incorporated in the detector used by Cochrane and Wilson[52] for the analysis of some nitrogen-containing herbicides, e.g. triazines, substituted-ureas and carbamates.

A possible procedure for the analysis of nitrosamines in various foods and beverages was suggested by Rhoades and Johnson[53]. The CCD was converted to an amine detector by the use of an inert gas, such as argon; an empty quartz tube was substituted for the catalyst reactor tube; and a furnace temperature of 400–600°C was used. In this temperature range ammonia was obtained as a degradation product of many amines and nitrosamines, whereas other organic nitrogen-containing compounds produced virtually none.

The response of such detectors is linear over a range of approximately two orders of magnitude. There are several factors which constitute disadvantages in this method of detection: the detector is destructive; quantitative results can only be obtained if qualitative information is available; and the selectivity of the detector is low, being increased only by indirect methods.

I. Microcoulometric detector (MCD)

Coulometry can be used for quantitative detection in GC, e.g. a non-selective detector could be produced by the pyrolysis of effluent material to carbon dioxide, which could then pass into a coulometric cell. More specific coulometric detectors basically consist of three parts, viz. a destructive unit where compounds from the GC column are either oxidised or reduced into simple inorganic compounds, a titration cell and a coulometer.

Coulson and Cavanagh[54] developed an automatic combustion and coulometric titration method for determining halogens in organic compounds, in which a titration cell connected to the exit of a combustion unit absorbed the halide produced from combustion in oxygen at 800°C, and automatically titrated it with silver ions generated by a coulometric system. The concentration of silver ions in the titration cell electrolyte was kept at a constant value by a variable electric current. The total silver ion concentration produced in the course of titration was directly proportional to the amount of halogen in the sample, and if the current was integrated as a function of time, an indication of the halogen content was automatically obtained from the integral, i.e. from the resulting chromatogram. For chlorine-containing compounds the minimum detectable amount was 2×10^{-9} g as chloride, but with some slight modification this has been lowered to about 1 ng. Bromine-containing compounds only gave half the expected response, since they were hydrolysed to HOBr in addition to HBr, and only the latter was precipitated by silver ions.

In addition to halogen, the oxidation mode was used by Coulson et al.[55] for the determination of sulphur in organic compounds by combustion to sulphur dioxide at 800°C. The silver ion cell used for hydrogen halides was only sensitive to compounds which could precipitate silver and included hydrogen sulphide and phosphine, but not sulphur dioxide. This fact enabled hydrogen halides, excluding fluoride, to be measured selectively. However, a modification was made in which the silver ion cell was replaced by a cell selective to sulphur, which had electrodes made of noble metal, and enabled the sulphur dioxide entering the cell to be automatically titrated by iodine to sulphate.

Although the main use of the MCD was originally in the analysis of organo-halogen and organosulphur compounds, it was modified and extended to determine phosphorus, by Burchfield and his co-workers[56-58], for application in pesticide residue analysis. In this system the detector was utilised in the reductive mode using molecular hydrogen at 950°C, thus forming phosphine from organophosphorus compounds, and hydrogen sulphide and hydrogen chloride from organosulphur and organochlorine compounds, respectively. All of these were titratable in the silver ion cell, and to avoid simultaneous examination, a tube of alumina was inserted after the furnace to remove H_2S and HCl when determining phosphorus, whereas a tube of silica gel removed HCl and separated PH_3 and H_2S. Thus the amounts of phosphorus and sulphur in one compound could be determined simultaneously.

The MCD is an extremely useful detector in residue analysis, as it is completely specific, being totally insensitive to other components, including stationary phase bleed, and yet its incorporation in other types of GC analysis appears to be rare. Its practical limits of detection are generally of the order of $0.01-0.1$ μg of the ions.

J. Electron capture detector (EC detector)

The EC detector is a non-destructive detector that utilises the ability of compounds to capture free electrons. Many types of compounds have a high electron-capturing capacity (electron affinity), which enables the EC detector to detect them at very low levels. It is this factor which accounts for the detector's excellent selectivity.

An ionisable carrier gas passes into an ionisation chamber containing a source of β-radiation, i.e. a copper or stainless steel foil coated on one surface usually with a thin layer of titanium in which tritium has been occluded (50–250 mC activity). A more recent commercial detector uses a source composed of tritium occluded to scandium. Alternatively, ^{63}Ni and ^{147}Pm may be used as the source[59]. The ionisation chamber contains electrodes across which is applied a potential. The potential chosen is just sufficient to collect all the free electrons that are produced as the result of the ionisation of the carrier gas molecules by the β-particles. The migration of the electrons to the anode produces a steady baseline current, sometimes referred to as the standing current, of about $10^{-8} - 10^{-9}$ A. When an electron-capturing solute enters the chamber, it reacts with the electrons to form stable negative molecular ions or charged particles, with the net result that electrons are removed from the system and the current flowing in the chamber is reduced, which is shown as a negative peak on the recorder trace.

Lovelock[60] demonstrated that the ionisation chamber could be used either at a fixed potential (polarisation voltage) or by applying a short pulse of potential to the electrodes (pulsed mode). Operation in the pulsed mode was found to increase the reproducibility, sensitivity and stability of the detector, being less affected by changes in flow rate and temperature. Consequently, it was more suitable for programmed techniques.

The carrier gases suitable for use in the pulsed mode are oxygen-free nitrogen and various combinations of gases, of which argon–methane combinations are preferred. Pure argon is unsuitable because it is readily converted to the metastable form (compare the argon ionisation detector, p. 58), but the addition of methane, hydrogen or carbon dioxide inhibits this formation. A combination of 5% methane in argon gives the best results. For the fixed potential (d.c. method), nitrogen is the preferred carrier gas.

In order to obtain the highest sensitivity for the EC detector, the operating conditions must be chosen carefully. For instance, it is necessary to consider the influence on the detector response of the nature of the stationary phase, the column temperature, the purity of the carrier gas, the polarising voltage (d.c. method), the pulse period (pulse method), the detector temperature and the baking time of the detector. In the d.c. method the flow rate through the detector must also be optimised with the help of a scavenger flow added to the carrier gas after the column in order to obtain the best stability as well as maximum sensitivity. All of these factors were considered in considerable detail by Devaux and Guiochon[61-63], with a view to obtaining the ultimate sensitivity from the EC detector.

Unfortunately, the EC detector has a very small linear dynamic range (approximately 10^3), and therefore it is not particularly suited to quantitative work, except in specialised fields, e.g. the analysis of halogen-containing pesticide residues. The linear dynamic range of the detector with a [63]Ni source is slightly less than that of the detector with a tritium source. The addition of a scavenger flow (purge gas) into the detector extends the linear dynamic range by diluting the sample, but this does, however, increase the noise level of the detector and reduces the sensitivity.

Contamination of the surface of the radioactive foil decreases the low-energy β-radiation available to ionise the carrier gas, thereby reducing the standing current produced at a given operating voltage and, hence, reducing the detector sensitivity. Contamination of the detector is also indicated by an increase in electrical noise, a decrease in the linear response range and the generation of false or erroneous responses. This contamination usually arises from stationary phase bleed and the non-volatile degradation material which deposits on to the active surface of the foil and other internal surfaces of the detector cell.

Various methods have been suggested for cleaning these detectors, including solvent washing with 5% alcoholic potash in an ultrasonic bath, and the use of a metal cleaning paste, e.g. Solvol—Autosol nickel chrome cleaner, followed by rinsing in hexane, as described by Holden and Wheatley[64]. This mild abrasive is applied to both the foil and the other parts of the detector, including the detector body.

In addition to its more obvious applications for the analysis of halogen-containing compounds, e.g. pesticide residues, the EC detector has been applied to the analysis of phosphorus-, oxygen-, sulphur- and nitrogen-containing compounds. Hoffsommer[65] recently used a [63]Ni detector for the analysis of nitro-compounds in the μg to pg range.

K. Other radioactive ionisation detectors

In addition to the EC detector, the other more frequently used radioionising detectors are the argon detector, the helium detector and the cross-section detector. All are non-destructive detectors.

The argon ionisation (AI) detector was at one time a very popular detector, but the advent of the EC detector and FID has considerably reduced the number of its applications. Argon is the carrier gas used with this detector, and, on entering the ionisation chamber, it is excited to a metastable state by high-energy β-particles from a ^{90}Sr source. When a solute whose ionisation potential is less than the energy of the metastable argon atoms enters the chamber, collisions result in the transfer of energy to the solute molecules, and lead to the formation of ions. This causes an increase in the current flowing across the detector, which is amplified and recorded. The applied potential in the ionisation chamber is about 500 V. Most organic compounds have ionisation potentials lower than the energy of metastable argon (11.5 eV), but water vapour, methane, oxygen, carbon dioxide, nitrogen, carbon monoxide, ethane, acetonitrile and fluorocarbons all have ionisation potentials greater than 11.5 eV, and therefore do not respond to this detector. Consequently, performance is impaired by the presence of water vapour or air in the carrier gas, and this is one of the disadvantages of this detector. The detector also has a small linear dynamic range, and calibration of all substances at any concentration level is necessary for quantitative work. It is very easy to overload the detector, although the upper limit of detection can be increased by using a mixed carrier gas of nitrogen and argon.

Lovelock made two improvements to the basic detector, a micro version[66], which has a smaller dead volume, suitable for use with capillary columns, and a triode detector[67] in which a third collecting electrode enabled the sensitivity to be increased a thousandfold.

Helium metastable atoms have a higher energy (19.6 eV) than those of argon, so that, by using helium as the carrier gas, the helium detector is able to sense virtually all compounds with very good sensitivity. In practice, the detector is usually used only for the trace analysis of permanent gases, where no other detector is adequate.

The cross-section detector consists of a radioionising chamber containing two electrodes across which a potential gradient is applied, and the carrier gas within the chamber is irradiated by a β-ray source. The passage of radiation through the gas, which is usually hydrogen, produces ion pairs which are collected by the applied potential across the electrodes. Increasing the polarising voltage increases the current to a saturation value, i.e. no ions recombine, so all ions are collected at the electrodes. The current is, however, significantly changed by the presence of a sample constituent from the column. The detector is very satisfactory for quantitative analysis, as it gives a linear response with respect to concentration over its entire operating range. The upper limit of detection is high, although the lower limit does not approach that of the argon ionisation detector. The detector is made by several manufacturers, and the construction of micro cross-section detectors has been described by Lovelock, Shoemake and Zlatkis[68, 69] and Abel[70].

A universal radioionising detector was described by Lasa, Owsiak and Kostewicz[71] which could be used as a cross-section, argon or EC detector, up to temperatures of 400°C.

References

1. Krejci, M. and Dressler, M., *Chromatog. Rev.,* **13**, 1 (1970)
2. Gudzinowicz, B.J., in *The Practice of Gas Chromatography* (Ettre and Zlatkis, Eds.), Wiley, New York, 239 (1967)
3. Lawson, A. E. and Miller, J. M., *J. Gas Chromat.,* **4**, 273 (1966)
4. Camin, D.L., King, R.W. and Shawhan, S.D., *Anal. Chem.,* **36**, 1175 (1964)
5. Petrocelli, J.A., *Anal. Chem.,* **35**, 2220 (1963)
6. Preau, G. and Guiochon, G., *J. Gas Chromat.,* **4**, 343 (1966)
7. Farre-Rius, F. and Guiochon, G., *J. Gas Chromat.,* **1** (November), 33 (1963)
8. Farre-Rius, F. and Guiochon, G., *J. Chromatog.,* **13**, 382 (1964)
9. Swingle, R.S., *Industrial Res.,* **14**, 40 (1972)
10. Guillemin, C.L. and Auricourt, F., *J. Gas Chromat.,* **2**, 156 (1964)
11. Guillemin, C.L. and Auricourt, F., *J. Gas Chromat.,* **4**, 338 (1966)
12. Guillemin, C.L., Auricourt, F. and Vermont, J., *Chromatographia,* **1**, 357 (1968)
13. Walsh, J.T. and Rosie, D.M., *J. Gas Chromat.,* **5**, 232 (1967)
14. Bevan, S.C. and Thorburn, S., *J. Chromatog.,* **11**, 301 (1963)
15. Bevan, S.C., Gough, T.A. and Thorburn, S., *J. Chromatog.,* **42**, 336 (1969)
16. Bevan, S.C., Gough, T.A. and Thorburn, S., *J. Chromatog.,* **43**, 192 (1969)
17. Bevan, S.C., Gough, T.A. and Thorburn, S., *J. Chromatog.,* **44**, 14 (1969)
18. Bevan, S.C. and Thorburn, S., *Chem. Brit.,* **1**, 206 (1965)

19. Bevan, S.C., Gough, T.A. and Thorburn, S., *J. Chromatog.*, **44**, 241 (1969)
20. Walker, B.L., *J. Gas Chromat.*, **4**, 384 (1966)
21. Dressler, M., *J. Chromatog.*, **42**, 408 (1969)
22. Hoffmann, R.L. and Evans, C.D., *J. Gas Chromat.*, **4**, 318 (1966)
23. Hoffmann, R.L. and Evans, C.D., *Science, N.Y.*, **153**, 172 (1966)
24. Speakman, F.P. and Waring, C., *Column*, **2** (3), 2 (1968)
25. Swan, D.F.K , *Column*, No. 14, 9 (1972)
26. Brazhnikov, V.V., Gurev, M.V. and Sakodynskii, K.I., *Chromatog. Rev.*, **12**, 1 (1970)
27. Lakota, S. and Aue, W.A., *J. Chromatog.*, **44**, 472 (1969)
28. Gerhardt, K.O. and Aue, W.A., *J. Chromatog.*, **52**, 47 (1970)
29. Dressler, M. and Janak, J., *J. Chromat. Sci.*, **7**, 451 (1969)
30. Aue, W.A., Gehrke, C.W., Tindle, R.C., Stalling, D.L. and Ruyle, C.D., *J. Gas Chromat.* **5**, 381 (1967)
31. Hartmann, C.H., *J. Chromat. Sci.*, **7**, 163 (1969)
32. Craven, D.A., *Anal. Chem.*, **42**, 1679 (1970)
33. Juvet, R.S. and Durbin, R.P., *Anal. Chem.*, **38**, 565 (1966)
34. Brody, S.S. and Chaney, J.E., *J. Gas Chromat.*, **4**, 42 (1966)
35. Fetzer, L.E., *Bull. Environ. Contam. Toxicol.*, **3**, 227 (1968)
36. Moye, H.A., *Anal. Chem.*, **41**, 1717 (1969)
37. Bowman, M.C. and Beroza, M., *Anal. Chem.*, **40**, 1448 (1968)
38. Grice, H.W., Yates, M.L. and David, D.J., *J. Chromat. Sci.*, **8**, 90 (1970)
39. Bowman, M.C. and Beroza, M., *J. Chromat. Sci.*, **7**, 484 (1969)
40. Bowman, M.C., Beroza, M. and Nickless, G., *J. Chromat. Sci.*, **9**, 44 (1971)
41. Versino, B. and Rossi, G., *Chromatographia*, **4**, 331 (1971)
42. McCormack, A.J., Tong, S.C. and Cooke, W.D., *Anal. Chem.*, **37**, 1470 (1965)
43. Moye, H.A., *Anal. Chem.*, **39**, 1441 (1967)
44. Bache, C.A. and Lisk, D.J., *Anal. Chem.*, **37**, 1477 (1965)
45. Bache, C.A. and Lisk, D.J., *Anal. Chem.*, **39**, 786 (1967)
46. Bache, C.A. and Lisk, D.J., *Anal. Chem.*, **38**, 1757 (1966)
47. Bache, C.A. and Lisk, D.J., *Anal. Chem.*, **38**, 783 (1966)
48. Bache, C.A. and Lisk, D.J., *Anal. Chem.*, **43**, 950 (1971)
49. Coulson, D.M., *J. Gas Chromat.*, **3**, 134 (1965)
50. Coulson, D.M., *J. Gas Chromat.*, **4**, 285 (1966)
51. Patchett, G.G., *J. Chromat. Sci.*, **8**, 155 (1970)
52. Cochrane, W.P. and Wilson, B.P., *J. Chromatog.*, **63**, 364 (1971)
53. Rhoades, J.W. and Johnson, D.E., *J. Chromat. Sci.*, **8**, 616 (1970)
54. Coulson, D.M. and Cavanagh, L.A., *Anal. Chem.*, **32**, 1245 (1960)
55. Coulson, D.M., Cavanagh, L.A., de Vries, J.E. and Walther, B., *J. Agric. Food Chem.*, **8**, 399 (1960)
56. Burchfield, H.P., Johnson, D.E., Rhoades, J.W. and Wheeler, R.J., *J. Gas Chromat.*, **3**, 28 (1965)
57. Burchfield, H.P., Rhoades, J.W. and Wheeler, R.J., *J. Agric. Food Chem.*, **13**, 511 (1965)
58. Burchfield, H.P. and Wheeler, R.J., *J. Ass. Off. Anal. Chem.*, **49**, 651 (1966)
59. Lubkowitz, J.A. and Parker, W.C., *J. Chromatog.*, **62**, 53 (1971)
60. Lovelock, J.E., *Nature*, **189**, 729 (1961)
61. Devaux, P. and Guiochon, G., *J. Gas Chromat.*, **5**, 341 (1967)
62. Devaux, P. and Guiochon, G., *J. Gas Chromat.*, **7**, 561 (1969)
63. Devaux, P. and Guiochon, G., *J. Chromat. Sci.*, **8**, 502 (1970)
64. Holden, A.V. and Wheatley, G.A., *J. Gas Chromat.*, **5**, 373 (1967)
65. Hoffsommer, J.C., *J. Chromatog.*, **51**, 243 (1970)
66. Lovelock, J.E., *Nature*, **182**, 1663 (1958)
67. Lovelock, J.E., *Anal. Chem.*, **33**, 162 (1961)
68. Lovelock, J.E., Shoemake, G.R. and Zlatkis, A., *Anal. Chem.*, **35**, 460 (1963)
69. Lovelock, J.E., Shoemake, G.R. and Zlatkis, A., *Anal. Chem.*, **36**, 1410 (1964)
70. Abel, K., *Anal. Chem.*, **36**, 954 (1964)
71. Lasa, J., Owsiak, T. and Kostewicz, D., *J. Chromatog.*, **44**, 46 (1969)

4

Sampling and Sample Derivatisation

A. Introduction

Apart from the conventional method of introducing a sample to a GC, viz. the injection of a solution with a syringe, many other sampling techniques are used to cope with the variety of samples often encountered. These include the use of such devices as solid samplers and gas sampling valves as well as procedures for the sampling of head space vapours. There are sampling techniques that involve the use of precolumns to remove co-extractives, which may otherwise cause contamination of the column or interfere with the analysis. In addition, there are many other techniques that are used to isolate components prior to GC analysis.

Besides the chemical reactions involved in the extraction, and isolation and clean-up of sample components, there are many other ways in which reactions can be employed advantageously in GC. Chemical reactions can be used to modify a compound to increase its volatility, to increase the detector's sensitivity towards it, to reduce its activity towards the column packing material and to establish its identity. This requires the preparation of a suitable derivative of the compound, the procedure being referred to as 'derivatisation', examples of which are reviewed in the section on sample derivatisation.

A volatile compound, via its functional group, may be caused to react with a reagent to produce a non-volatile derivative. Consequently, a chromatogram obtained after derivatisation would indicate the loss of a peak, in comparison with that obtained before derivatisation, and substantiate the position of the volatile compound. This procedure is often referred to as a 'subtractive technique'. Alternatively, separated compounds emerging from a GC column can be caused to react with suitable reagents to show characteristic visual changes, e.g. precipitation or colour change.

B. Sampling

1. Gas sampling

Gases are usually flushed directly into the GC column via some type of sampling loop system, which often incorporates a multiport valve arrangement, frequently

61

referred to as a gas sampling valve. Such a valve was described by Hamilton[1], in which either small or large samples could be introduced by changing the position of the valve to connect either one or two sampling loops into the carrier gas flow.

There are many different types of sampling loop system, varying considerably in their structure and complexity, and the choice often depends on the specific problem. Some gas sampling loops, which are attached to the GC, are used to sample *in situ*, while others are used to collect the sample elsewhere and then brought to the GC system for analysis.

Besides the gas sampling loops, gases may be sampled with the aid of large-volume gas-tight syringes, and specially toughened syringe needles are available for direct insertion into the head space of food cans, etc., without opening the container. Such a procedure was used by Novak, Gelbicova-Ruzickova and Wicar[2] for taking representative liquid or gaseous samples from moderately pressurised containers, e.g. canned beverages.

The direct analysis of head space vapours over food products by GC eliminates the artefact formation that can occur with concentration procedures. Consequently, it is the easiest and most convenient way to study the volatiles of flavour and aroma. Normal procedures involve the use of a container fitted with a rubber stopper or septum which is penetrated with a syringe needle for sampling[3-5].

The head space technique was used by Sinclair *et al.*[6] for the determination of dimethyl sulphide in beer. The sample was chilled at 0°C and mixed with NaCl, and n-butanol was added as an internal standard. The mixture was set aside at 30°C for 1.5 h before sampling the head space vapour.

Fore, Rayner and Dupuy[7] and Dupuy, Rayner and Fore[8] used head space technique for the determination of residual isopropanol and acetone in oilseed meals and flour.

For the detection of trace components some advantage may be obtained by using the cryogenic injection technique of Rushneck[9], in combination with large head space vapour samples[10]. In this procedure successive injections are made into the GC, and the sample is condensed at subambient temperatures in the initial portion of the column. After temperature programming, the trace components are thus accumulated in amounts sufficient for detection.

Commercial electropneumatic injection systems are available for use in automatic head space analysis in food control.

2. Solid sampling

Samples, both solid and liquid, are subjected to GC after dissolving the material in a suitable solvent and injecting an aliquot with a microsyringe. The solvent is generally chosen with a boiling point much lower than that of the sample being analysed, so that it is eluted completely before the sample components appear in the chromatogram. The chromatogram frequently consists of a large solvent peak together with considerably smaller peaks of the sample components, such that the tail of the solvent peak may interfere with the early sample component peaks and may even obscure them, thus preventing examinations at high sensitivities. Many techniques have been used for reducing or eliminating the solvent peak, including those of solid injection and the use of a non-volatile solvent, such as silicone oil, to dissolve the sample[11].

Solid samplers often shorten total analysis time, eliminating the need for such isolation techniques as steam distillation and solvent extraction. Generally, these samplers overcome the problems that arise from the direct injection of samples containing non-volatile components, which gradually accumulate on the GC column and change the character of the stationary phase.

Solid injectors or solid samplers occur in numerous forms, and can be grouped according to their mode of operation. Many are modifications of injection syringes[12-19], some use glass inserts[20, 21], and other use metal gauzes which are dropped or moved by a magnet into the injection area[22-25]. Encapsulation of samples in polythene or indium, which are subsequently melted with the release of the volatiles, have also been used[26, 27]. In addition, there are several commercial units available, although not all of these are suitable for every chromatograph, e.g. some are not applicable to instruments with horizontal injection.

Wong and Schwartz[28] described a versatile injector which was suitable for either placing and removing solid samples in the carrier gas stream, or through which liquid samples could be injected. It had the advantages that it did not require shut-down between samples, produced no disturbance of the baseline in operation and could be used with the majority of GC instruments. Another facet of this injector was its potential in reaction chromatography.

Darbre and Islam[29] used a precolumn packed with support, to retain the solutes after evaporating the solvents *in vacuo*. The precolumn was then placed in the heated inlet of the GC. The method has a potentially greater sensitivity than conventional syringe injection, as most of the sample solution can be applied to the precolumn, thus enabling more solute to be chromatographed.

Another method concerning the elimination of the solvent peak in chromatograms was described by Teuwissen and Darbre[30]. It was used in the preparation of trifluoroacetylated amino acid methyl esters, and consisted of an apparatus specifically designed to enable the derivatisation to take place on a platinum support, viz. a small ball of platinum wire. Two alternative methods were then described for dropping the platinum balls into the heated inlet zone of the chromatograph.

Precolumns in the form of packed injection port liners have been used by several analysts for GC analysis of volatiles in the presence of non-volatile material. A large sample is diffused into a packed liner, which is inserted in the heated injection port of a GC, such that the volatiles are immediately swept on to the GC column. It is therefore essential that the GC injection port should have a separate heater to enable a temperature to be selected that will rapidly volatilise the constituents. Injection port liners packed with sand have been used to remove non-volatile material, and to prevent the contamination of the injection block and column by the pyrolysis of this material. Dupuy, Fore and Goldblatt[31] used an injection port liner packed with glass wool for the analysis of volatiles in vegetable oils. Brown, Dollear and Dupuy[32] modified the packing of the liner, such that it could be used as a solid sampler for the examination of volatiles from peanuts.

3. Other sampling procedures

Many sampling techniques are extremely involved and are often specific to a particular problem, e.g. the extraction and isolation of flavour volatiles prior to GC examination, by such techniques as vacuum distillation. Most of these techniques are not directly related to the GC, and are therefore considered beyond the scope of this book.

The limitation of head space analysis can be the requirement of exceedingly high volumes of head space vapours in order to detect trace amounts of many components. The minor components may be so diluted with carrier gas that they escape detection. Low-temperature GC precolumns are frequently used to concentrate head space components[33,34], but nevertheless the quantity of head space vapour that can be used is severely limited because of diffusion. Such a method for concentration was chosen by Morgan and Day[35] for the analysis of flavour volatiles, in which they modified the on-column trapping procedure for volatiles due to Hornstein and Crowe[36]. The entrainment apparatus, which was rigidly supported at the side of the GC oven, is shown in *Figure 4.1*.

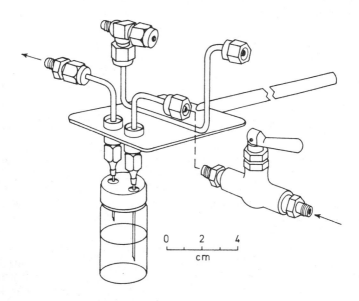

Figure 4.1. Details of entrainment apparatus (after Morgan and Day[35])

Screw-capped vials containing sufficient anhydrous Na_2SO_4 to saturate the aqueous samples were placed in an oven at 105°C, overnight. The sample was then added to the vial, a drilled cap containing a silicone rubber liner was fitted and the vial was attached to the apparatus. The vial was heated to the desired entrainment temperature and agitated intermittently. Carrier gas was passed through the apparatus into the GC column, where the volatiles were condensed in the initial portion of the column, which had been cooled with dry ice in a Dewar flask. At the end of the entrainment period, the column was disconnected from the apparatus and coupled up to the GC injection port, and the GC oven was heated to the analysis temperature to obtain the chromatogram in the usual manner.

Other means of concentrating volatiles from head space vapours have been carried out, including the use of activated charcoal and similar adsorbents prior to GC analysis[34,37,38].

Hartman, Rose and Vandaveer[39] isolated and concentrated volatiles from vegetable oil by bubbling purified helium through the samples, which were heated to 350°C in an oil-bath. The volatile components were collected in specially prepared activated charcoal collection tubes over a period of 2 h and then extracted with CS_2 containing an internal standard. A 400-fold concentration of the volatile components was achieved with this procedure, which gave good reproducibility.

C. Sample derivatisation

1. Alkylation

Alkylation is used for the derivatisation of many organic compounds in GC, including carboxylic acids, lipids and glycerides. Methyl, propyl, butyl and amyl esters are the alkyl esters generally used, the most popular of which is the methyl ester. It has been used as a derivative for fatty acids, amino acids, hydroxy acids and sulphonic acids.

A great deal of attention has been given to the methylation of fatty acids, and many methods and reagents have been recommended, including methanol alone in micro-autoclaves, methanol—sulphuric acid, methanol—hydrochloric acid, methanol with catalysts such as p-toluenesulphonic acid and boron trihalide complexes, dimethyl sulphate and diazomethane.

Methylation with diazomethane is usually carried out by treating the acids in methanol with a solution of diazomethane in ether for up to several hours [40,41], and until a yellow colour persists. The solvent and excess reagent are removed *in vacuo* at room temperature, and the esters are dissolved in methanol for GC. Diazomethane usually gives excellent quantitative recoveries, but it has the disadvantages of a short shelf life and an explosive nature, together with a generally long preparation time.

Dimethyl sulphate in methanol is an efficient methylating reagent, but is highly toxic[42]. It has been used in conjunction with 2,2-dimethoxy-propane[43]. Scoggins and Fitzgerald[44] used this reagent for the methylation of chlorophenoxyacetic acid herbicides. In comparison with acid-catalysed reactions, dimethyl sulphate was found to be quicker and more quantitative, and approached the quantitativity achieved with diazomethane.

In addition to the use of diazomethane for methylation, diazoethane, diazopropane and diazobutane have been used to prepare their corresponding esters, usually in the presence of catalytic amounts of BF_3[45].

Two frequently used methylation methods involve the use of HC1—methanol [46] and H_2SO_4—methanol[47-49]. These are particularly useful for the derivatisation of triglycerides, as they convert the triglyceride to its fatty acid methyl esters without going through the saponification stage. The carboxylic acids are refluxed with the reagent for 2 h, the esters are extracted with petrol and the extract is concentrated for GC. The methylation of non-volatile acids with methanol—HCl and thionyl chloride was studied by Gee[50], and applied to amino acids[51]. The acids were refluxed with the reagents for 10—30 min and the excess reagents were removed *in vacuo*. In the case of the amino acids, these esters were in the form of hydrochlorides and no loss occurred on evaporation. The methyl esters of the amino acids were then trifluoroacetylated with trifluoroacetic anhydride.

Another common reagent for the formation of methyl esters is BF_3—methanol [52-54], particularly for general fatty acid analysis. The conditions necessary for the methylation of various lipids were extensively studied by Morrison and Smith[55]. Many investigators have used different conditions such as long reaction times and high reaction temperatures[56] and different reagent concentrations[57].

The use of the reagents BF_3—propanol and BF_3—butanol forms the corresponding propyl and butyl esters. Staruszkiewicz, Bond and Salwin[58] and Staruszkiewicz and Starling[59] used BF_3—propanol to form the propyl ester of β-hydroxybutyric acid in the determination of this acid in eggs. The method was quantitative and recoveries of greater than 90% of the acid were obtained with good precision. The procedure was a modification of the one used by Salwin and Bond[60] for the analysis of lactic and succinic acids in eggs.

Klopfenstein[61] carried out the methylation of unsaturated fatty acids using both BF_3 in methanol and BCl_3 in methanol. BCl_3 in methanol was shown to be the reagent of choice to avoid the destruction of highly unsaturated fatty acids under prolonged or harsh reaction conditions.

A simple procedure for the quantitative preparation of methyl esters of fatty acids from glyceride fats and oils was described by Luddy *et al.*[62] The procedure involved the alkaline-catalysed reaction of the oil with potassium methylate in anhydrous methanol. For oils containing a high content of free fatty acids, the procedure was modified to include treatment with the acidic catalyst, BF_3, in methanol after initial reaction with the potassium methylate catalyst. A similar procedure was used for methylating the fatty acids of peanut oil[63] and milk fat[64].

The propyl esters and higher esters have been used mainly as derivatives for short-chain fatty acids[65-68]. Appleby and Mayne[69] carried out the n-propyl esterification of mono- and dibasic fatty acids using the BF_3—propanol complex as the reagent. The short-chain fatty acids in butter oil and cheese were analysed by Iyer *et al.*[70] using butylation with butanol—H_2SO_4, although it was suggested that BF_3 might be a better catalyst.

Many alkyl esters of long-chain fatty acids can be quantitatively prepared by causing the acids to react with dimethylformamide dialkyl acetals[71] at 60°C for 10—15 min in a suitable solvent. GC is carried out directly on the mixture, once a complete solution is obtained. The method is as successful as the diazomethane procedure, and it would appear to be far more convenient.

Hydroxyl groups can be derivatised to methyl ethers by reaction with methyl iodide and silver oxide in dimethylformamide[72]. This method has been used for the derivatisation of sugars[73]. However, the methylation of a mixture of sugars gives rise to a complicated chromatogram due to their various isomers and therefore the procedure is only efficient for simple mixtures.

Methylation is frequently used for the derivatisation of phenols[74,75] to the corresponding ethers, which are less adsorptive and more volatile. Stark[76] methylated an extract of pentachlorophenol residues from fish with diazomethane, before analysis by GC with EC detection. Nitrosomethyl urea was used as the precursor for the preparation of the diazomethane as suggested in the simplified method of Arndt used by Hartman, Rose and Vandaveer[39]. This gave a somewhat impure reagent, but avoided the hazardous distillation of the diazomethane.

2. Silylation

The importance of silylation in GC, i.e. substitution of the active hydrogen in HO, HS and HN functions by the TMS group–$Si(CH_3)_3$, has increased since its introduction for the preparation of volatile derivatives. TMS derivatives are prepared relatively easily and rapidly. Reactive species yield derivatives which are more volatile, less polar and thermally stable than most other derivatives. Nevertheless, the usual silylation procedures are not generally applicable to compounds mixed with reactive solvents, e.g. water. Either aqueous solutions must be freeze dried or solutes extracted into some other solvent prior to derivatisation.

A number of methods have been employed for TMS derivatisation, including procedures for alcohols and amines[77], phenols[78, 79], sugars[80, 81], amino acids [82], vitamins[83] and polyols[84]. The quantitative aspects of this technique have been investigated by Mason and Smith[85]. A detailed account of silylation is given in the excellent book by Pierce[86].

The most common reagents for silylation are hexamethyldisilazane (HMDS), trimethylchlorosilane (TMCS) and bis-(trimethylsilyl)acetamide (BSA). In addition to these, there are other silylating reagents that are not so frequently encountered, e.g. trimethylsilyldiethylamine (TMSDEA) and bis-trimethylsilyltrifluoroacetamide (BSTFA).

Silylation of the sample with HMDS:TMCS, 3:1, in pyridine is frequently used, and the mixture is available commercially under the name 'Tri-Sil'. Generally, the mixture is heated at 60–80°C for a few hours, but difficult samples are refluxed until the reaction is complete. NH_4Cl is formed as a precipitate, but does not interfere with the analysis, and, usually, the reaction solution can be directly injected into the GC. Pyridine can be replaced with any other suitable solvent that does not react with the reagents; e.g. dimethylformamide, dimethyl sulphoxide, acetonitrile, dioxan, tetrahydrofuran and carbon disulphide are frequently used. The procedure for the reagent BSA is similar, with the exception that it is often sufficient to let the mixture react at room temperature. BSTFA is a more volatile reagent than BSA, and has the advantage that it produces HF in the FID, which reacts with the silicone to form the volatile SiF_4, thus keeping the detector free of silicone deposits.

N-trimethylsilylimidazole (TSIM), in dry pyridine, obtained commercially as 'Tri-Sil Z', is an excellent silylating reagent for polyhydroxy compounds, even in aqueous solution[87]. Silylation can be completed effectively within a few minutes at 60°C.

Pyridine is almost universally accepted as the preferred solvent for the silylating medium but does sometimes cause extensive tailing on the chromatogram, especially when HMDS and TMCS are used as reagents. Ellis[88] has claimed that a mixture of dimethylformamide and dimethyl sulphoxide offers certain advantages in this respect as the solvent. Alternatively, pyridine can be effectively removed by evaporation under vacuum, and replaced with dry hexane as described by Lehrfeld[89], or by extracting the derivatised mixture with chloroform before application to the GC column[90].

In a review of the TMS derivatives of food carbohydrates, Birch[91] suggested that the presence of water in the silylation of wet samples could lead to variability in column characteristics, in addition to the probable instability of the resulting derivatives on the column. This could be overcome by using an on-column silylation technique.

Silylation in aqueous solution can be carried out with both qualitative and quantitative reliability, provided there are sufficient reagents to react with all the water present. This fact was illustrated by the procedure of Weiss and Tambawala [92] in which excess HMDS, TMCS and pyridine were caused to react for 3 h at 35–40°C with an aqueous solution of polyols. The derivatised sample was extracted with $CHCl_3$ to eliminate the pyridine, and the $CHCl_3$ extract was water-washed before GC.

The silylation of amino acids was carried out by Ruehlmann and Giesecke[82] and then extensively investigated by Gehrke, Nakamoto and Zumwalt[93]. This technique offered certain advantages, in that silylation of the 20 protein amino acids was completed in a single reaction medium and they could be separated on a single chromatographic column. BSTFA was used as the silylating agent, as suggested by Stalling, Gehrke and Zumwalt[94].

Gehrke and Leimer[95,96] investigated the effects of various solvents on derivatisation with BSTFA, and concluded that polar solvents, e.g. acetonitrile, give different derivatisation characteristics with certain amino acids from those given by non-polar solvents, e.g. methylene chloride. In the derivatisation procedure the water was removed from the amino acids with a stream of dry nitrogen at 75°C, and methylene chloride was added and evaporated to form an azeotrope of any remaining water. An excess of BSTFA was then added in a solvent, and the sample tube was closed and heated at 150°C for 2.5 h in an oil-bath. It was concluded that the method was complementary to the *N*-TFA butyl ester technique for the GC analysis of protein amino acids (p. 70).

3. Trifluoroacetylation

The TFA derivatives of many compounds are frequently used for GC, since they are usually very volatile and simple to prepare, and can give good sensitivities with EC detection, owing to the high electron affinity of the fluorine atoms. In particular, trifluoroacetylation has found considerable application in the analysis of the NH_2 functional group.

The direct GC analysis of free amines is often impaired by peak tailing caused by partial adsorption on to the column material and also by the high volatility of the lower amines. The conversion of the amines to suitable derivatives, such as the corresponding trifluoroacetamides, offers an alternative procedure which avoids the adsorption problem. In addition, the boiling range of the amines is decreased, since the TFA derivatives of the higher amines are more volatile, while those of the lower amines are less so.

Several workers have reported the separation of a number of TFA derivatives of alkylamines on both packed[97-99] and capillary columns[100,101]. The derivatives were prepared by dissolving the amine in a suitable organic solvent and adding a slight excess of TFAA. The mixture was allowed to stand for a few minutes, washed with sodium bicarbonate solution and dried, and the solvent was evaporated.

Ethyl trifluoroacetate (ETFA) was used by Lubkowitz[102] as an alternative reagent to the conventional TFAA for the preparation of TFA derivatives of amines. ETFA reacted with amines to produce the trifluoroacetamides and ethyl alcohol. The reaction proceeded rapidly at 60–70°C, but at room temperature the reaction took place slowly. The reaction yields were improved if small quantities of ammonia were bubbled through the reagent solution before storage. This was believed to neutralise any traces of trifluoroacetic acid, which might be produced by the hydrolysis of the ETFA and thus prevent the production of the corresponding amine salt during the derivatisation of the amine. EFTA overcame the drawbacks of TFAA as a derivatising reagent for amines in that the reaction was not so violent, and trifluoroacetic acid was not produced as a reaction by-product. Trifluoroacetic acid is highly corrosive and can damage metallic oven and detection components.

Many methods of identifying sugars by GC utilise silylation procedures, even though each simple sugar produces at least two anomeric TMS derivatives, and many of the derivatives have long elution times. As with silylation, the direct trifluoroacetylation of individual sugars[103,104] gives anomeric peaks. However, the reduction of sugars to polyols and subsequent derivatisation to acetate esters – in particular, the TFA esters, which are more volatile than the TMS derivatives – gives much better results. This procedure was described by Shapira[105].

Since the low volatility of amino acids has prevented their direct analysis by GC, suitable volatile derivatives were sought. Trifluoroacetic anhydride (TFAA) has been the reagent used by many analysts for the preparation of such derivatives, viz. the TFA amino acid alkyl esters.

N-TFA methyl esters of amino acids Some of the TFA methyl esters of amino acids are extremely volatile and losses have been reported when concentrating a solution containing these derivatives. Consequently, many analysts have now chosen to utilise the corresponding n-butyl ester derivatives.

Islam and Darbre[106] and Darbre and Islam[107] successfully used the TFA methyl ester derivatives by carrying out a thorough investigation of the various methods of evaporating the solvent. Protein samples were hydrolysed with 6N hydrochloric acid in sealed tubes at 105°C for 20 h and the acid removed by rotary evaporation. The protein hydrolysates were then methylated with methanol–hydrochloric acid and acylated with TFAA in methylene chloride at room temperature for 20 min. The excess reagent was then rotary evaporated with an oil pump at 0°C for a maximum time of 4 min. The derivatives were maintained under rigorous anhydrous conditions to avoid hydrolysis before GC[108].

N-*TFA n-butyl esters of amino acids* Zomzely, Marco and Emery[109] investigated the N-TFA n-butyl esters as possible derivatives for amino acids. Lamkin and Gehrke[110] described a method for preparing these derivatives in which they formed the butyl ester via the methyl ester of the amino acids and then carried out the trifluoroacetylation with TFAA in methylene chloride. Detailed experimental conditions for the quantitative derivatisation and chromatographic separation were given by Gehrke and Stalling[111] and Roach and Gehrke[112].

A method for the direct esterification of amino acids to yield their butyl ester derivatives before trifluoroacetylation was carried out by Roach, Gehrke and Zumwalt[113] and Roach and Gehrke[114]. This considerably simplified the derivatisation procedure, and was both rapid and precise. Two methods were described, one applicable to 1–20 mg total amino acids and the other to 1–200 μg total amino acids, and both formed the butyl ester derivatives by heating with n-butanol–3N HCl at 100°C for 15–30 min, and enabled a complete analysis of protein amino acids to be carried out in less than an hour. The solubility problems which had been associated with cysteine and some of the other amino acids in the earlier methods were removed. The procedure was used by Gehrke and Leimer[115] with some modification and a simple apparatus for carrying out the derivatisation was described by Mee and Brooks[116].

Several accounts of the GC of protein amino acids were presented by Zumwalt and his co-workers[117-119]. Isolation of the acids was achieved by ion exchange methods and N-TFA n-butyl ester derivatives were prepared using the methods referred to previously. They showed that the N-TFA n-butyl esters were better than the N-TFA methyl esters for amino acid derivatives, as they are less volatile

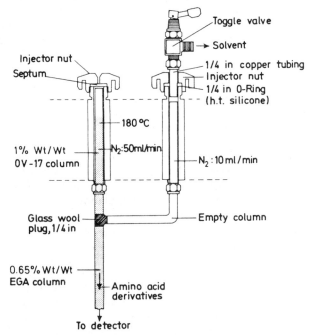

Figure 4.2. Injection port solvent vent device (after Zumwalt et al.[119])

and no evaporation losses occurred on concentration of the latter, particularly at the microgram and submicrogram level. A modification of the method was described[119] which was suitable for the analysis of samples containing nanogram to picogram amounts of amino acids and gave recoveries of amino acids greater than 80%. The high sensitivity was achieved by the use of a 'solvent-vent chromatographic device', illustrated in *Figure 4.2*, which allowed injection of up to 100 μl of derivatised sample on a standard analytical column, and thus eliminated the need for sample concentration. In addition, it removed the large solvent and reagent interference peaks from the chromatograms. Further increase in sensitivity was obtained by the use of EC detection.

4. Halogeno-esterification

In addition to the use of TFA derivatives many other halogeno-esters have now been used successfully as derivatives for GC, particularly with EC detection.

Chloroacetate derivatives A rapid procedure for the chloroacetylation of microgram quantities of phenols and their derivatives was carried out by Argauer[120]. Aqueous sodium hydroxide solutions of the phenols were treated with chloroacetic anhydride in benzene to produce chloroacetates that were sensitive to EC detection. The derivatives were chromatographed directly without any further treatment.

Retention times for the derivatives of 32 phenols were given but the method was found to be unsatisfactory for several nitrophenols and polyhydroxybenzenes. Some phenols would require more controlled conditions than others for quantitative analysis.

The method was subsequently applied to the determination of several carbamate pesticide residues after hydrolysis[121].

Trichloroethyl esters Derivatives for GC with EC detection of carboxylic acids are few. Smith and Tsai[122] prepared their trichloroethyl esters by the reaction with 2,2,2-trichloroethanol in TFAA, and found them to be sensitively measured with a [63]Ni EC detector. The ester derivatives of ten model benzenoid acids were prepared using this rapid and quantitative esterification technique. Ten milligrams of the carboxylic acids was caused to react for 10 min on a steam-bath with 1 ml of 10% 2,2,2-trichloroethanol in TFAA. The excess TFAA was evaporated in a stream of dry air, and the residual trichloroethanol was removed by passing a solution of it in ethyl acetate through a silica gel column. Percentage recoveries for most of the acids studied were close to 100, and the detection limits were subnanogram, in general.

Pentafluorobenzyl esters Kawahara[123, 124] carried out the analysis of phenols, mercaptans and organic acids at the nanogram level after the preparation of pentafluorobenzyl derivatives. The compounds were derivatised by refluxing under alkaline conditions for 3 h with a 2% ethanolic solution of α-bromo-2,3,4,5,6-pentafluorotoluene and an alkyl halide. A solvent extract of the derivatives was chromatographed using EC detection, and the chromatograms exhibited virtually no tailing. Very good recoveries are possible with these derivatives if a favourable extraction procedure is used to isolate the compounds from the foodstuff. Nevertheless, the method is regarded by many analysts as cumbersome, and requires the use of an exceedingly strong lachrymator.

Pentafluoropropionyl esters These esters have been used as derivatives for amines and alcohols by heating an extract of them in a suitable solvent with pentafluoropropionic anhydride at $60°C$ for 10 min.

Masuda and Hoffmann[125,126] used pentafluoropropionic anhydride to form the pentafluoropropionamides of primary amines of polynuclear aromatic hydrocarbons. The amines were dissolved in dry ether and caused to react with an excess of pentafluoropropionic anhydride at room temperature for 18 h. The amides produced were partially separated on a Florisil column and finally separated and identified using GC with EC detection.

Heptafluorobutanoyl esters These derivatives have been particularly useful for the picogram detection of steroids. The steroid is dissolved in tetrahydrofuran and an excess of heptafluorobutyric anhydride is added. The solution is heated at $60°C$ for 30 min and then evaporated to dryness with a stream of nitrogen. The residue is dissolved in a small amount of acetone and injected into the GC.

Wilson, Lawson and Kodicek[127] suggested that a few steroids were broken down in the presence of heptafluorobutyric anhydride at elevated temperatures, and they carried out the derivatisation of vitamins D_2 and D_3 at $0°C$. The reaction mixture was left for 1 h and then water was added to terminate the reaction. The ester was extracted with petrol for application to the GC. This procedure enabled vitamins D_2 and D_3 to be determined at levels which occur in natural sources.

5. Dinitrophenylation

The 2,4-dinitrophenyl derivatives of amines are amenable to GC, and possess strong electron-capturing properties. Consequently, they are suitable for the detection of amines at nanogram levels.

These derivatives were chosen by Day, Golab and Koons[128] for the detection of low molecular weight aliphatic amines, using the reagent 1-fluoro-2,4-dinitrobenzene. The same reagent was used by Cohen and Wheals[129] for the determination of substituted urea and carbamate herbicides after hydrolysis to their amine moieties. The herbicides were separated on a silica gel chromatoplate and hydrolysed by spraying with 5% hydrochloric acid, and then the 2,4-dinitrophenyl derivatives of the amines were prepared by spraying the plate with a solution of 1-fluoro-2,4-dinitrobenzene in acetone. The derivative was then eluted from the adsorbent and examined by GC incorporating EC detection.

The determination of phenols at the nanogram level necessitates the development of a sensitive method and is usually approached by the preparation of a suitable derivative to enable the use of more specific and sensitive detectors. The 2,4-dinitrophenyl ether derivatives of phenols prepared from 1-fluoro-2,4-dinitrobenzene were shown by Cohen *et al.*[130] to display strong electron-capturing properties. Three methods were described for preparing the derivatives from microgram quantities of phenols. Higher yields were generally obtained with the method referred to as the 'sandwich layer' technique, in which an acetone solution of the phenols was spotted on to a Whatman silica gel loaded paper, SG 81. It was then sprayed with a saturated solution of sodium methoxide in methanol, followed by a solution of 4% v/v 1-fluoro-2,4-dinitrobenzene in acetone. The paper was sandwiched between two glass plates clipped tightly together and then heated in an oven at 190°C for 40 min. When cool, the area of interest was removed by cutting out the spot and the derivatives were extracted with 10 ml of acetone and two drops of water. The method was applied to the detection and determination of certain pesticide residues, such as carbamates, which yield phenols on hydrolysis. This was the subject of a subsequent paper by these authors[131], in which the hydrolysis of the carbamate to the phenol and formation of the corresponding 2,4-dinitrophenyl ether was effected concurrently.

6. Formation of 2,4-dinitrophenyl hydrazones

The 2,4-dinitrophenyl hydrazones have been used as derivatives for many carbonyl compounds, including formaldehyde[132,133]. Volatile carbonyl compounds are important flavour components in foodstuffs, accounting for both many off-flavours and pleasant odours, but as their concentrations are generally very low, they are difficult to identify or determine. Fifteen carbonyl compounds that were known to be flavour components were analysed as their 2,4-dinitrophenyl hydrazones by Kallio, Linko and Kaitaranta[134]. The derivatives were prepared by shaking 100 μl of each carbonyl compound with 100 ml of a saturated solution of 2,4-dinitrophenyl hydrazine in aqueous 2N hydrochloric acid and allowing the mixture to stand at room temperature overnight. The precipitate that formed was isolated by filtration, washed with hydrochloric acid and water, and then dried in a vacuum desiccator. This precipitate was then dissolved in a solvent, e.g. benzene or ethyl acetate, and chromatographed using either FID or EC detection. It proved an advantage to silanise the GC column before use to increase its ability to resolve the dinitrophenyl hydrazones. The analysis of the carbonyls using this method showed that the use of an EC detector led to essentially greater sensitivity and selectivity. Whereas nanogram amounts could be analysed with FID, it was possible to analyse subnanogram amounts with EC detection. The linear dynamic range with the latter was, however, very narrow, e.g. 20–500 pg for 2-butanone, and rendered it undesirable for precise quantitative work. Nevertheless, the EC detector was extremely well suited to the trace analytical problems associated with the identification of carbonyl compounds as their dinitrophenyl hydrazone derivatives in the investigation of natural and artificial flavourings.

7. Oxidation

Oxidation methods for distinguishing unsaturated and saturated fatty acids were used by James and Webb[135] and Gunstone and Sykes[136].

Sampling and Sample Derivatisation

The formation of derivatives by oxidation was used by Althorpe et al.[137] for the conversion of nitrosamines to nitramines, which were found to give 200 times greater response to EC detection than to FID and enabled nitrosamines to be determined at the picogram level. One millilitre of a solution of the nitrosamine in dichloromethane was placed in a stoppered tube, 0.1 ml of peroxytrifluoroacetic acid (PTFA) was added and the mixture was allowed to stand for 3.5 h. One or two drops of water was then added, followed by an excess of calcium carbonate, and anhydrous sodium sulphate. The solution was then examined by GC incorporating EC detection on a column of Carbowax 20M at 140°C. Table 4.1 gives the efficiency of conversion of nitrosamine to nitramine, and the retention times of the nitrosamines and corresponding nitramines.

A similar procedure was used by Sen[138] for the determination of dimethylnitrosamine as dimethylnitramine also at picogram levels. The method was applied successfully to the estimation of dimethylnitrosamine isolated from nitrite-treated fish.

Table 4.1 Retention data and efficiency of conversion of nitrosamines to nitramines (after Althorpe et al.[137])

Nitrosamine	Conversion/%	R_t Nitrosamine/min	R_t Nitramine/min
Dimethyl	86	4.0	9.8
Methyl ethyl	85.5	4.9	11.25
Diethyl	84.5	5.5	12.4
Methyl-isopropyl	76	5.8	12.65
Di-isopropyl	72	7.15	15.05
Di-n-propyl	84	9.5	20.5
Di-isobutyl	82.5	10.45	20.65
Di-n-butyl	83	20.2	42.9
N-nitrosopiperidine	25	22.5	39.6
N-nitrosopyrrolidine	85	25.25	57.65

References

1. Hamilton, L. H., J. Gas Chromat., 2, 302 (1964)
2. Novak, J., Gelbicova-Ruzickova, J. and Wicar, S., J. Chromatog., 60, 127 (1971)
3. Bassette, R., Ozeris, S. and Whitnah, C. H., Anal. Chem., 34, 1540 (1962)
4. Nelson, P. E. and Hoff, J. E., Food Technol., 22 (11), 61 (1968)
5. Davis, P. L., J. Chromat. Sci., 8, 423 (1970)
6. Sinclair, A., Hall., R. D., Thorburn Burns, D. and Hayes, W. P., J. Sci. Food Agric., 21, 468 (1970)
7. Fore, S. P., Rayner, E. T. and Dupuy, H. P., J. Amer. Oil Chem. Soc., 48, 140 (1971)
8. Dupuy, H. P., Rayner, E. T. and Fore, S. P., J. Amer. Oil Chem. Soc., 48, 155 (1971)
9. Rushneck, D. R., J. Gas Chromat., 3, 318 (1965)
10. Heins, J. T., Maarse, H., ten Noever de Brauw, M. C. and Weurman, C., J. Gas Chromat., 4, 395 (1966)
11. Cramers, C. A. and van Kessel, M. M., J. Gas Chromat., 6, 577 (1968)
12. Lurie, A. O. and Villee, C. A., J. Gas Chromat., 4, 160 (1966)
13. McComas, D. B. and Goldfien, A., Anal. Chem., 35, 263 (1963)
14. Renshaw, A. and Biran, L. A., J. Chromatog., 8, 343 (1962)
15. Zahuta, J., J. Chromatog., 12, 404 (1963)
16. Kroman, H. S., King, M. O. and Bender, S. R., J. Chromatog., 15, 92 (1964)

17. Carson, L. M. and Uglum, K. L., *J. Gas Chromat.*, **3**, 208 (1965)
18. Stahl, W. H., Sullivan, J. H. and Voelker, W. A., *J. Ass. Off. Agric. Chem.*, **46**, 819 (1963)
19. Yannone, M. E., *J. Gas Chromat.*, **6**, 465 (1968)
20. Levins, R. J. and Ikeda, R. M., *J. Gas Chromat.*, **6**, 331 (1968)
21. Dean, A. C., Bradford, E., Hubbard, A. W., Pocklington, W. D. and Thomson, J., *J. Chromatog.*, **44**, 465 (1969)
22. Collins, W. P. and Sommerville, I. F., *Nature*, **203**, 836 (1964)
23. Ros, A., *J. Gas Chromat.*, **3**, 252 (1965)
24. Ruchelman, M. W., *J. Gas Chromat.*, **4**, 265 (1966)
25. Bailey, R. E., Under, O. M. and Grettie, D. P., *J. Gas Chromat.*, **6**, 340 (1968)
26. von Rudloff, E., *J. Gas Chromat.*, **3**, 390 (1965)
27. Hudy, J. A., *J. Gas Chromat.*, **4**, 350 (1966)
28. Wong, N. P. and Schwartz, D. P., *J. Chromat. Sci.*, **7**, 569 (1969)
29. Darbre, A. and Islam, A., *J. Chromatog.*, **49**, 293 (1970)
30. Teuwissen, B. and Darbre, A., *J. Chromatog.*, **49**, 298 (1970)
31. Dupuy, H. P., Fore, S. P. and Goldblatt, L. A., *J. Amer. Oil Chem. Soc.*, **48**, 876 (1971)
32. Brown, D. F., Dollear, F. G. and Dupuy, H. P., *J. Amer. Oil Chem. Soc.*, **49**, 81 (1972)
33. Mendelsohn, J. M., Steinberg, M. A. and Merritt, C., *J. Food Sci.*, **31**, 389 (1966)
34. Heinz, D. E., Sevenants, M. R. and Jennings, W. G., *J. Food Sci.*, **31**, 63 (1966)
35. Morgan, M. E. and Day, E. A., *J. Dairy Sci.*, **48**, 1382 (1965)
36. Hornstein, I. and Crowe, P. F., *Anal. Chem.*, **34**, 1354 (1962)
37. Dhont, J. H. and Weurman, C., *Analyst*, **85**, 419 (1960)
38. Jennings, W. G. and Nursten, H. E., *Anal. Chem.*, **39**, 521 (1967)
39. Hartman, K. T., Rose, L. C. and Vandaveer, R. L., *J. Amer. Oil Chem. Soc.*, **48**, 178 (1971)
40. Quin, L. D. and Hobbs, M. E., *Anal. Chem.*, **30**, 1400 (1958)
41. Schlenk, H. and Gellerman, J. L., *Anal. Chem.*, **32**, 1412 (1960)
42. Martin, H. F. and Driscoll, J. L., *Anal. Chem.*, **38**, 345 (1966)
43. Simmonds, P. G. and Zlatkis, A., *Anal. Chem.*, **37**, 302 (1965)
44. Scoggins, J. E. and Fitzgerald, C. H., *J. Agric. Food Chem.*, **17**, 156 (1969)
45. Wilcox, M., *Anal. Biochem.*, **32**, 191 (1969)
46. Stoffel, W., Chu, F. and Ahrens, E. H., *Anal. Chem.*, **31**, 307 (1959)
47. Archibald, F. M. and Skipski, V. P., *J. Lipid Res.*, **7**, 442 (1966)
48. Litchfield, C., Farquhar, M. and Reiser, R., *J. Amer. Oil Chem. Soc.*, **41**, 588 (1964)
49. Rogozinski, M., *J. Gas Chromat.*, **2**, 136 (1964)
50. Gee, M., *Anal. Chem.*, **37**, 926 (1965)
51. Gee, M., *Anal. Chem.*, **39**, 1677 (1967)
52. Metcalfe, L. D. and Schmitz, A. A., *Anal. Chem.*, **33**, 363 (1961)
53. Metcalfe, L. D., Schmitz, A. A. and Pelka, J. R., *Anal. Chem.*, **38**, 514 (1966)
54. van Wijngaarden, D., *Anal. Chem.*, **39**, 848 (1967)
55. Morrison, W. R. and Smith, L. M., *J. Lipid Res.*, **5**, 600 (1964)
56. Hyun, S. A., Vahouny, G. V. and Treadwell, C. R., *Anal. Biochem.*, **10**, 193 (1965)
57. Lough, A. K., *Biochem. J.*, **90**, 4c (1964)
58. Staruszkiewicz, W. F., Bond, J. F. and Salwin, H., *J. Chromatog.*, **51**, 423 (1970)
59. Staruszkiewicz, W. F. and Starling, M. K., *J. Ass. Off. Anal. Chem.*, **54**, 773 (1971)
60. Salwin, H. and Bond, J. F., *J. Ass. Off. Anal. Chem.*, **52**, 41 (1969)
61. Klopfenstein, W. E., *J. Lipid Res.*, **12**, 773 (1971)
62. Luddy, F. E., Barford, R. A., Herb, S. F. and Magidman, P., *J. Amer. Oil Chem. Soc.*, **45**, 549 (1968)
63. Barnes, P. C. and Holaday, C. E., *J. Chromat. Sci.*, **10**, 181 (1972)
64. Christopherson, S. W. and Glass, R. L., *J. Dairy Sci.*, **52**, 1289 (1969)
65. Sampugna, J., Pitas, R. E. and Jensen, R. G., *J. Dairy Sci.*, **49**, 1462 (1966)
66. Craig, B. M., Tulloch, A. P. and Murty, N. L., *J. Amer. Oil Chem. Soc.*, **40**, 61 (1963)
67. Jones, E. P. and Davison, V. L., *J. Amer. Oil Chem. Soc.*, **42**, 121 (1965)
68. Bezard, J. and Bugaut, M., *J. Chromat. Sci.*, **7**, 639 (1969)
69. Appleby, A. J. and Mayne, J. E. O., *J. Gas Chromat.*, **5**, 266 (1967)

70. Iyer, M., Richardson, T., Amundson, C. H. and Boudreau, A., *J. Dairy Sci.,* **50,** 285 (1967)
71. Thenot, J. P., Horning, E. C., Stafford, M. and Horning, M. G., *Anal. Lett.,* **5,** 217 (1972)
72. Walker, H. G., Gee, M. and McCready, R. M., *J. Org. Chem.,* **27,** 2100 (1962)
73. Kircher, H. W., *Anal. Chem.,* **32,** 1103 (1960)
74. Kirkland, J. J., *Anal. Chem.,* **33,** 1520 (1961)
75. Boggs, H. M., *J. Ass. Off. Anal. Chem.,* **49,** 772 (1966)
76. Stark, A., *J. Agric. Food Chem.,* **17,** 871 (1969)
77. Langer, S. H., Connell, S. and Wender, I., *J. Org. Chem.,* **23,** 50 (1958)
78. Friedman, S., Jahn, C., Kaufman, M. L. and Wender, I., *U. S. Bur. Mines Bull.,* No. 609, 27 (1963)
79. Dallos, F. C. and Koeppl, K. G., *J. Chromat. Sci.,* **7,** 565 (1969)
80. Sweeley, C. C., Bentley, R., Makita, M. and Wells, W. W., *J. Amer. Chem. Soc.,* **85,** 2497 (1963)
81. Flynn, C. and Wendt, A. S., *J. Ass. Off. Anal. Chem.,* **53,** 1067 (1970)
82. Ruehlmann, K. and Giesecke, W., *Angew. Chem.,* **73,** 113 (1961)
83. Fisher, A. L., Parfitt, A. M. and Lloyd, H. M., *J. Chromatog.,* **65,** 493 (1972)
84. Smith, B. and Carlsson, O., *Acta Chem. Scand.,* **17,** 455 (1963)
85. Mason, P. S. and Smith, E. D., *J. Gas Chromat.,* **4,** 398 (1966)
86. Pierce, A. E., *Silylation of Organic Compounds,* Pierce Chemical Co., Illinois (1968)
87. van Ling, G., *J. Chromatog.,* **44,** 175 (1969)
88. Ellis, W. C., *J. Chromatog.,* **41,** 325 (1969)
89. Lehrfeld, J., *J. Chromat. Sci.,* **9,** 757 (1971)
90. Partridge, R. D. and Weiss, A. H., *J. Chromat. Sci.,* **8,** 553 (1970)
91. Birch, G. G., *J. Food Technol.,* **8,** 229 (1973)
92. Weiss, A. H. and Tambawala, H., *J. Chromat. Sci.,* **10,** 120 (1972)
93. Gehrke, C. W., Nakamoto, H. and Zumwalt, R. W., *J. Chromatog.,* **45,** 24 (1969)
94. Stalling, D. L., Gehrke, C. W. and Zumwalt, R. W., *Biochem. Biophys. Res. Commun.,* **31,** 616 (1968)
95. Gehrke, C. W. and Leimer, K., *J. Chromatog.,* **53,** 201 (1970)
96. Gehrke, C. W. and Leimer, K., *J. Chromatog.,* **57,** 219 (1971)
97. Dove, R. A., *Anal. Chem.,* **39,** 1188 (1967)
98. Morrissette, R. A. and Link, W. E., *J. Gas Chromat.,* **3,** 67 (1965)
99. McCurdy, W. H. and Reiser, R. W., *Anal. Chem.,* **38,** 795 (1966)
100. Irvine, W. J. and Saxby, M. J., *Phytochemistry,* **8,** 473 (1969)
101. Irvine, W. J. and Saxby, M. J., *J. Chromatog.,* **43,** 129 (1969)
102. Lubkowitz, J. A., *J. Chromatog.,* **63,** 370 (1971)
103. Tamura, Z. and Imanari, T., *Chem. Pharm. Bull., Japan,* **15,** 246 (1967)
104. Luke, M. A., *J. Ass. Off. Anal. Chem.,* **54,** 937 (1971)
105. Shapira, J., *Nature,* **222,** 792 (1969)
106. Islam, A. and Darbre, A., *J. Chromatog.,* **43,** 11 (1969)
107. Darbre, A. and Islam, A., *Biochem. J.,* **106,** 923 (1968)
108. Makisumi, S. and Saroff, H. A., *J. Gas Chromat.,* **3,** 21 (1965)
109. Zomzely, C., Marco, G. and Emery, E., *Anal. Chem.,* **34,** 1414 (1962)
110. Lamkin, W. M. and Gehrke, C. W., *Anal. Chem.,* **37,** 383 (1965)
111. Gehrke, C. W. and Stalling, D. L., *Separation Sci.,* **2,** 101 (1967)
112. Roach, D. and Gehrke, C. W., *J. Chromatog.,* **43,** 303 (1969)
113. Roach, D., Gehrke, C. W. and Zumwalt, R. W., *J. Chromatog.,* **44,** 269 (1969)
114. Roach, D. and Gehrke, C. W., *J. Chromatog.,* **52,** 393 (1970)
115. Gehrke, C. W. and Leimer, K., *J. Chromatog.,* **53,** 195 (1970)
116. Mee, J. M. L. and Brooks, C. C., *J. Chromatog.,* **62,** 138 (1971)
117. Zumwalt, R. W., Roach, D. and Gehrke, C. W., *J. Chromatog.,* **53,** 171 (1970)
118. Zumwalt, R. W., Kuo, K. and Gehrke, C. W., *J. Chromatog.,* **55,** 267 (1971)
119. Zumwalt, R. W., Kuo, K. and Gehrke, C. W., *J. Chromatog.,* **57,** 193 (1971)
120. Argauer, R. J., *Anal. Chem.,* **40,** 122 (1968)
121. Argauer, R. J., *J. Agric. Food Chem.,* **17,** 888 (1969)
122. Smith, R. V. and Tsai, S. L., *J. Chromatog.,* **61,** 29 (1971)
123. Kawahara, F. K., *Anal. Chem.,* **40,** 1009 (1968)

124. Kawahara, F. K., *Anal. Chem.,* **40,** 2073 (1968)
125. Masuda, Y. and Hoffmann, D., *Anal. Chem.,* **41,** 650 (1969)
126. Masuda, Y. and Hoffmann, D., *J. Chromat. Sci.,* **7,** 694 (1969)
127. Wilson, P. W., Lawson, D. E. M. and Kodicek, E., *J. Chromatog.,* **39,** 75 (1969)
128. Day, E. W., Golab, T. and Koons, J. R., *Anal. Chem.,* **38,** 1053 (1966)
129. Cohen, I. C. and Wheals, B. B., *J. Chromatog.,* **43,** 233 (1969)
130. Cohen, I. C., Norcup, J., Ruzicka, J. H. A. and Wheals, B. B., *J. Chromatog.,* **44,** 251 (1969)
131. Cohen, I. C., Norcup, J., Ruzicka, J. H. A. and Wheals, B. B., *J. Chromatog.,* **49,** 215 (1970)
132. Soukup, R. J., Scarpellino, R. J. and Danielczik, E., *Anal. Chem.,* **36,** 2255 (1964)
133. Leonard, R. E. and Kiefer, J. E., *J. Gas Chromat.,* **4,** 142 (1966)
134. Kallio, H., Linko, R. R. and Kaitaranta, J., *J. Chromatog.,* **65,** 355 (1972)
135. James, A. T. and Webb, J., *Biochem. J.,* **66,** 515 (1957)
136. Gunstone, F. D. and Sykes, P. J., *Chem. & Ind.,* 1130 (1960)
137. Althorpe, J., Goddard, D. A., Sissons, D. J. and Telling, C. M., *J. Chromatog.,* **53,** 371 (1970)
138. Sen, N. P., *J. Chromatog.,* **51,** 301 (1970)

5
Reaction Chromatography

A. Introduction

Derivatisation is frequently accomplished prior to GC, following the extraction, isolation and clean-up of the sample components. However, there are instances when it is possible to carry out the derivatisation *in situ*, i.e. within the GC. This technique is known as 'reaction chromatography', and this term encompasses any structural change of a compound occurring within the GC. Reaction sites in this type of GC may be at several points in the chromatographic system: they may be ahead of the injection port in a unit referred to as a microreactor[1]; they may occur in the injection port, precolumn or chromatographic column, or in a short column preceding the detector. The high-temperature pyrolysis of a compound to obtain a fingerprint chromatogram is the most common example of reaction chromatography, but has few applications in food analysis.

Reaction chromatography has been employed to carry out such procedures as hydrogenation, dehydrogenation, hydrolysis, oxidation, bromination, esterification and saponification. Where these have been used to confirm data on the qualitative identification of various components in a mixture, the chromatogram produced is called a discrimination chromatogram.

B. Catalytic reaction chromatography

1. Hydrogenation

Hydrogenation is a technique widely used for determining the structure of unsaturated compounds. Mixtures of hydrocarbons and of fatty acids have been chromatographed before and after hydrogenation to determine, by difference, the proportions of their unsaturated and saturated components. Hydrogenation techniques in reaction chromatography have also been applied to other classes of unsaturated compounds, including esters, ketones, aldehydes, amines and halides. The reaction products depend on the amount and type of catalyst and the reaction temperature. Exhaustive hydrogenation to methane is obtained at about $1000°C$, and the methane produced gives a measure of the total carbon present[2].

78

Several different procedures and apparatus were described by Beroza and Sarmiento[3] for carrying out the direct hydrogenation of many organic compounds. One of these, which was based on the method of Mounts and Dutton[4], was particularly useful for the quantitative hydrogenation of fatty acid esters of glyceride oils. The apparatus, which could be made easily at little expense, was located in the oven of the GC, and, consequently, operated at the same temperature as the analytical column. Another apparatus utilised the GC injection port, so that the catalyst could be maintained at one temperature while the analytical column was at another. The catalyst used in both these hydrogenators was a small amount of neutral palladium (1% Pd on 60/80-mesh Gas Chrom P). Hydrogen was used as the carrier gas. Results showed that the hydrogenation procedures were sufficiently powerful to saturate multiple bonds in straight-chain, ring and substituted compounds, yet most functional groups remained unchanged. Hence, the procedures were useful for confirming the identification of unsaturated compounds by providing additional retention data. If the precolumn is provided with a by-pass[5], so that part of the sample escapes hydrogenation, the hydrogenated sample and unhydrogenated sample both appear on the same chromatogram. Consequently, the change in retention caused by hydrogenation can be observed directly.

The hydrogenation of unsaturated fatty acids, using a GC column filled with palladium catalyst precipitated on Celite, was described by Koman[6].

Figure 5.1 illustrates the results of the GC separations of some natural fatty acids on the column with and without palladium, and indicates complete conversion of unsaturated fatty acids into the corresponding saturated ones.

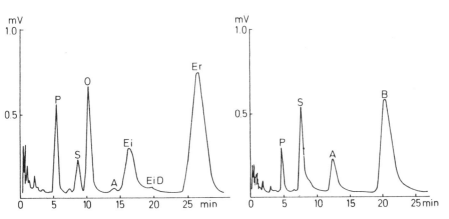

Figure 5.1. Chromatograms of mixtures of fatty acids separated on columns without (left-hand side) and with palladium (right-hand side). P = palmitic ($C_{16:0}$); S = stearic ($C_{18:0}$); O = oleic ($C_{18:1}$); A = arachidic ($C_{20:0}$); Ei = eicosenoic ($C_{20:1}$); EiD = eicosadienoic ($C_{20:2}$); Er = erucic ($C_{22:1}$); and B = behenic ($C_{22:0}$) acid. (After Koman[6])

Beroza and his co-workers[7-9] devised a method for determining the carbon skeleton of microgram quantities of organic compounds by hydrogenation. It has been called the Beroza Carbon Skeleton Determinator. The sample, in a stream of hydrogen, was passed through a heated tube packed with a palladium catalyst. The reaction products were then swept into a GC and separated. Identification of the products was made by matching retention time with known hydrocarbons or by collecting samples for MS. This procedure has been utilised and modified by other workers[10].

Stransky, Gruz and Ruzicka[11] used a microreactor for determining the parent hydrocarbons of compounds by the reduction with zinc dust and hydrogen as the carrier gas. All aliphatic and aromatic halogen derivatives, alcohols, ketones and aldehydes were fully converted into their corresponding parent hydrocarbons by this method.

2. Dehydrogenation

In reaction gas chromatography, dehydrogenation has been used for the identification of hydrocarbons[12], and for some alcohols and amines[13], using the dehydrogenation equipment of Franc and Kolouskova[5]. Keulemans and Voge[14] and Rowan[15] found that a platinum–alumina–halogen catalyst used at 350°C was best for dehydrogenation, using hydrogen or helium carrier gas.

Mizrahi and Nigam[16] described the dehydrogenation of terpenes, using a platinum catalyst in a precolumn, and studied the reactions by the identification of the aromatics formed in the reaction.

C. Reagent reaction chromatography

1. Derivative formation

Many techniques have been described in which a precolumn or microreactor, containing a reagent, reacts with a sample to produce volatile components that are more readily separated by GC than the components of the original sample. Such a technique may be applicable to a non-volatile sample which may be converted to a volatile derivative for GC after injection into the apparatus.

Flash exchange chromatography One of the first applications of derivative formation in reaction chromatography was that of 'flash exchange' analysis introduced by Ralls[17-19] and applied to aldehydes, ketones and carboxylic acids. Aldehydes and ketones were isolated in the form of their 2,4-dinitrophenylhydrazones, and these derivatives were injected into a precolumn containing α-ketoglutaric acid, where they were heated for 10 s at 250°C. The involatile ketonic acid displaced the aldehydes and ketones, which were chromatographed in the usual way. For acids, their potassium or sodium salts were injected into the reactor, which contained potassium or sodium hydrogen sulphate to regenerate the volatile acids, but these determinations were not quantitative. Hunter[20] improved and extended the technique for application to carboxylic acids up to C_{18}. He formed the ethyl esters of the organic acids in an exchange reaction using potassium ethyl sulphate at 275°C. The resulting chromatogram of esters could be used to confirm the identities of carboxylic acids, previously detected in their free state by GC.

Acid and alkali reactions Salts of fatty acids in aqueous solution were converted by Thompson and Smith[21] to free fatty acids in a precolumn packed with an acid-coated support. In a similar manner, it would be expected that an alkaline packed precolumn would liberate amines from their salts prior to GC analysis.

Kato and Yamaura[22] used a precolumn of Chromosorb W mixed with either sodium or potassium hydroxide at 200°C for the determination of the fatty acid composition of edible oils and fats. The oil was injected into the precolumn in methanolic solution to form the methyl esters of the fatty acids, which were then separated by GC. Subsequently[23], a precolumn of Chromosorb W treated with 10% sulphuric acid was used for the esterification.

A precolumn of KOH on a suitable support may also be suitable for the saponification of relatively non-volatile esters, and thus liberate the alcohol for GC analysis. Saponification was used by Lundquist and Meloan[24] for the determination of polysorbates in food products. After extraction and clean-up, the esters were saponified on a reaction precolumn of soda-lime beads, the acid salt being retained and the polyol being separated and determined on a Carbowax 20M column at 120°C. The complete analysis was carried out in about 2.5 h, which was considerably shorter than many alternative methods. The reaction precolumn was changed every ten to twelve samples. Although the method was easy to make quantitative, all the polysorbates gave peaks of similar retention time, so the identification of individual polysorbates was not possible.

Esterification Considerable effort has been made to develop techniques for esterification in reaction chromatography, since esters chromatograph far more easily than their parent acids or alcohols. The conversion of acids to their methyl esters has been the subject of extensive study, particularly the derivatisation of fatty acids.

Bitner, Lanser and Dutton[25] carried out the esterification of fatty acids with tetramethylammonium hydroxide in a microreactor. The procedure consisted of taking up the tetramethylammonium hydroxide and sample fat into a microsyringe and introducing it into the microreactor. The reaction mixture from the microreactor was injected into the port of a GC at 250–300°C with helium gas flow. As the procedure involved the use of thermal degradation, it is similar to the examples found in the section on pyrolysis (see p. 84).

The microreactor apparatus of Davison and Dutton[26] was adapted by these workers for use in the transesterification of triglyceride oils[27]. The fatty acid compositions of soyabean, linseed and safflower oils were determined and compared with those obtained by the more usual procedures. The sample of oil for analysis was drawn into a syringe, together with a solution of sodium methoxide in methanol, and injected into the microreactor, without mixing. Excess methanol was evaporated off with the reactor at 50°C for approximately 30 s. Formic acid was then injected and the reactor temperature raised to 100°C. When no indication of acid was shown at the reactor outlet, the needle of the reactor was inserted into the GC injection port and the temperature was raised to 250°C to vaporise the methyl esters on to the GC column.

Moye[28] described a method for the on-column transesterification of *N*-methylcarbamate pesticide residues with methanol and NaOH. The reaction took place in the injection port, which was packed with untreated glass microbeads and held at a temperature of 220°C. The methyl *N*-methylcarbamate produced was chromatographed on a column of Porapak P and detected with a Rb_2SO_4 pellet AFID. This method was used by Van Middelem, Moye and Janes[29] for the determination of carbofuran residues in lettuce.

Silylation The application of reaction chromatography to the silylation of various classes of compounds was described by Esposito[30] and Esposito and Swann[31]. The technique had the ability to accommodate aqueous and alcoholic solutions, and its application was illustrated with fatty acids and polyhydric alcohols. A solution containing the sample was injected on to the chromatographic column, followed, immediately, by an injection of the TMS reagent. TMS derivatives were formed as the volatile reagents swept through the zones occupied by the reactive compounds, and were then chromatographed as they continued through the column. The TMS reagent was Silyl-8 (a mixture of BSA, TMSDEA and HMDS), and was chosen because of the absence of HCl release. Liquid phases, with active hydrogen sites, could not be used for the method, and the analyses were carried out on silicone oil columns.

2. Subtractive processes

The reaction of a sample with a reagent by chemical or physical means to selectively remove a component from the sample within the GC is an example of a reaction chromatography process referred to as a subtractive technique or subtractive process. A typical example is the chemical reaction between a functional group and a reagent to selectively remove the component by conversion to a non-volatile derivative. An example, by physical means, is the use of a molecular sieve to selectively remove a component such as water.

When combined with retention data, this technique can often be used to conclusively identify sample components. The sample would be chromatographed twice, once using the subtractive technique and once without it. Components which react with the reagent in the GC would be absent from the first chromatogram, and can thus be separated. The technique may also be used to resolve two components which are not separated on a particular GC column, by chromatographing the sample both with and without the use of a reagent that removes one of the components.

Subtractive techniques usually entail the use of a precolumn containing the reagent, although analytical columns containing a section coated with the reagent have also been used.

The technique has found a particular application in food analysis in the examination of food flavour volatiles, since these are generally rich in relatively low-boiling alcohols and carbonyl compounds, which can be selectively removed relatively easily by using an appropriate reagent.

Table 5.1 gives a list of selective reagents which have been used together with the classes of compounds which they remove. Usually the reagent is suspended on an inert material such as Celite, and is prepared in a similar manner to the GC column packings. It should be noted that several of the reagents have not been extensively used, as they appear to have poor reliability. Molecular

Table 5.1 Reagents for the selective removal of compounds in reaction chromatography

Compounds removed	Reagent	Ref.
Acids	potassium hydroxide	21, 32
Acids	zinc oxide	26, 33
Alcohols	boric acid	33, 34
Alcohols	boric acid + Carbowax	35
Alcohols	3-nitrophthalic anhydride	34
Aldehydes	o-dianisidine	33, 34
Aldehydes	FFAP	34, 36
Aldehydes	Sodium bisulphite + ethylene glycol	37, 38
n-Alkanes	molecular sieve	39
Amines	phosphoric acid	21
Carbonyl compounds	benzidine	33, 34
Carbonyl compounds	semicarbazide	34
Carbonyl compounds	sodium bisulphite	37, 40
Sulphur-containing compounds	mercuric chloride	37
Water	calcium carbide	41
Water	Drierite + PEG 400	42
Water	molecular sieve	43

sieves have been shown to remove only some straight-chain alkanes, and Withers[44] demonstrated that FFAP is unreliable for aldehydes.

The use of precolumn reactions for the selective removal of alcohols and carbonyl compounds in the GC analysis of mixtures was investigated by Cronin[34]. He used 3-nitrophthalic anhydride and semicarbazide as reagents for selectively subtracting alcohols and carbonyl compounds, respectively, from mixtures containing these compounds, and compared these reagents with a number of other compounds, viz. boric acid for alcohols, FFAP and o-dianisidine for aldehydes, and benzidine for aldehydes and ketones. The reagents were deposited on diatomaceous earth support, packed into a short precolumn and heated in the injection block of a GC. Details of the construction of the precolumns were given together with methods for attachment to PLOT columns. Each reagent was found to have advantages and disadvantages over the other for a specific functional group; e.g. in the case of alcohols, boric acid was reactive over a wide temperature range of 50–200°C but gave irregularities in the baseline on temperature programming, whereas 3-nitrophthalic anhydride was reactive only at moderately high temperatures but gave a stable baseline. Similarly, semicarbazide produced less peak broadening of non-carbonyl compounds than benzidine, but the former had a rather limited temperature range.

Bierl, Beroza and Ashton[33] introduced the term 'reaction loops' for the subtractor column containing suitable reagents located between the GC column and the detector. They found that a 6 in column of 20% benzidine on Chromosorb P effectively subtracted aldehydes, epoxides and most ketones. However, Haken, Ho and Withers[40] found this reagent and also o-dianisidine to be inferior to bisulphite for the removal of carbonyl compounds. They used a 10 in subtractor column packed with 10% sodium metabisulphite on Celite at 60°C, and obtained useful results with carbonyl compounds of boiling points up to 200°C.

Reaction Chromatography

The use of metal hydrides in GC precolumn reactions for the formation of non-volatile derivatives of alcohols, aldehydes, ketones, esters and epoxides was reported by Regnier and Huang[45]. Lithium borohydride and lithium aluminium hydride were found to be of equivalent reactivity, both subtracting all the functional groups studied. Sodium borohydride was effective in differentiating between aldehydes and ketones, and sodium trimethoxyborohydride in differentiating between ketones and all esters except acetates.

D. Pyrolysis

The best-known example of reaction chromatography is pyrolysis gas chromatography, where the samples are usually pyrolysed at high temperatures and the resulting fragments examined by GC. It is a particularly useful technique for the analysis of non-volatile materials. The pyrolysis gas chromatogram (pyrogram) provides a 'fingerprint' for the compound which may then serve to establish its identity.

Pyrolysis causes the thermal fragmentation of molecules in the compound and, under ideal conditions, the primary pyrolysis products are characteristic of the structure of this parent compound. Unfortunately, in the commonly used pyrolysis techniques the primary rupture of bonds is frequently followed by a complex series of competitive reactions and further secondary pyrolysis of the primary products. These serve to complicate the final fragmentation pattern, and are often the cause of poor reproducibility. Consequently, the ultimate aim of any pyrolyser design is to reduce secondary reactions to a minimum. Certain conditions are necessary in order to minimise the extent of secondary reactions: the heating rate must be as high as possible; the quantity of material to be pyrolysed must be minimal; and the primary fragments must diffuse quickly into the stream of carrier gas for dilution and cooling, in order to increase their stability.

Many different systems and methods have been described to produce the fragmentation pattern, and these were reviewed in great detail by Levy[46].

The choice of pyrolyser may depend to a large extent upon the particular application, but is usually one of three types: metal filaments in the form of a wire or ribbon resistively heated to the appropriate temperature[46]; a microreactor consisting of a sample boat enclosed in a furnace[46]; a ferromagnetic wire heated to its Curie temperature with an induction heater. The latter was described by Simon et al.[47] and Buehler and Simon[48], and is often referred to as the Curie point pyrolyser.

Several other fragmentation techniques have been reported, including the use of an electron discharge[49] and lasers[50]. The use of pulsed lasers for analytical pyrolysis is rapidly increasing. Although relatively expensive, they possess many advantages over other pyrolysers. For further details the reader is referred to the work of Biscar[51] and Ristau and Vanderborgh[52]. One considerable advantage is that they produce a fragmentation pattern close to ideal, i.e. secondary reactions at a minimum. A pyrolyser of this type was described by Kojima and Morishita[53].

In a pyrolyser the extent of degradation depends on the temperature and time of pyrolysis. Low-temperature pyrolysis or thermal degradation is used

84

to cause simple chemical changes to form another compound, the molecular weight of which does not differ greatly from the original; hence, the rupture of carbon—carbon bonds is minimal. This technique has been applied to the preparation of certain derivatives, and can often be performed successfully in the heated injection port of the GC.

Methyl esters can be prepared by heating the tetramethylammonium salts of carboxylic acids, which expels trimethylamine and leaves the ester. The reaction was applied to the direct production of a number of aryl and alkyl acid esters in the injection port of a GC by Robb and Westbrook[54]. Although a quantitative yield of the ester could be obtained, the temperature of the injection port was critical, and Downing[55] modified the procedure so that it could be applied to most simple organic acids, viz. saturated fatty acids. The essential feature of his modification was the removal of water before pyrolysis, which had the advantage that the efficiency of the conversion to methyl esters was unaffected by sample size or concentration. The yields were also largely insensitive to reagent concentration or to variation, within the wide limits, of injection port temperature.

Bailey[56], working with non-aqueous conditions, showed that the technique could be applied to the determination of mono- and dibasic aromatic and aliphatic acids when these are present as major components in mixtures containing non-acidic compounds.

Downing's method has been incorporated in the analysis of aqueous solutions of mono- and dicarboxylic acids[57], and in the analysis of polyunsaturated fatty acids[58]. When applied to polyunsaturated fatty acids, it was necessary to bring the solution of the tetramethylammonium salts to pH 7.5—8.0 with 5% acetic acid before taking it up in a capillary injection probe.

Generally, high-temperature pyrolysis has limited applications in food analysis, and few examples are cited in the literature. Dhont[59] used pyrolysis at 600°C to identify several organic compounds in food odours, and Karmen, Walker and Bowman[60] used it for the microdetermination of lipids.

References

1.   Bitner, E.D., Davison, V.L. and Dutton, H.J., *J. Amer. Oil Chem. Soc.*, **46**, 113 (1969)
2.   Huyten, F.H. and Rijnders, G.W.A., *Z. Anal. Chem.*, **205**, 244 (1964)
3.   Beroza, M. and Sarmiento, R., *Anal. Chem.*, **38**, 1042 (1966)
4.   Mounts, T.L. and Dutton, H.J., *Anal. Chem.*, **37**, 641 (1965)
5.   Franc, J. and Kolouskova, V., *J. Chromatog.*, **17**, 221 (1965)
6.   Koman, V., *J. Chromatog.*, **45**, 311 (1969)
7.   Beroza, M. and Acree, F., *J. Ass. Off. Agric. Chem.*, **47**, 1 (1964)
8.   Beroza, M., *Nature*, **196**, 768 (1962)
9.   Beroza, M. and Sarmiento, R., *Anal. Chem.*, **35**, 1353 (1963)
10.   Brownlee, R.G. and Silverstein, R.M., *Anal. Chem.*, **40**, 2077 (1968)
11.   Stransky, Z., Gruz, J. and Ruzicka, E., *J. Chromatog.*, **59**, 158 (1971)
12.   Klesment, I., *J. Chromatog.*, **31**, 28 (1967)
13.   Pacakova, V. and Smolkova, E., *J. Gas Chromat.*, **6**, 426 (1968)
14.   Keulemans, A.I.M. and Voge, H.H., *J. Phys. Chem.*, **63**, 476 (1959)
15.   Rowan, R., *Anal. Chem.*, **33**, 658 (1961)

16. Mizrahi, I. and Nigam, I.C., *J. Chromatog.*, **25**, 230 (1966)
17. Ralls, J.W., *Anal. Chem.*, **36**, 946 (1964)
18. Ralls, J.W., *J. Agric. Food Chem.*, **8**, 141 (1960)
19. Ralls, J.W., *Anal. Chem.*, **32**, 332 (1960)
20. Hunter, I R., *J. Chromatog.*, **7**, 288 (1962)
21. Thompson, G.F. and Smith, K., *Anal. Chem.*, **37**, 1591 (1965)
22. Kato, A. and Yamaura, Y., *Chem. & Ind.*, 1260 (1970)
23. Kato, A., Tomita, H. and Yamaura, Y., *Chem. & Ind.*, 302 (1971)
24. Lundquist, G. and Meloan, C.E., *Anal. Chem.*, **43**, 1122 (1971)
25. Bitner, E.D , Lanser, A.C. and Dutton, H.J., *J. Amer. Oil Chem. Soc.*, **48**, 633 (1971)
26. Davison, V.L. and Dutton, H.J., *Anal. Chem.*, **38**, 1302 (1966)
27. Davison, V.L. and Dutton, H.J., *J. Lipid Res.*, **8**, 147 (1967)
28. Moye, H.A., *J. Agric. Food Chem.*, **19**, 452 (1971)
29. Van Middelem, C.H., Moye, H.A. and Janes, M.J., *J. Agric. Food Chem.*, **19**, 459 (1971)
30. Esposito, G.G., *Anal. Chem.*, **40**, 1902 (1968)
31. Esposito, G.G. and Swann, M.H., *Anal. Chem.*, **41**, 1118 (1969)
32. Sato, T., Shinriki, N. and Mikami, Y., *Bunseki Kagaku*, **14**, 223 (1965)
33. Bierl, B.A., Beroza, M. and Ashton, W.T., *Mikrochim. Acta*, 637 (1969)
34. Cronin, D.A., *J. Chromatog.*, **64**, 25 (1972)
35. Ikeda, R.M., Simmons, D.E. and Grossman, J.D., *Anal. Chem.*, **36**, 2188 (1964)
36. Allen, R.R., *Anal. Chem.*, **38**, 1287 (1966)
37. Bassette, R. and Whitnah, C.H., *Anal. Chem.*, **32**, 1098 (1960)
38. Kerr, J.A. and Trotman-Dickenson, A.F., *Nature*, **182**, 466 (1958)
39. Whitham, B.T., *Nature*, **182**, 391 (1958)
40. Haken, J.K., Ho, D.K.M. and Withers, M.K., *J. Chromat. Sci.*, **10**, 566 (1972)
41. Kung, J.T., Whitney, J.E. and Cavagnol, J.C., *Anal. Chem.*, **33**, 1505 (1961)
42. Cohen, E.N. and Brewer, H.W., *J. Gas Chromat.*, **2**, 261 (1964)
43. McKinney, R.W. and Jordan, R.L., *J. Gas Chromat.*, **5**, 501 (1967)
44. Withers, M.K., *J. Chromatog.*, **66**, 249 (1972)
45. Regnier, F.E. and Huang, J.C., *J. Chromat. Sci.*, **8**, 267 (1970)
46. Levy, R.L , *Chromatog. Rev.*, **8**, 48 (1966)
47. Simon, W., Kriemler, P., Voellmin, J.A. and Steiner, H., *J. Gas Chromat.*, **5**, 53 (1967)
48. Buehler, C. and Simon, W., *J. Chromat. Sci.*, **8**, 323 (1970)
49. Barlow, A , Lehrle, R.S. and Robb, J.C., *Polymer*, **2**, 27 (1961)
50. Folmer, O.F. and Azarraga, L.V., *J. Chromat. Sci.*, **7**, 665 (1969)
51. Biscar, J.P., *Anal. Chem.*, **43**, 982 (1971)
52. Ristau, W.T. and Vanderborgh, N.E., *Anal. Chem.*, **42**, 1848 (1970)
53. Kojima, T. and Morishita, F., *J. Chromat. Sci.*, **8**, 471 (1970)
54. Robb, E.W. and Westbrook, J.J., *Anal. Chem.*, **35**, 1644 (1963)
55. Downing, D.T., *Anal. Chem.*, **39**, 218 (1967)
56. Bailey, J.J., *Anal. Chem.*, **39**, 1485 (1967)
57. Downing, D.T. and Greene, R.S., *Lipids*, **3**, 96 (1968)
58. Downing, D.T., and Greene, R.S., *Anal. Chem.*, **40**, 827 (1968)
59. Dhont, J.H., *Nature*, **200**, 882 (1963)
60. Karmen, A., Walker, T. and Bowman, R.L., *J. Lipid Res.*, **4**, 103 (1963)

6
Interpretation of Results

A. Qualitative interpretation

If a GC column is sufficiently specific, the components of the sample being analysed are separated completely and are indicated as individual peaks on the chromatogram. Frequently this is not the case, and components are not separated, or are only partially so, on the particular column. Thus it is very difficult to establish whether a chromatographic peak represents one or more components. Even if complete separation is obtained, it is unlikely that a study of the chromatogram alone will conclusively establish the identity of a particular component, and the analyst may have to make use of other identification techniques.

Generally, the identification of unknown components in a sample is attempted by an investigation and comparison of retention data, which are derived from the measurement of the retention times of the components on one or more columns. The retention time is the time spent by a component in the column between injection and detection and, on a given column, is specific for a particular component. However, there are several limitations in using absolute values of retention times for comparison, since retention data depend on both the reproducibility of the column and the analytical conditions.

Both the retention time, t, and the adjusted retention time, t', where $t' = t - t_A$ (see *Figure 2.3*), are adequate for normal use, but in some cases, particularly for physical measurements, use is made of retention volume, V, which is the volume of carrier gas flowing through the column during the period of retention. It can be deduced from the equation:

$$V = tF \qquad (6.1)$$

where F is the carrier gas flow rate. Retention volumes are less dependent on column and operating characteristics than retention times, and yet their use has not been generally accepted.

Retention times frequently vary slightly from one analysis to another, so a direct comparison of the retention time of a sample component with that of a reference compound, under the same chromatographic conditions, is usually unsatisfactory for qualitative identification. It is also likely that more than one sample component may have the same retention time, so the comparative technique is qualitatively insufficient and frequently inconclusive. This problem can be overcome, to some extent, by evaluating the sample component and the reference compound under the same conditions on several columns containing different stationary phases, preferably of varying polarities. The difficulty of variable retention times can be effectively overcome by the use of relative retention times, *r*, instead of those which are absolute. In this case the retention times of the components are expressed relative to a certain standard component chromatographed under the same conditions. The standard may be a component of the sample, but, if not, it will then be added to the sample prior to GC analysis. In both instances the standard is referred to as an internal standard, and is preferred to an external one, where it is injected into the GC prior to the sample. In general, the choice of standard should be made with regard to its availability and its retention time, which ideally should fall near the middle of the chromatogram. A comparison of the relative retention time of an unknown component with that of a reference compound, chromatographed under the same conditions and related to the same standard, makes the identification of the component more feasible. In dubious cases relative retention data on two columns of different polarities should be obtained.

The literature is full of the retention data of compounds quoted on different columns, at different temperatures and with various internal standards, but because of these variable parameters such reported values are rarely of immediate value to other analysts. Unfortunately, there has been little standardisation in reporting this type of information.

The retention index system proposed by Kováts[1] was an attempt to standardise quoted retention data and ensured that the retention time of a component was expressed in a uniform scale compiled from a series of closely related standards. These standards, which were based on data from n-paraffins, provided fixed points on a linear scale and the retention index, *I*, of the component could be calculated from the equation:

$$I = 100 \left[\frac{\log t'(C_x) - \log t'(C_z)}{\log t'(C_{z+1}) - \log t'(C_z)} + z \right] \qquad (6.2)$$

where C_x is an unknown component, C_z and $C_z + 1$ are normal paraffins with carbon number z and $z + 1$, respectively, and t' is the adjusted retention time or distance. Here the retention index for the n-paraffins is 100 times the carbon number, e.g. 200 for ethane. The n-paraffins chosen for use as standards are those that are found to straddle the unknown component on the chromatogram, and consequently the retention index for the unknown component will fall between the retention indices of these particular paraffins. The retention data for the two adjacent n-paraffins can be determined by mixing them with the unknown component or determining them individually under identical conditions. Although it is preferable to use n-paraffins as the standards, it may not always be possible and other standards can be used that have had their retention indices previously measured relative to the n-paraffin scale.

Retention indices thus offer a means of reporting retention data on a relative basis and it is rapidly becoming accepted as the method of choice for reporting chromatographic data, being especially useful for qualitative identification. The concept has a considerable literature, and Ettre[2] has given a detailed summary of every aspect of the system. McReynolds[3] has listed retention indices for many substances on 77 different liquid phases.

An important characteristic of the Kováts system is that a plot of retention index against column temperature is usually linear for a particular component on a given column. Such plots enable retention indices at different temperatures to be established by interpolation and they assist in the assessment of elution sequences for sample components on a particular column at different temperatures.

The Kováts system has enabled the relationship between retention indices and molecular structure to be investigated, and this aspect is discussed in detail in the literature[1,4].

Even though the Kováts system was designed to minimise errors in the reporting of retention data from different columns, large ranges of values have sometimes been reported for solutes on supposedly similar columns operated under identical conditions[5]. These may be attributed to effects of sample size, nature of the liquid phase, polarity and temperature. In particular, care must be taken in applying retention indices in analyses of trace components. Errors in the determination of retention indices and the effects of various factors on these errors have been discussed in several papers[6-8].

The Kováts system is applied mainly in isothermal analysis. However, van den Dool and Kratz[9] showed that for normal linear temperature programmed GC the relationship between retention time and carbon number was nearly linear within a homologous series, such that the Kováts equation could be simplified by eliminating its logarithmic nature. This equation was equivalent to the one given by Douglas[10], as proposed by Harbourn for general use. Harbourn suggested that the analyst's unwillingness to adopt Kováts' retention index, universally, was due to its being a logarithmic relationship, and therefore time-consuming to measure.

The term 'retention master' was introduced by Rasquinho[11] in an attempt to standardise chromatographic data. It was, simply, a visual presentation of the Kováts index under linear temperature programmed conditions to assist with the characterisation of essential oils.

Other retention systems have been proposed, including one suitable for temperature programmed GC by Schmit and Wynne[12], in which a retention measurement was developed based on the ratio of the elution temperature of an internal standard and the sample component.

B. Quantitative interpretation

It is well known to analysts engaged in GC that the quantitative analysis of a sample involves the measurement of the area under a chromatographic peak, as this is directly related to the amount of component present. If the identity of the component is known, a quantitative assessment can be made by comparison with standard amounts of the pure compound under the same chromatographic conditions.

89

Interpretation of Results

1. Peak area calculation

Peak areas are generally determined by using one of the following methods.

(1) Multiplying the peak height at the maximum by the width of the peak at half its height. For a Gaussian peak, this multiple gives 94% of the total area obtained with an integrator. Since this applies equally to both the component peak and the standard peak, it does not reduce the accuracy of the calculation. The method is widely used, but has limitations, viz. half-widths are difficult to measure for sharp peaks and errors occur in the case of non-Gaussian peaks. Hawkes and Russell[13] evaluated this method for peaks which have a sloping baseline. Ball, Harris and Habgood[14] evaluated the effect of fractional height, i.e. the fraction of the height at which the width is measured, with respect to the precision of peak area measurement. The optimum fractional height was shown to depend on peak shape. In general, sharp Gaussian peaks should be measured for width close to the baseline, and flat broad peaks should be measured at the half-height. For practical purposes a single fractional height of a quarter would give the best compromise for the full range of Gaussian peaks normally encountered.

(2) Triangulation of the peak. A triangle is formed by drawing two tangents to the sides and through the inflection points of the peak, with the base taken as the intersection of the tangents with the baseline. The area is then calculated as one-half of the product of the height of the triangle and the base. The calculated area represents about 97% of the integrated area, but large errors occur with tailing peaks.

(3) Cutting out the recorded peaks from the chart and weighing the paper on an analytical balance. Although fairly accurate, the method is time-consuming, requires care in cutting and, consequently, is rarely used.

(4) Tracing the peak curve with a planimeter. The method is time-consuming and has no greater accuracy than alternative manual methods.

(5) Automatic integration, which may be achieved on integrators ranging from the semi-automatic ball and disc type to the electronic digital integrator and the computer integrator. These can be costly, but the expense may be justified on account of their greater accuracy.

Other methods of determining peak areas, or area values relative to them, have been used by analysts, including the product of the peak height and the retention time[15], and the product of the time taken between the deflection of the recorder pen from the baseline to the top of the peak and the peak height[16]. In the latter example it was claimed that the value is proportional to the amount of the component causing the peak, and is unaffected by peak shape.

The measurement of peak areas is tedious and sometimes difficult. The peaks at the beginning of a chromatogram are frequently tall and narrow; consequently, the errors in measuring the width are large. On the other hand, peak heights are easily and accurately measured, so, in some such cases, peak heights are used instead of areas for quantitative evaluation. Calculations based on peak height are generally preferable for analyses in which the peak width is small and also when the concentration of the component is high. Peak heights are very sensitive to instrument conditions such as changes in column temperature and flow rate, column efficiency and sample injection, and so they should only be used when these variations are eliminated.

Ball, Harris and Habgood[17] showed that the relative indeterminate error was always smaller if only peak height was used and not peak area. This supported the general conclusion often reached in practical work that with good experimental control and stability of the GC system maximum quantitative precision could be obtained from measurements of peak height. The potential gain in precision that would be achieved by the use of height measurements justified an intensive effort towards an improvement of experimental performance and instrumental design.

2. Partially overlapping peaks

When overlapping peak areas in chromatograms are quantitatively measured, interpretation is extremely difficult and large errors frequently result. Sometimes the problem can be overcome experimentally by using columns that provide better resolution, but this often results in excessive retention times. The exact form of the individual peaks cannot be determined easily, and approximations are necessary. Usually, approximations are made using one of two methods. In the first method the two peaks may be divided by a straight line drawn as perpendicular to the baseline at the minimum point between two peaks. Inherent errors in this method were investigated by Kishimoto, Miyauchi and Musha[18]. In the second method the areas of the two peaks are obtained by calculating the areas of the triangles formed by dropping the tangents to the two peaks at the inflection points on the baseline; and this method is generally used if the resolution between the peaks is very poor.

A method for evaluating partially resolved peaks was proposed by Proksch, Bruneder and Granzner[19], which consisted of measuring the areas at both sides of the perpendicular through the minimum of the two peaks with an electronic integrator and multiplying these areas by appropriate correction factors to obtain the true area of the two component peaks. A set of correction factors was tabulated for universal application to partially resolved Gaussian peaks.

3. Quantitative evaluations

The quantification of samples may be carried out by the selection of one of the following methods, in combination with an appropriate method of calculating the peak area.

Internal normalisation In this method the concentration of each component in the sample is calculated by establishing the peak area of each component and then expressing these as a percentage of the sum of the individual peak areas. The values obtained are referred to as the relative peak areas, and in certain cases would give, directly, the concentration of the components in the sample. This is particularly true for most detectors if a relatively small range of closely

91

related compounds is analysed, that have similar molecular weights, e.g. many free fatty acids or fatty acid methyl esters.

Comparison with a separate standard This method is only suitable where the chromatogram does not permit the use of an internal standard, e.g. when there are too many components in the chromatogram. Each peak is compared with the corresponding peak of the standard mixture. The accuracy of the method depends on the reproducibility of the analytical conditions and on the reproducibility of the amount of sample injected during the analysis of both the sample and the external standard.

Internal standard The internal standard method involves the addition of a known amount of a selected compound to a known amount of sample. This mixture is then chromatographed and evaluated by comparison with several reference mixtures containing the same amount of selected compound added to various amounts of the component under evaluation in the sample. The selected compound is referred to as an internal standard, and is chosen such that it is located in the chromatogram close to the peaks under evaluation, without any overlap and such that its concentration produces a peak height similar to those peaks of interest. Obviously, the compound chosen must not react chemically with the sample, and must not be present in it. Calibration curves can readily be established because, in this case, the ratio of the area of the sample component peak to the area of the standard peak in the reference compound is directly proportional to the concentration of the respective component in the sample.

The principal advantages of this method are that the exact reproducibility of sample injection volumes is not so important and the method can accommodate samples containing a multiplicity of peaks.

References

1. Kováts, E., *Helv. Chim. Acta,* **41,** 1915 (1958)
2. Ettre, L. S., *Anal. Chem.,* **36,** (8), 31A (1964)
3. McReynolds, W. O., *Gas Chromatographic Retention Data,* Preston Technical Abstracts Co., Evanston, Ill. (1966)
4. Kováts, E. and Strickler, H., *J. Gas Chromat.,* **3,** 244 (1965)
5. Kaiser, R., *Chromatographia,* **3,** 127 (1970)
6. Takacs, J., Rockenbauer, M. and Olacsi, I., *J. Chromatog.,* **42,** 19 (1969)
7. Takacs, J. and Kralik, D., *J. Chromatog.,* **50,** 379 (1970)
8. Erdey, L., Takacs, J. and Szalanczy, E., *J. Chromatog.,* **46,** 29 (1970)
9. van den Dool, H. and Kratz, P. D., *J. Chromatog.,* **11,** 463 (1963)
10. Douglas, A. G., *J. Chromat. Sci.,* **7,** 581 (1969)
11. Rasquinho, L. M. A., *J. Gas Chromat.,* **3,** 340 (1965)
12. Schmit, J. A. and Wynne, R. B., *J. Gas Chromat.,* **4,** 325 (1966)
13. Hawkes, S. J. and Russell, C. P., *J. Gas Chromat.,* **3,** 72 (1965)
14. Ball, D. L., Harris, W. E. and Habgood, H. W., *Anal. Chem.,* **40,** 1113 (1968)
15. Bartlet, J. C. and Smith, D. M., *Can. J. Chem.,* **38,** 2057 (1960)
16. Tamura, H. and Hozumi, K., *Bunseki Kagaku,* **20,** 149 (1971)
17. Ball, D. L., Harris, W. E. and Habgood, H. W., *Anal. Chem.,* **40,** 129 (1968)
18. Kishimoto, K., Miyauchi, H. and Musha, S., *J. Chromat. Sci.,* **10,** 220 (1972)
19. Proksch, E., Bruneder, H. and Granzner, V., *J. Chromat. Sci.,* **7,** 473 (1969)

7
Other Identification and Related Techniques

A. Introduction

GC is essentially a technique of separation, and additional evidence is usually required for the identification of individual components as they are eluted. A considerable amount of retention data is available, a large part of which is in the form of retention indices[1] (see Chapter 6, p. 88). Although it would be of great value, unambiguous identification using retention measurements from a single GC column is not possible. Further evaluation on other types of GC column may help or additional information may be obtained from the use of other chromatographic techniques, e.g. TLC[2], PC and HPLC. Nevertheless, other means of positive identification should be sought, and may entail collecting the column eluants from the GC using fraction collecting techniques. The isolated components may then be analysed by such spectroscopic techniques as IR, UV, NMR, laser Raman and MS. Sometimes the sample may not be eluted from the GC in a form suitable for subsequent examination by these techniques, in which case alternative methods of identification such as chemical reactions using micro-techniques and subtractive processes (see Chapter 5, p. 82) can be used.

Many problems arise in GC, where the component of interest is a trace constituent of the sample. Unless an analytical preparative GC is used, the injection of large amounts of the sample usually reduces the column efficiency, and so spectroscopic methods capable of examining microgram or submicrogram amounts are essential. Consequently, the trapping techniques discussed in the text have been subdivided into those applicable for the collection of milligram quantities and those specifically intended for isolation at the microgram level. The latter often involve special techniques, many of which are specific to the chosen type of spectroscopic examination.

The direct combination of a GC with an MS or with a rapid-scan IR spectrophotometer provides a very versatile analytical instrument. The former is expensive, requiring considerable expertise and lengthy interpretation. The latter is generally restricted in sensitivity to approximately 100 μg of compound, although greater sensitivity has been achieved. Thermal fragmentation of the GC column eluant in a controlled manner, using pyrolysis techniques, is another aid to identification and is less expensive than MS techniques.

93

B. Fraction collection

Many methods have been used for collecting fractions from GCs for further instrumental or chemical analysis, and were the subject of a review by Littlewood[3].

1. Milligram and submilligram quantities

Some methods that have been used are relatively simple and effective, while others are elaborate or too specific in their applications.

Three basic methods are frequently encountered. The first method involves the collection of the components in a tube containing GC packing, which gives good trapping efficiency but is often in an inconvenient form to subsequently obtain spectra. The second method involves the condensation of the components in trapping vessels or tubing, but this is usually less efficient than the first method owing to the number of parameters involved. These parameters include the temperature and geometry of the trapping vessel, the carrier gas flow rate and the vapour pressure of the components. Also, many high-boiling components tend to undergo aerosol formation and are not trapped successfully, especially if the cooling conditions are too severe. In general, compounds with a high vapour pressure require more extreme cooling for efficient trapping. The third method involves the adsorption of the components by contact with a solvent at or below ambient temperature.

A number of GC traps of varying complexity, designed to prevent the formation of aerosols, have been described in the literature. Thermal gradient[4-6], electrostatic[7-10] and centrifugal traps[11] have all been successful for the trapping of high-boiling compounds, but most were designed for preparative work and were too large for routine analysis. Total eluant traps have also been described, in which the carrier gas was condensed with the sample in a trap surrounded by liquid nitrogen. Argon[12,13] or carbon dioxide[14,15] was used as the carrier gas, carbon dioxide being preferred because of the ease of condensation. These methods were useful when dealing with low-boiling compounds, and quantitative recoveries were obtained in the milligram range. Other approaches for overcoming aerosol formation have been the co-condensation and solvent entrainment[16-18] and filter methods[19,20] using a filtration medium such as potassium bromide[21].

2. Microgram and submicrogram quantities

Many authors have described a variety of techniques for the collection and transfer of microgram amounts of samples eluted from packed[22-26] and, to a lesser extent, from capillary columns[27].

A technique which is frequently used is the collection of fractions in cooled capillary tubes[26,28] or in capillaries packed with adsorbent material. Various adsorbents have been used, including support material[29], column packing material[30,31], molecular sieve[32], activated charcoal[33] and porous glass[34].

A simple quantitative method for trapping both high- and low-boiling compounds from PLOT columns down to levels of 20 ng or lower was described by Cronin[35]. The fractions were collected in short U-traps of PLOT tubing coated with a liquid phase, which were cooled in a small Dewar flask containing liquid nitrogen. The technique was developed particularly for use in the study of food volatiles.

Curry *et al.*[36] and De Klein[37] collected samples down to a few micrograms by a cold trap technique. The collected material was transferred to potassium bromide powder and was examined by IR spectroscopy after the preparation of a microdisc, followed by the use of a refracting beam condenser to increase the sample transmission.

A technique for collecting microgram fractions in cooled glass tubes containing a solid support and transferring them into an IR cell by condensation with CCl_4 or CS_2 was described by Copier and Schutte[38], and found to be equally suitable for high- and low-boiling compounds.

Chang[39] described a trap suitable for direct insertion into a septum at the exit port of a chromatograph. It was constructed from capillary glass tubing bent into a U-shape with restrictions drawn every 1 cm to prevent aerosol formation. This was then immersed in a Dewar flask containing dry ice and after collection of the fraction it was centrifuged so that the sample was forced into the lower U-shaped section of the trap. It was then ready for washing out with a suitable solvent or for direct removal with a syringe. This trap was used successfully for the collection of volatile flavour compounds in food. A micro-cavity cell of 0.1 mm path length was used for the IR spectra. With solvent compensation and a beam condenser, an identifiable spectrum was obtained with 0.05 mg of sample.

C. MS

The analytical potential that has arisen as the result of the combination of MS with GC (GC–MS) has led to the development of various interface devices, known as molecular separators. These have been reviewed in a publication by Littlewood[40]. The function of the separator is to diminish the total mass of the carrier gas entering the MS to a level that will enable the maintenance of an adequate vacuum sufficient to prevent losses in MS resolution. In addition, it should be possible to achieve this function without giving rise to any appreciable loss of sample.

The technique of molecular separation commenced with the jet orifice separator[41] and the fritted glass separator[42, 43], but many other separators have since been suggested. Most separators now rely on a membrane device through which the carrier gas, usually helium, and sample components are separated, as the result of their different rates of permeation. The separator is usually enclosed in a constantly evacuated chamber, to enable the carrier gas to be pumped away efficiently.

The Watson–Biemann separator[42] consists of a heated fritted-glass tube, the exterior of which is exposed to a vacuum (1 mmHg) to preferentially remove the carrier gas. The separator is constructed of glass for chemical inertness, and yet, even at normal operating temperatures, there has been a history of adsorption, resulting in peak distortion, peak delay and low yields. MacLeod and Nagy[44] reported a significant loss of certain organic compounds, particularly polar compounds, even after washing the separator with acid. Deactivation of the porous glass surface by silanisation *in situ* with BSA using a microsyringe to inject it through a septum positioned in the transfer line proved to be satisfactory. Any excess reagent and by-products were pumped out of the MS after a few hours' baking. Other silanisation reagents have been used successfully for this purpose, including TMCS.

The efficiency or yield of a separator is defined as the ratio of the amount of sample surviving the enriching process to the amount which enters the separator, expressed as a percentage:

$$\% \text{ efficiency} = \frac{\text{sample out}}{\text{sample in}} \times 100 \qquad (7.1)$$

From an experimental point of view, the efficiency value alone is insufficient to describe the operating parameters of an enrichment device, since a value of 100% would be obtained if no separation was used at all. Consequently, the enrichment factor (separation factor) is used to give a measure of the degree of enrichment that has taken place by comparing the amount of sample component in the separator effluent with the amount in the carrier gas, i.e. solute enrichment:

$$\frac{\text{enrichment}}{\text{factor}} = \frac{\text{sample out}}{\text{carrier out}} \times \frac{\text{carrier in}}{\text{sample in}} = \text{efficiency}\left(\frac{\text{carrier in}}{\text{carrier out}}\right) \quad (7.2)$$

Krueger and McCloskey[45] constructed two interface systems for GC, consisting of silanised porous stainless steel tubing. They were similar in design to the Watson—Biemann separator[43]. The principal advantages of this porous stainless steel type interface were given as rugged construction, high efficiency, ability to bake at high temperatures, ease of disassembly for cleaning and adaptability by design to any instrument.

The ability of palladium and a few of its alloys to selectively diffuse hydrogen has led to their use as molecular separators, as they are potentially completely efficient in the removal of this gas. These separators are unique, as, in effect, it is possible to remove all the carrier gas and therefore the calculated value for the enrichment factor approaches infinity. Consequently, the performance of palladium separators cannot be strictly evaluated in the terms used for other types of separators.

Lucero and Haley[46] discussed the theory of the palladium—silver alloy hydrogen separator, and examples of its use in GC—MS were reported by Lovelock, Charlton and Simmonds[47].

Simmonds, Shoemake and Lovelock[48] developed an efficient interface of palladium—silver tubing which was both simple and rugged and gave a quantitative delivery of most compounds. It diffused hydrogen with an optimum efficiency close to 100% at temperatures in excess of 200°C, if it was placed in an oxidising atmosphere such as air. They reported an investigation of the extent of catalytic changes that may arise in the structure or composition of certain compounds as the result of using this type of separator. Potentially sensitive functional groups, such as aldehydes and nitriles, did not undergo reductions, and, in general, catalytic reduction only occurred in compounds with conjugated unsaturation.

The use of chemical ionisation MS in GC–MS[49] enables the effluent from the GC to be used directly without a molecular separator, if the carrier gas is suitable as a chemical ionisation reactant gas. Chemical ionisation spectra (CI–MS) are frequently less complex and easier to interpret than electron impact spectra (EI–MS). Frequently methane is used as the carrier gas and often the MS is of the quadrupole type. A MS with two ion sources, one EL and one CI, has been described[50], which allows the simultaneous EI–CI MS of the GC effluent.

D. Thermal fragmentation

The use of thermal fragmentation to identify components trapped at the outlet of an analytical GC has been investigated by many analysts[51–54].

Levy and Paul[55] described a two-unit GC system, consisting of a conventional analytical GC, a flow delay-line trap, a pyrolyser and a second GC. It was shown that the pyrolysis of several types of organic compounds was reproducible and analogous to their MS.

A similar unit was described by Sternberg, Krull and Friedel[56] which used a high-pressure electrical discharge between the chromatographs. The discharge caused fragmentation of the eluted compounds from the first chromatograph, and the stable fragments were identified on the second chromatograph, which produced a fragmentation spectrum.

A bibliography of thermal fragmentation was compiled by Sarner[57].

E. Rapid scan IR spectrometry

Several IR spectrometers have been adapted to record the spectra of GC fractions. Bartz and Ruhl[58] and Wilks and Brown[59] modified instruments that gave satisfactory spectra for scan speeds of 5–40 s with minimum sample amounts of 100–1000 μg depending upon the transmission characteristics of the compound. A multiple scan interference spectrometer was used by Low and Freeman[60] to record the effluents from a GC column. Although the instrument could scan the 2500–250 cm^{-1} region in 1s, multiple scans greater than 100 were needed to obtain good sensitivity.

Brown *et al.*[61] described a rapid scan instrument which had been developed for operation with SCOT and packed columns. The spectra were measured as the sample flowed through a heated, windowless, multiple internal reflectance cell of 1 ml volume. The scan speeds of the instrument for the 3700–750 cm^{-1} region were variable between 5 and 20 s. Good-quality spectra were obtained with as little as 20 μg of sample, and functional group bands were observed at 3 μg level.

F. Chemical microtechniques

Microreactors are frequently employed between the injection point and the GC column and, occasionally, they can be used to some advantage at the exit of the column.

97

Other Identification and Related Techniques

For the examination of food flavour volatiles, GC–MS utilising capillary columns is frequently used, and, in the case of fully resolved components, the MS and retention data usually provide sufficient evidence for identification. Often, unresolved or trace components require additional structural evidence to ensure identification, but rarely is there sufficient sample for other spectrometric examinations to be made.

However, the use of a micro-reactor at the exit of the GC column, which could cause the catalytic reaction with hydrogen, would give hydrogenation or hydrogenolysis products suitable for further examination by GC, or preferably GC–MS. Teeter et al.[62] and Issenberg, Kobayashi and Mysliwy[63] carried out such a method, placing the micro-reactor 'on-line' between the GC column exit and the MS inlet.

To avoid any misinterpretation of the spectra resulting from incomplete hydrogenation, Stanley and Murray[64] preferred to use an 'off-line' method, collecting the resolved components from a stream-splitter at the GC–MS interface[65], then causing them to react with hydrogen and subsequently re-examining the products by GC–MS.

Many other analysts have employed the microhydrogenation technique for the identification of GC effluents[66-71]. It has been applied to sulphur-, halogen-, oxygen-, nitrogen- and phosphorus-containing compounds, often to microgram quantities.

References

1. Kováts, E., in *Advances in Chromatography*, Vol. 1 (J.C. Giddings and R.A. Keller, Eds.), Marcel Dekker, New York (1965)
2. Humphrey, A.M., *J. Chromatog.*, **53**, 375 (1970)
3. Littlewood, A.B., *Chromatographia*, **1**, 223 (1968)
4. Schlenk, H. and Sand, D.M., *Anal. Chem.*, **34**, 1676 (1962)
5. Teranishi, R., Corse, J.W., Day, J.C. and Jennings, W.G., *J. Chromatog.*, **9**, 244 (1962)
6. Stevens, R.K. and Mold, J.D., *J. Chromatog.*, **10**, 398 (1963)
7. Kratz, P., Jacobs, M. and Mitzner, B.M., *Analyst*, **84**, 671 (1959)
8. Thompson, A.E., *J. Chromatog.*, **6**, 454 (1961)
9. Ross, W.D., Moon, J.F. and Evers, R.L., *J. Gas Chromat.*, **2**, 340 (1964)
10. Fish, D.W. and Crosby, D.G., *J. Chromatog.*, **37**, 307 (1968)
11. Wehrli, A. and Kováts, E., *J. Chromatog.*, **3**, 313 (1960)
12. Swoboda, P.A.T., *Nature*, **199**, 31 (1963)
13. Howlett, M.D.D. and Welti, D., *Analyst*, **91**, 291 (1966)
14. Hornstein, I. and Crowe, P., *Anal. Chem.*, **37**, 170 (1965)
15. Bache, C.A. and Lisk, D.J., *J. Chromat. Sci.*, **7**, 296 (1969)
16. Jones, J.H. and Ritchie, C.D., *J. Ass. Off. Agric. Chem.*, **41**, 753 (1958)
17. Hardy, R. and Keay, J.N., *J. Chromatog.*, **17**, 177 (1965)
18. Tsuda, T. and Ishii, D., *J. Chromatog.*, **47**, 469 (1970)
19. Witte, K. and Dissinger, O., *Z. Anal. Chem.*, **236**, 119 (1968)
20. Schluter, E.C., *Anal. Chem.*, **41**, 1360 (1969)
21. Leggon, H.W., *Anal. Chem.*, **33**, 1295 (1961)
22. Badings, H.T. and Wassink, J.G., *J. Chromatog.*, **18**, 159 (1965)
23. Teranishi, R., Flath, R.A., Mon, T.R. and Stevens, K.L., *J. Gas Chromat.*, **3**, 206 (1965)
24. Beroza, M., *J. Gas Chromat.*, **2**, 330 (1964)

25. Bierl, B.A., Beroza, M. and Ruth, J.M., *J. Gas Chromat.*, **6**, 286 (1968)
26. Woolley, W.D., *Analyst*, **94**, 121 (1969)
27. Odland, R.K., Glock, E. and Bodenhamer, N.L., *J. Chromat. Sci.*, **7**, 187 (1969)
28. Burson, K.R. and Kenner, C.T., *J. Chromat. Sci.*, **7**, 63 (1969)
29. Shearer, D.A., Stone, B.C. and McGugan, W.A., *Analyst*, **88**, 147 (1963)
30. Murray, K.E., Shipton, J., Robertson, A.V. and Smyth, M.P., *Chem. & Ind.*, 401 (1971)
31. Dandoy, J. and Delvaux, J., *Chem. & Ind.*, 592 (1971)
32. Cartwright, M. and Heywood, A., *Analyst*, **91**, 337 (1966)
33. Damico, J.N., Wong, N.P. and Sphon, J.A., *Anal. Chem.*, **39**, 1045 (1967)
34. Kane, D.M. and Karasek, F.W., *J. Chromat. Sci.*, **10**, 501 (1972)
35. Cronin, D.A., *J. Chromatog.*, **52**, 375 (1970)
36. Curry, A.S., Read, J.F., Brown, C. and Jenkins, R.W., *J. Chromatog.*, **38**, 200 (1968)
37. De Klein, W.J., *Z. Anal. Chem.*, **246**, 294 (1969)
38. Copier, H. and Schutte, L., *J. Chromatog.*, **47**, 464 (1970)
39. Chang, S.S., *Food. Technol.*, **27** (4), 27 (1973)
40. Littlewood, A.B., *Chromatographia*, **1**, 37 (1968)
41. Ryhage, R., *Anal. Chem.*, **36**, 759 (1964)
42. Watson, J.T. and Biemann, K., *Anal. Chem.*, **36**, 1135 (1964)
43. Watson, J.T. and Biemann, K., *Anal. Chem.*, **37**, 844 (1965)
44. MacLeod, W.D. and Nagy, B., *Anal. Chem.*, **40**, 841 (1968)
45. Krueger, P.M. and McCloskey, J.A., *Anal. Chem.*, **41**, 1930 (1969)
46. Lucero, D.P. and Haley, F.C., *J. Gas Chromat.*, **6**, 477 (1968)
47. Lovelock, J.E., Charlton, K.W. and Simmonds, P.G., *Anal. Chem.*, **41**, 1048 (1969)
48. Simmonds, P.G., Shoemake, G.R. and Lovelock, J.E., *Anal. Chem.*, **42**, 881 (1970)
49. Schoengold, D.M. and Munson, B., *Anal. Chem.*, **42**, 1811 (1970)
50. Arsenault, G.P., Dolhun, J.J. and Biemann, K., *Anal. Chem.*, **43**, 1720 (1971)
51. Dhont, J.H., *Nature*, **200**, 882 (1963)
52. Dhont, J.H., *Analyst*, **89**, 71 (1964)
53. Weurman, C., *Chem. Weekbl.*, **59**, 489 (1963)
54. Cramers, C.A.M.G. and Keulemans, A.I.M., *J. Gas Chromat.*, **5**, 58 (1967)
55. Levy, E.J. and Paul, D.G., *J. Gas Chromat.*, **5**, 136 (1967)
56. Sternberg, J.C., Krull, I.H. and Friedel, G.D., *Anal. Chem.*, **38**, 1639 (1966)
57. Sarner, S.F., *J. Chromat. Sci.*, **10**, 65 (1972)
58. Bartz, A.M. and Ruhl, H.D., *Anal. Chem.*, **36**, 1892 (1964)
59. Wilks, P.A. and Brown, R.A., *Anal. Chem.*, **36**, 1896 (1964)
60. Low, M.J.D. and Freeman, S.K., *Anal. Chem.*, **39**, 194 (1967)
61. Brown, R.A., Kelliher, J.M., Heigl, J.J. and Warren, C.W., *Anal. Chem.*, **43**, 353 (1971)
62. Teeter, R.M., Spencer, C.F., Green, J.W. and Smithson, L.H., *J. Amer. Oil Chem. Soc.*, **43**, 82 (1966)
63. Issenberg, P., Kobayashi, A. and Mysliwy, T.J., *J. Agric. Food Chem.*, **17**, 1377 (1969)
64. Stanley, G. and Murray, K.E., *J. Chromatog.*, **60**, 345 (1971)
65. Murray, K.E., Shipton, J., Whitfield, F.B., Kennett, B.H. and Stanley, G., *J. Food Sci.*, **33**, 290 (1968)
66. Thompson, C.J., Coleman, H.J., Ward, C.C. and Rall, H.T., *Anal. Chem.*, **32**, 424 (1960)
67. Thompson, C.J., Coleman, H.J., Ward, C.C. and Rall, H.T., *Anal. Chem.*, **34**, 151 (1962)
68. Sevenants, M.R., Ward, C.C. and Rall, H.T., *Anal. Chem.*, **34**, 154 (1962)
69. Thompson, C.J., Coleman, H.J., Hopkins, R.L., Ward, C.C. and Rall, H.T., *Anal. Chem.*, **32**, 1762 (1960)
70. Thompson, C.J., Coleman, H.J., Hopkins, R.L. and Rall, H.T., *J. Gas Chromat.*, **5**, 1 (1967)
71. Sevenants, M.R. and Jennings, W.G., *J. Food Sci.*, **31**, 81 (1966)

Part 2
Recommendations

8

Recommended Methods and Stationary Phases

A. Introduction

This chapter brings together recommended GC methods for the analysis of foods in quality assessment and for the determination of food additives and contaminants. A small library of GC columns is suggested which should enable the analyst to perform the majority of analyses required in a general food laboratory.

In ensuing chapters, criticisms have been made of some of the methods outlined, but because it would be impossible for the authors to practically evaluate all the methods cited in this book, they have decided on a positive line of recommendation rather than criticism. One, and sometimes two, methods per analysis are recommended, based on a combination of factors which include the authors' general practical experience, other chromatographers' experience and commonsense. Good recoveries, sensitivity, simplicity of operation, and the ease and speed of any chemistry involved in the analysis have been taken into account in these recommendations.

Recommended methods help the gas chromatographer who is under pressure to produce results as soon as possible, from an unfamiliar method, when normally he would prefer to explore the literature for himself and then try a selection of methods before choosing the most appropriate one for his needs.

The choice of stationary phase for a particular analysis is probably the biggest issue to the gas chromatographer. As new phases have been developed, being given manufacturers' code-names and code-numbers, chromatographers can easily purchase a new phase, unaware that they already possess one which has almost the same polarity characteristics and which would perform equally as efficiently as the new phase, given a simple adjustment of temperature or flow rate.

For those gas chromatographers who are concerned with general food analysis, it is important to have a library of GC columns. It is impossible and unnecessary to have such a library which contains every stationary phase at different loadings on every support and *Table 8.1* lists nine stationary phases together with the classes of compounds which can be separated and determined on these phases. This small collection of stationary phases covers most of the applications of GC in food analysis.

Table 8.1 Stationary phases and the compounds which have been analysed using them, as cited in the text

Recommended stationary phase	Type	Compounds determined
one of the following: DC-200, E-301, JXR, OV-1, OV-101, SE-30, SF-96	methyl silicone	acetic acid and isobutyric acid (decyl esters); aconitic acid (TMS derivative); aliphatic acids; aliphatic acids (TMS derivatives); benzoic acid; benzoic acid (TMS derivative); BHA and BHT; BHA, BHT, NDGA, PG and TBHQ (TMS derivatives); binapacryl; brominated vegetable oils (TMS derivatives); butyric acid (ethyl ester); caffeine; capsaicin (TMS derivative); carbamates; coumarin; cyclamic acid (methyl ester); dinitrophenols (methyl esters); diphenyl; flavanols (TMS derivatives); glycerides (TMS derivatives); glycyrrhetic acid (TMS derivative); hydroxy benzoates (TMS derivatives); indene; nicotinamide and nicotinic acid (TMS derivatives); nicotine; N-nitrosamines; OC pesticides; OC pesticides (TMS derivatives); ON pesticides; OP pesticides; pantothenic acid and pantothenol (TMS derivatives); patulin chloroacetate; PCBs; phenolic acids (TMS derivatives); o-phenyl phenol; piperonyl butoxide; polyols (TMS derivatives); propoxur (TMS derivative); pyrethrins; saccharin (methyl ester); sorbic acid; sorbitol hexa-acetate; sterols; sterol acetates; sterol butyrates; sterols (TMS derivatives); sugars (TFA derivatives); sugars (TMS derivatives); sugar alcohols (TMS derivatives); thiabendazole; tocopherols (TMS derivatives); tocopheryl butyrates; triacetin, triethyl citrate; triglycerides; triterpene alcohols; vitamin A (TMS derivative); vitamin B_6 (acetyl derivatives); vitamins D_2 and D_3; vitamin E (see tocopherol); vitamin K_5 (TMS derivative)
one of the following: OV-17, DC-710	methyl phenyl silicone	aliphatic acids; carbon disulphide; chloroform; glutamic acid (TFA derivative of the butyl ester); inositol hexa-acetate; OC fumigants; stilboestrol (diethyl ester); vitamins D_2 and D_3; vitamins D_2 and D_3 (TMS derivatives); vitamin K_3
XE-60	nitrile silicone	BHA, BHT, DLTDP, Ionox-100 and NDGA; binapacryl; captafol, captan, folpet; carbamates (2,4-dinitroaniline derivatives); cylohexylamine (2,4-dinitrophenyl derivative); gallates and

Recommended stationary phase	Type	Compounds determined
		THBP (TMS derivatives); OC pesticides; TDPA (dimethyl ester)
one of the following: Apiezon L, Apiezon M	hydro-carbon	aliphatic acids (methyl esters); aliphatic aldehydes; aliphatic sulphides; benzaldehyde; BHA and BHT; cyclohexene; dichlorvos and malathion; lactones; 2-phenyl ethanol; thioctic acid (methyl ester); tocopherols (TMS derivatives); vitamin K (TMS derivative)
PEG 20M (Carbowax 20M)	poly-ethylene glycol	acetaldehyde; aliphatic alcohols; aliphatic esters; aliphatic sulphides; benzoic acid; BHA and BHT; butyric acid; cyclohexylamine; heptanol; isosorbide; methyl mercuric chloride; methyl mercuric iodide; N-nitrosa-mines; pentane; polyols; propionic acid; pyribenzamine; saccharin (methyl ester); sorbic acid; terpenes, sesquiter-penes; vanillin, ethyl vanillin; vicinal diketones
one of the following: PEG 1500 (Carbowax 1500) PEG 1540 (Carbowax 1540)	poly-ethylene glycol	aliphatic acids (methyl esters); aliphatic aldehydes; cyclohexene; diethyl carbonate; N-nitrosamines; pectins (pyrolysis products)
DEGS	polyester	aliphatic acids (butyl esters); aliphatic acids (ethyl esters); aliphatic acids (methyl esters); aliphatic acids (propyl esters); BHT; malathion and malaoxon; terpenes
one of the following: ethylene glycol adipate (EGA) butanediol succinate (BDS)	polyester	aliphatic acids (methyl esters); aliphatic aldehydes; amino acids (TFA derivatives of the butyl esters); methyl mercuric bromide; sorbic acid; vitamins D_2 and D_3 (TMS derivatives)
Porapak Q	porous polymer	acetone; chlorinated aliphatic hydro-carbons; ethanol; glycerol; hexane; isopropanol; nitrous oxide

The selection of support materials depends on the particular analysis and, for this reason, no one material has been specifically recommended, but Appendix 2 gives guidance in this respect.

In the recommended methods which follow, stationary phases and their loadings, support materials, their treatment and mesh size, temperatures or temperature programmes, and detection methods, are quoted from the literature. However, it would not always be necessary, or practical, to use the exact GC conditions as quoted, and by reference to Appendix 1, Appendix 2 and *Table 2.1* the chromatographer should be able to select an approximately equal set of conditions, using the materials already at his disposal. The authors have taken the liberty of suggesting improvements to some of the older recommended methods in order to bring them into line with present-day technology, e.g. the replacement of firebrick and Celite 545 by support material specifically developed for GC.

B. Recommended methods for the analysis of dairy products

In the analysis of dairy products methods are recommended which include those concerned with their quality control, e.g. fatty acid composition of the glycerides, sterols and volatile constituents.

1. Milk

GC methods are used to determine total fat and fatty acid composition of milk. The technique has particular value when the sample undergoing analysis is too small for the application of the normal gravimetric or volumetric methods, and the recommended method is that of Glass, Lohse and Jenness[1]. Samples of the order of 30—60 mg of milk are mixed with an extraction solvent comprising 3:2:3 methanol:dimethyl carbonate:benzene and containing methyl tridecanoate as internal standard. The mixture is centrifuged and an aliquot of the supernatant is treated with 2N sodium methoxide, followed after 5 min by 10% anhydrous hydrochloric acid in anhydrous methanol. Centrifugation removes sodium chloride and the methyl esters of the fatty acids in the supernatant are determined using a column of 10% DEGS on 80/90 Anakrom F6, using temperature programming between 90 and 210°C at 6°/min, and using FID. Total fat and fatty acid composition are determined by relating peak area measurement of the methyl esters of the fatty acids to the internal standard. The method is quick and gives good agreement with that of Rose-Gottlieb.

Although most GC methods for fatty acid analysis rely on the formation and determination of methyl esters, one drawback is that in the analysis of milk, owing to the wide span of volatility of its various fatty acids, the chromatogram has normally to be produced under two sets of operating conditions in order to include both the short-chain and the long-chain fatty acids. Butyl esterification, on the other hand, has the advantage that only one set of GC conditions need obtain for the complete analysis of milk fatty acids, and the recommended method of Sampugna, Pitas and Jensen[2] uses this rapid procedure. Milk fat is placed in a test-tube and the butyl esters of the fatty acids of the glycerides are formed by heating the fat with di-n-butyl carbonate and sodium butoxide. GLC is performed using a column of 18% DEGS on Anakrom ABS with temperature programming between 50 and 215°C at three different rates, and using FID. These conditions separate the fatty acids, C_4 to C_{18}.

The volatile constituents of milk are interesting from several points of view, not the least the off-flavours associated with either staleness or rancidity. The method of Arnold, Libbey and Day[3], developed for the examination of sterilised, concentrated milk for stale flavour constituents, is recommended for the detection of ketones, aldehydes and other classes of compounds found in milk. The method depends on the on-column trapping of volatiles as described by Morgan and Day[4] (see Chapter 4, p. 64). GLC is carried out using a column of 20%

Apiezon M on 80/100 Celite 545, using a short precolumn of Apiezon M containing 5% of sodium hydroxide to remove fatty acids. FID is used and temperature programming is performed between 100 and 200°C. It is suggested that Apiezon L be used in place of Apiezon M, which may not be available, and that a Chromosorb or equivalent be used in place of Celite 545.

In the heat treatment of milk some biacetyl is formed, and if its concentration rises above the flavour threshold, the milk becomes unpalatable. The recommended method for the determination of biacetyl in milk is that of Scanlan and Lindsay[5]. The sample of milk, with added tetradecanol to prevent foaming, is purged with nitrogen and the biacetyl is collected by on-column trapping[4]. GLC is performed at 70°C, using EC detection on a column of 20% TCEP coated on 80/100 Celite 545. It is suggested that a Chromosorb or equivalent be used in place of Celite.

Although the sugar composition of milk is of interest to the dairy industry, GC was little used for its determination until the arrival of TMS derivatisation. The method of Reineccius, Kavanagh and Keeney[6] is recommended for the determination of lactose, glucose and galactose in milk, using TMS derivatives. The free carbohydrates are isolated from milk by dialysis and, after derivative preparation, GLC is performed using a column of 5% SE-30 on 80/100 AW, DMCS Chromosorb G, using FID and also MS to identify isomers. Temperature programming between 130 and 270°C at 6°/min is used and the internal standard is *N*-acetyl-D-glucosamine.

2. Butter

The main interest in the manufacture of butter is probably its quality, and the focal point of law enforcement analysis is the admixture of butter with hardened vegetable oils.

In the early days of GC, the presence of vegetable oils in butter was detected by using various ratios of fatty acids and deducing the extent of admixture after mathematical interpretation of the chromatographic results. It is easier to concentrate on one particular acid, viz. butyric, for the quantitative assessment of butter, when it is present in other foods as a minor constituent. Two methods are recommended for the determination of butter content. In Withington's[7] method, the sample is heated with a solution of sodium hydroxide in ethanol at 80°C in a water-bath, and this transesterification produces ethyl butyrate from the butyric acid, which is produced by the hydrolysis of those glycerides containing it. GLC is performed using a column of 5% SE-30 on 60/80 DMCS Chromosorb W, and using FID. Methyl hexanoate is incorporated in the analysis as the internal standard and temperature programming is used between 70 and 200°C at 8°/min. Withington uses 4.79% as the average ethyl butyrate figure for commercial butter, and his method is recommended for the determination of up to 10% butter in margarine.

In the Phillips and Sanders method[8] the free butyric acid is determined by using a column of 5% Carbowax 20M containing 0.5% terephthalic acid on 100/120 AW Supasorb, operated at 125°C and using FID. In the sample preparation the butter fat is saponified and the product is acidified and filtered. The filtrate contains the water-soluble fatty acids, including butyric, which is found to be 3.6% in average butter. This semimicro method, which uses only 0.1g fat, is recommended for the determination of butter in such foods as milk chocolate coatings, cream soups, fish and meat pastes with butter, butter and milk biscuits and butter sugar confectionery.

The main sterol in vegetable oils is sitosterol, but it is not present in butter. An examination of the sterol content of butter for sitosterol will therefore indicate whether there has been any addition of vegetable oil. The method recommended for such an examination is that of Eisner *et al.*[9] The fat is saponified and the unsaponified matter is extracted and cleaned up on a column of Florisil. The cleaned-up extract is chromatographed using a column of 2% SE-52 on 80/100 DMCS Chromosorb P, which at 212°C separates β- and γ-sitosterols and sigmasterol. These workers used an AI detector, but FID might be an improvement. The β-sitosterol peak is evaluated against that produced by standard mixtures of vegetable oils, e.g. margarine, in butter, and as little as 1% margarine in butter gives a measurable peak for β-sitosterol.

The principal contributors to butter flavour appear to be lactones, but the subject is of academic interest, and Chapter 9 contains references to methods for the determination of lactones and other butter volatiles.

3. Cheese

There is little interest in the composition of cheese fat, probably because cheese is rarely adulterated by the addition of other fats or oils. The prime analytical involvement is with the flavour constituents, which are important in differentiating varieties of cheese.

The flavour constituents of cheese may be detected by GC methods, and two are recommended, which encompass most volatile flavour compounds encountered in the different varieties.

The method of Langler, Libbey and Day[10] involves a low-pressure, low-temperature distillation of cheese fat, recovering the volatile constituents by the on-column trapping technique of Morgan and Day[4] (see Chapter 4, p. 64) using a column consisting of 20% TCEP on 60/80 AW Celite 545. GLC is carried out at 50°C, using FID. It is suggested that a Chromosorb or equivalent be used in place of Celite 545.

The second method is that of Liebich *et al.*[11], in which no clean-up stage is incorporated. The cheese oil is centrifuged and injected on to a type of precolumn held at 190°C for 10 min to allow the volatiles to enter the GC capillary column, which is coated with Dowfax 9N15. The column is temperature programmed between room temperature and 125°C and FID is used. If this stationary phase is unavailable, it is suggested that Igepal CO 880 be used.

Both of these methods[10,11] attempt to prevent the formation of artefacts by keeping the isolation temperature as low as possible.

C. Recommended methods for the analysis of fats and oils

In the analysis of fats and oils methods are recommended which help to differentiate one species from another, as well as providing some knowledge of quality.

1. Animal fats

GC methods are used to determine the fatty acid constitution of animal fats with a view to differentiating species. In this context, it is not necessary to determine every fatty acid present in an animal fat, but to concentrate the analysis on those fatty acids which are qualitatively significant.

The method of Castledine and Davies[12] is recommended for the differentiation of animal fats, such as pork from lamb, pork from beef, and also for horse, dog, cat and duck from other fats. It does not differentiate beef and lamb fats. The fat is separated from the sample by conventional methods and a portion is transesterified using 0.6% methanolic sodium hydroxide at 70°C for 5 min. The resulting methyl esters are extracted with diethyl ether, propyl palmitate is added as internal standard, and GLC is carried out using a column of 10% PEGA on 100/120 Celite, operated at 190°C and using FID. It is suggested that the Celite be replaced by a Chromosorb or equivalent support. Qualitative evaluation of unsaturated fatty acids indicates the presence or absence of certain animal fats and also vegetable oils or fats. Quantitative evaluation of myristic, palmitic, stearic and oleic acids and a calculation of the ratios of these is an important aid to the differentiation of some animal fats and to their admixture.

In the second recommended method, of Hubbard and Pocklington[13], fat is extracted from meat and dissolved in benzene, prior to formation of the methyl esters of fatty acids by heating under reflux with methanolic 0.2N hydrochloric acid for 30–40 min. The esters are extracted with diethyl ether and GLC is carried out using a column of 5% Apiezon L on 100/120 acid- and alkali-washed Celite, at a temperature between 185 and 200°C, and using AI detection. For the specific separation of some unsaturated esters, 8% BDS is used. It is suggested that Celite be replaced by a Chromosorb or equivalent, and that FID be used in place of the AI detector. In this method the acids, both saturated and unsaturated, from C_{10} to C_{20} are separated, and a close examination of certain acid concentrations will indicate some small differences between beef and lamb fats, but it is impossible to differentiate these fats in admixture.

The flavour reversion of beef and mutton tallow can be measured according to the 2,6-nonadienal plus 4-heptenal content using the recommended method of Hoffmann and Meijboom[14]. The volatiles are collected from the tallow by sweeping them out with nitrogen at 160°C under vacuum and trapping them in an ethanol/dry ice-cooled trap at $-$ 80°C and a liquid nitrogen-cooled trap at $-$ 196°C. The volatiles are dissolved in petrol and the free acids removed with sodium bicarbonate solution. GLC is performed using a column of 5% PEGA on Celite at 104°C, and using FID.

2. Vegetable oils and fats

The two major considerations in the analysis of vegetable oils and fats are adulteration and refinement quality. Many cheap oils have been used to sophisticate more expensive ones and also there are often marked differences in the quality of oils according to their state of refinement.

The analytical factors which help to establish the identity and quality of vegetable oils are the balance of the glyceride fatty acids and the identities and

109

proportions of sterols, triterpene alcohols and tocopherols which occur in the unsaponifiable matter of these oils.

Like animal fats, vegetable oil glycerides contain fatty acids which lie within certain ranges according to the variety. There is a greater proportion of unsaturated fatty acids in vegetable oils than in animal fats, a factor used to aid differentiation of these two groups.

The recommended method for the fatty acid analysis of vegetable oils is that of Grieco and Piepoli[15]. This entails extraction of the oil with petrol and the preparation of methyl esters by transesterification. GLC is carried out using 20% PEGS on Chromosorb W at 213°C, using TC detection. An FID would probably be the chosen detector nowadays. This method can be used to determine the fatty acids between C_6 and C_{24}, including the unsaturated ones. The ratios of certain acids give an indication of adulteration and can also be used for quality control purposes.

After normal saponification procedure, the unsaponifiable matter may be examined for the sterol, tocopherol or triterpene alcohol composition, since the ratios of the individual compounds within these three groups can yield useful information concerning possible adulteration or of the refinement quality.

The recommended method for the determination of triterpene alcohols is that of Fedeli *et al.*[16] The terpene fraction of the unsaponifiable matter of vegetable oils is analysed using a column of 1% SE-30 on 100/120 silanised Gas Chrom P at 230°C, and using FID. From a study of the ensuing fingerprint chromatogram, together with chromatograms from standard materials, much can be deduced as to the identity and quality of the oil.

It is possible to determine sterol and tocopherol fractions direct, but it is recommended that TMS derivatives of these be utilised. The method of Amati, Carraro Zanitaro and Ferri[17] is recommended for the analysis of sterols in vegetable oils. The oil is saponified with methanolic potassium hydroxide, and the unsaponifiable matter is extracted with diethyl ether. A preliminary separation of sterols from other compounds is achieved using silica gel TLC, and the sterol fraction is dissolved in diethyl ether for conversion to TMS derivatives. GLC is performed using a column of 3% JXR on 100/120 acid-washed, silanised Gas Chrom P at 260°C, and using FID.

The method of Losi and Piretti[18] is recommended for the determination of tocopherols in vegetable oils. This entails saponification, TLC fractionation and TMS derivatisation, similar to those described above[17], with GLC carried out using a column of 1% SE-30 on Gas Chrom P at 240°C, and using FID.

3. Fish oils

The fatty acid composition of fish oil glycerides is best determined using Ackman's [19] method. Fish oils, in comparison with vegetable oils, contain fatty acids of greater unsaturation and also some of greater chain length. After saponification and formation of methyl esters by conventional methods, the latter are chromatographed using a column of 3% EGSP-Z (an organosilicone polyester) on 100/120 Gas Chrom Q at a temperature between 180 and 200°C, and using FID. This stationary phase has the advantage of separating almost all the fish oil glyceride fatty acids in one chromatogram.

D. Recommended methods for the analysis of meat, fish and eggs

In the analysis of meat and fish, apart from the determination of their lipid constitution already described in the previous section, the importance of GC methods lies in the determination of their volatile flavour components. The determination of the simple aliphatic acids associated with deterioration of egg covers the most important use of GC methods for that food.

1. Meat

Although the GC conditions are important in methods applied to the determination of flavour volatiles, the method of isolation is also vital and can have considerable influence on the components found, i.e. some very volatile compounds can be lost while artefacts might be formed, the net result being the portrayal of the wrong flavour picture.

The first recommended method, although originally applied to cooked chicken volatiles, could equally be used in a similar examination of other cooked or uncooked meats. In this method of Nonaka, Black and Pippen[20], cooked chicken slurry is distilled and the aqueous distillate is extracted with isopentane. The condensed isopentane extract is chromatographed on a capillary column coated with Apiezon L. FID is used and temperature programming is carried out between 75 and 200°C at 2°/min. In the initial study MS ought to be used to give positive identification of the complex array of peaks on the chromatogram. The method will separate aldehydes, ketones, sulphur-containing compounds, and aromatic and aliphatic hydrocarbons.

The second recommended method is that of Cross and Ziegler[21]. Two methods for the extraction of the volatile components are used, both being relatively simple, yet effective. In the first extraction the meat sample is held at 60°C and purged with nitrogen for 20 min, passing the effluent vapour on to a precolumn consisting of 1% DC-550 on glass microbeads, being cooled in dry ice. Before GC, the precolumn is warmed and nitrogen is used to sweep the volatile components on to the analytical column.

In the second extraction the meat sample is mixed with water in a flask and heated to 85°C, and nitrogen is bubbled through the solution for 4–5 h. The effluent vapour is passed through a series of trapping solutions, viz. 2N hydrochloric acid solution to trap bases, 2,4-dinitrophenylhydrazine solution to trap carbonyls and mercuric chloride solution to trap sulphur-containing compounds. Following this procedure, the various solutions are analysed separately, the carbonyls being regenerated from their 2,4-dinitrophenylhydrazones prior to GC.

GLC is performed using a column of 1% SE-30 on glass microbeads at 40°C, using FID. This work was published in 1965 and it is suggested that capillary column GC, using a non-polar stationary phase coating, would be more appropriate nowadays.

2. Fish

Although the flavour volatiles of fish are important from the point of view of the differentiation between species, it is the off-flavour and particularly the

111

high aliphatic amine content, caused by spoilage, which has stimulated most GC consideration.

The first recommended method for the determination of fish volatiles is that of Wong, Damico and Salwin[22]. Low-temperature vacuum distillation at 35°C is used to isolate the volatiles, these being collected in traps cooled by dry ice and liquid nitrogen, respectively. GLC is carried out using a column of 5% $\beta\beta'$-oxydipropionitrile on 60/80 Gas Chrom P. Cryogenic temperature programming is used between − 60°C and + 50°C at 1°/min. FID is used, but MS is incorporated, at least in the first instance, to positively identify the compounds, which include amines, aldehydes, ketones, hydrocarbons and sulphur-containing compounds.

The second method, which is specifically recommended for the determination of di- and trimethylamine in fish, is that of Keay and Hardy[23]. The sample is blended with 0.6N perchloric acid and filtered and the filtrate is steam distilled under alkaline conditions into a hydrochloric acid solution. This receiving solution is diluted to give a final acidity of 0.06N and aliquots are taken for GC analysis after the addition of ethanol as internal standard. A specific column mixture for the separation of aliphatic amines is used, comprising two sections. In the first section there is a mixture of 20% Dowfax 9N9 and 2.8% potassium hydroxide on 80/100 Silocell C22, and in the second section a mixture of 17% Dowfax 9N9 and 14% potassium hydroxide on the same support material. FID is used and the column temperature is 100°C. If Dowfax 9N9 is unavailable, Igepal CO 880 is suggested as an alternative stationary phase. It is further suggested that one of the Chromosorbs, or equivalent, be used as support material in place of Silocell C22.

3. Eggs

Lactic, succinic and some other aliphatic acids are produced in food by microbial or enzymic action. Spoilage due to such action can occur in frozen or dried egg, and GC is used to quantitatively determine these acids as a guide to the freshness and quality of the egg.

The recommended method is that of Salwin and Bond[24], in which a mixture of sulphuric and phosphotungstic acids is used to liberate lactic and succinic acids from egg samples, these acids being extracted into diethyl ether. After evaporation of the ether extract, propyl derivatives are prepared by refluxing the residue with BF_3 and propanol for 10 min. Ammonium sulphate solution is added and also acetophenone as internal standard. The mixture is shaken with chloroform, this organic phase being ultimately used for GC. This is performed using a column of 10% DEGS on 100/120 Gas Chrom Z at 130°C, using FID. 3-Hydroxybutyric acid, which is also found in spoiled eggs, can also be determined by this method.

Johansen and Voris[25] have produced a method which is recommended for the determination of cholesterol of egg when used as an ingredient in baked foods. An aqueous slurry of the food, e.g. egg noodles, cake or hamburger, is saponified and the unsaponifiable matter is isolated and dissolved in dimethylformamide. 5-α-Cholestane is added as internal standard and the TMS derivatives are prepared. These derivatives are chromatographed using a column of

5% QF-1 on 80/100 Gas Chrom Q at 230°C, and using FID. This method is only applicable if it is certain that ingredients do not contain animal fat, e.g. butter or lard, which contain cholesterol.

E. Recommended methods for the analysis of essential oils

In the analysis of essential oils the problems include the separation of different classes of compound, as well as of individual members within each class. For this reason, GC analysis is facilitated by fractionation of the compound classes beforehand, using suitable techniques, e.g. molecular distillation and column chromatography.

Essential oils are produced from fruit, leaf, whole plant, flower and root, and in most cases it is the hydrocarbons which characterise them, the terpene family in particular. Some essential oils contain characteristic alcohols and carbonyls, and therefore these classes, too, have analytical significance.

It would be impossible to recommend GC methods for every essential oil, but the following ones, used for the analysis of citrus oils, are generally applicable and have been selected according to the versatility of their isolation procedure as well as their ultimate chromatography.

Hunter and Brogden[26] have separated terpenes and sesquiterpenes from oxygenated compounds in cold-pressed orange oil, and their method is recommended. The oil is centrifuged in two stages and the supernatant is fractionated in a molecular still. Under controlled conditions, the terpenes and oxygenated compounds of low molecular weight are distilled, leaving the sesquiterpenes in the residue. The sesquiterpene fraction is chromatographed using a column of 25% Carbowax 20M on 30/60 Chromosorb P at 175°C, and using FID.

The distillate, containing terpenes and the low molecular weight oxygenated compounds, is passed through a basic alumina column, the terpenes being subsequently eluted with hexane and leaving behind the oxygenated compounds. The terpene fraction is chromatographed using the same conditions as for the sesquiterpene fraction but with a column temperature of 135°C.

It is advisable to confirm identities of peaks, at least in the first instance, by MS and IR spectroscopy.

In the separation and determination of essential oil alcohols the method of Hunter and Moshonas[27], as applied to orange oil, is recommended. The feature of the separation of the alcohol fraction in the extraction procedure is a liquid–liquid partition step, using carbon tetrachloride and propylene glycol. The major portion of the oil, being hydrocarbon in nature, remains in the carbon tetrachloride, and the alcohols, which are retained in the propylene glycol, are cleaned up by alumina column chromatography. GLC is performed using a column of 17% Carbowax 20M on AW Chromosorb G, using FID and a temperature programme between 150 and 220°C at 1.1°/min. In the first instance MS and IR spectroscopy should be used to verify the GC findings.

In the examination of essential oils the terpene and sesquiterpene fractions are somewhat similar for oils extracted from the different varieties of citrus fruits. It is the oxygenated compounds, particularly carbonyls, which are responsible for their flavour differences. The recommended GC method for the analysis of carbonyls in essential oils is that of Moshonas[28]. The oil is vacuum distilled and molecular distilled, and the final distillate is cleaned up on deactivated Florisil. The eluate is analysed using a column of 20% Carbowax 20M on 60/80 Chromosorb P, using TC detection and a temperature programme between 100 and 225°C at 1°/min. MS and IR should be used, in the first instance, to positively identify the compounds.

For the specific determination of aldehydes in essential oils, the isolation procedure of Stanley et al.[29], as applied to citrus oils, is recommended. Water-soluble Girard T-aldehyde compounds are formed, which permit the removal of non-carbonyls by organic solvent extraction. The aldehydes are regenerated by reaction with aqueous formaldehyde and are extracted with isopentane. In this early GC work the column comprised BDS on firebrick, but one of the Chromosorbs or equivalent would be an improved support material.

If only the major components of an essential oil are required to be separated, regardless of class, then the GC conditions used by Cieplinski and Averill[30], as applied to the analysis of peppermint oil, are worthy of note. These include a capillary column coated with UCON 50-HB-2000 liquid phase and a temperature programme of 80°C for 3 min and 80 to 150°C at 2°/min, using FID.

F. Recommended methods for the analysis of fruits, vegetables and non-alcoholic beverages

In the analysis of fruits, apart from the interest in their essential oil composition (see Section E), GC can be used to help in the identification of a fruit in a fruit product by using parameters such as fixed aliphatic acids, amino acids and sugars. Otherwise, the main use of GC is in the qualitative and sometimes, quantitative evaluation of flavour volatiles for fruits, vegetables and non-alcoholic beverages, such as coffee, cocoa and tea.

1. Fruit

It would be impossible and unnecessary to recommend methods for the determination of compounds which are specific to every kind of fruit, and in recommending the following, general application has been considered.

The examination of fruit and fruit products for fixed aliphatic acids, amino acids and sugars, taken collectively, yields valuable information concerning the identification of fruit species. The recommended method for the determination of fixed fruit acids is that of Fernandez-Flores, Kline and Johnson[31]. Ethanolic extracts of the fruits are made and the acids, viz. glycolic, succinic, fumaric, malic, tartaric, citric, syringic and quinic, are precipitated as their lead salts. The lead salts are converted to TMS derivatives for GC, a column of 3.8% SE-30 on 60/80 silanised Diatoport S being used. TC detection is used and also temperature programming between 90 and 240°C at 6°/min.

The recommended method for the determination of amino acids in fruits and fruit juices is that of Fernandez-Flores et al.[32] An 80% aqueous ethanolic extract is made, the ethanol is evaporated and the resulting solution is passed through a column of Dowex 50W-X8 resin. The amino acids are retained on the column and then eluted with 1N ammonia solution. After evaporation, the amino acids in the residue are converted to their N-TFA butyl esters for GC analysis, using a column of 0.65% EGA on 60/80 AW Chromosorb W. FID is used and also temperature programming between 80 and 205°C at 4°/min.

The recommended method for the determination of sugars in fruits is that of Kline, Fernandez-Flores and Johnson[33]. An aqueous ethanolic extract is made of the fruit and Celite is added plus lead acetate solution in order to precipitate interfering acids. The TMS derivatives are made for GC, which is carried out using the same conditions as for the determination of fixed fruit acids[31], except that the temperature programme is from 160 to 270°C at 4°/min.

The flavour constituents of fruits are often simple volatile compounds. Thus simple esters, aldehydes and ketones characterise many fruit flavours. For this reason, the recommended technique for the extraction of flavour components of fruits is head space sampling (see Chapter 4, p. 62). As an example, the method of Romani and Ku[34] is typical of the type recommended for obtaining the simple volatile constituents of fruits or vegetables. Pears are placed in a sealed jar at 20°C and the volatiles are allowed to collect in the head space, which is sampled at intervals. If a study of the change of volatile composition with time is required, the head space is flushed out with water-saturated air and the samples are left to generate the next series of volatiles. For pear volatiles, which include aliphatic acetates, acetaldehyde and ethylene, GLC is performed using a column of 8% DEGS on 30/60 AW Chromosorb P at a temperature of 95°C, and using FID.

It is recommended that the simple sulphur-containing volatiles be sampled and analysed according to the method of Baerwald and Miglio[35], who used this method on hops and hop extracts. Samples are held under nitrogen in a sealed ampoule at 50°C for 30 min, and 20 ml of head space is taken for GC analysis. This is performed using a column of 10% Triton X-305 on Chromosorb G, with temperature programming between 40 and 150°C at 4°/min, and using FPD.

The quality of capsicum is largely dependent on the capsaicin content, and several GC methods have been employed for its determination. The recommended one is that of Mueller-Stock, Joshi and Buechi[36]. Powdered capsicum is extracted with methanol, the extract is evaporated and the resulting oil is dissolved in acetic acid. The acidic solution is extracted with petrol, the petrol solution extracted back into acetic acid and the capsaicin further extracted into dichloromethane. After clean-up on activated carbon, the TMS derivative is prepared for GC using a column of 3% JXR on 100/120 Gas Chrom Q. FID is used and a temperature programme between 150 and 230°C at 6°/min. Co-extracted dihydrocapsaicin and nordihydrocapsaicin may also be found on the chromatogram.

2. Vegetables

Vegetables have received little GC attention, other than in the analysis of flavour components. For the extraction of simple volatiles, suitable head space sampling techniques can be used at appropriate temperatures, according to the type of flavour components required to be analysed.

Sulphur-containing compounds are among the most important flavour constituents of vegetables. The recommended GC conditions for the determination of such constituents is that of Brodnitz, Pollock and Vallon[37], as used in the analysis of onion oil. GLC is carried out using a capillary column coated with Carbowax 20M, which is temperature programmed between 50 and 175°C at 2°/min, and using FID.

3. Non-alcoholic beverages

The dependence of the consistent quality of non-alcoholic beverages, such as cocoa, coffee and tea, has rested almost entirely on taste panels. In support of the organoleptic results from these panels, GC, together with MS, has added a more scientific approach to the analysis of aroma and flavour components of these beverages.

The recommended method for the determination of aroma constituents of roasted cocoa beans is that of van Praag, Stein and Tibbetts[38]. Two techniques are used for the extraction of the volatiles, viz. head space sampling and steam distillation. The distillate of the latter is divided into basic, acidic and neutral fractions. GC is performed using a capillary column coated with Carbowax 20M, and temperature programming is carried out between 30 and 175°C at 2°/ min, and using FID. Similar GC conditions could be used for the determination of aroma and flavour constituents of coffee or tea, and it is also recommended that MS be used, at least in the first instance, to verify the identities of compounds.

In the examination of the flavour of tea, flavanols are probably the most important compounds. The recommended method for the analysis of these compounds in tea is that of Pierce *et al.*[39] Tea is powdered in ice and the flavanols are extracted with pyridine. The TMS derivatives are prepared for GC, which is carried out using a column of 3% OV-1 on 60/80 Gas Chrom Q at 250°C, and using FID.

The determination of caffeine in tea may be conveniently carried out using Newton's method[40]. The sample extract is cleaned up by column chromatography and the eluate is extracted with chloroform. The chloroform extract is mixed with an equal volume of ethanol containing pentobarbitone as internal standard and this solution is chromatographed using a column of 10% DC-200 on 80/100 Gas Chrom Q using a temperature of 190°C and AFID.

G. Recommended methods for the analysis of alcoholic beverages and vinegar

In the analysis of beer, cider, wine and spirits, GC has been used to determine ethanol, methanol, fusel oil, flavour volatiles, sugars and acids.

1. Beer, wort and cider

There is considerable interest, both commercial and domestic, in the ethanol content of most alcoholic beverages. Many satisfactory chemical and physical methods have been used for the determination of ethanol, but these often lack specificity. GC methods have brought this required specificity, and the recommended method for the determination of ethanol in beer or wort is that of

116

Trachman[41]. The beer is decarbonated, 2-butanol is added as internal standard and an aliquot of the mixture is injected on to a column containing 50/80 Porapak Q. GSC is performed using FID, with temperature programming from 100 to 200°C at 10°/min.

The balance between the various sugars in beer and wort is very important, commercially, and the GC method of analysis according to Clapperton and Holliday[42] is recommended. Phenyl-β-D-glucopyranoside is added as internal standard to the beer or wort sample, and the solution is evaporated to dryness. The TMS derivatives are prepared and GLC is performed using a column of 1.5% silicone gum rubber on 80/100 AW, DMCS Chromosorb W. FID is used and also temperature programming between 100 and 350°C at 8°/min.

As with sugars, glycerol is an important ingredient in beer, and the recommended method for its determination is that of Feil and Marinelli[43]. It would be convenient to use the TMS derivative, but the dry conditions necessary for its formation are difficult to obtain, and so direct injection of the glycerol is preferred, using a column containing the macroreticular resin, Par 1. GSC is carried out using FID and at a temperature of 195°C.

Like sugars, amino acids are precursors of the flavour compounds of alcoholic beverages, and their analysis is important, commercially; the method of Kurosky and Bars[44] is recommended. The amino acids in beer or wort samples are separated on a Dowex 50W-X8 resin column and are eluted using 1N aqueous ammonia solution. The methyl esters are prepared and are treated with trifluoroacetic anyhydride to form the N-TFA methyl esters. These are dissolved in dichloromethane and chromatographed using a column mixture of 0.75% Carbowax 1540 and 0.25% 2,2-dimethylpropane-1,3-diol succinate on Chromosorb W. FID is used and also temperature programming between 80 and 200°C at 4°/ min.

The fusel oil constituents of alcoholic beverages, particularly wines and spirits, are among the main flavour components. These constituents, particularly 2-phenylethanol, also influence the flavour of beer, cider and perry, and the recommended method for the determination of 2-phenylethanol is that of Kieser et al.[45], as applied to cider. The method involves a three-part distillation, the 2-phenylethanol in the final distillate being extracted with ethylene chloride. GC is performed using a column of 0.2% Apiezon L on 60-mesh glass microbeads at 80°C, and using FID.

GC, often coupled to MS, has been the stock technique in the evaluation of the volatile flavour components of alcoholic beverages. Head space sampling technique has a lot to commend it and there are many methods which have different conditions for incubation, i.e. different incubation temperatures and times. No specific method is recommended, therefore, because this would depend on the class of volatiles being analysed.

In order to include the less volatile flavour components as well as the very volatile ones, the method of Powell and Brown[46] is recommended. The beer sample is saturated with salt and extracted with carbon disulphide, n-octanol being added as internal standard. GLC is performed on this solution using a column of 8% FFAP on 80/100 AW, DMCS Chromosorb G. FID is used and a temperature programme between 60 and 225°C at two different rates. This method, which can be applied to the analysis of hop essential oil as well as beer, covers the analysis of alcohols, esters and acids.

117

Sulphur-containing compounds are among the important flavour components of beer. The method of Richardson and Mocek[47] is recommended for their determination and applies to bottled beer, although it could be adapted for use with draught beer. The bottle of beer is chilled to 4°C, uncapped, de-foamed and recapped before incubation at 50°C for an hour. Head space samples are analysed on a column of 10% Triton X-305 on 80/100 AW, DMCS Chromosorb G. FPD, set in the sulphur mode, is used and temperature programming is carried out between 40 and 150°C at 6°/min.

It is suggested that cider be examined for ethanol, sugars, amino acids and general flavour components using the same methods as recommended for beer.

2. Wine

The most convenient method of determining ethanol in wine is to dilute the sample and use Trachman's[41] direct injection method as applied to beer.

Similarly, the GC conditions of Clapperton and Holliday[42] for the determination of sugars as their TMS derivatives is conveniently applied to wine sugar analysis after extraction by the method of Capella and Losi[48]. In this extraction the wine sample is cleaned up on an ion exchange resin prior to evaporation in ethanol in preparation for the formation of TMS derivatives.

The balance between sugars, fixed acids and tannins is very important to the vintner, and GC can be used to separate and determine the concentrations of the fixed wine acids; the method recommended for this is due to Brunelle, Schoeneman and Martin[49]. The acids are precipitated as their lead salts from the wine and these salts are converted to their TMS derivatives for GLC using a column of 3.8% SE-30 on Diatoport S at 130°C, and using FID.

Wine quality is dependent on its volatile flavour constituents, which include fusel oil and esters. It is recommended that the fusel oil concentration be determined after distillation by the method of Martin and Caress[50], as applied to spirits and described in the section which follows.

The extraction of wine esters can be achieved by solvents such as diethyl ether, isopentane and mixtures of the two and also by chlorinated hydrocarbons. The method recommended for the extraction and separation of wine esters is that of Hardy and Ramshaw[51], which uses trichlorofluoromethane as extractant. After concentration of the extract, it is shaken with propylene glycol, which removes the higher alcohols. The remaining trichlorofluoromethane extract is chromatographed using a capillary column coated with Ucon LB-550X, temperature programming being carried out between 32 and 165°C at 1.5°/min, and using FID. It is necessary, in the first instance, to positively identify the esters by MS examination.

It has sometimes been necessary to determine the carbon dioxide content of wines, and this is conveniently performed using the GSC method of Ashmead, Martin and Schmit[52]. The sample is directly injected on to a short column of 60/80 charcoal, operated at 40°C, and using TC detection. The method has the advantage over manometric methods in that there is no interference from sulphur dioxide.

3. Spirits

The determination of the ethanol content of spirits is important, legally as well as commercially. It is possible to use direct injection methods with appropriate aqueous ethanol standards and high attenuation, but it is recommended that, after suitable dilution, the method of Trachman[41] be used.

The determination of methanol in spirits is important from a toxicological standpoint, and the method recommended for its determination is that of the UK Research Committee on the Analysis of Potable Spirits[53]. Methyl formate is added as internal standard and samples are injected on to a 100/120 Porapak Q column, using FID and a temperature selected between 70 and 125°C, depending on the type of spirit undergoing analysis. Conditions are selected so that water, methanol, acetaldehyde and methyl formate are separated.

Ethyl acetate and fusel oil constituents of spirits are best determined by the method of Martin and Caress[50]. Butanol is added as internal standard to the spirit sample and the mixture is injected directly on to a long column of 3% Carbowax 400 on 100/120 Chromosorb W. GLC is performed at 82°C, using FID. In the first instance, MS is used to confirm identities of the compounds.

For the specific determination of ethyl esters in spirits, the method of Guymon and Crowell[54] is recommended. The sample is diluted so that the ethanol content is 20% v/v and is then extracted with four portions of dichloromethane. The concentrated, combined extract is chromatographed using a column of 10% FFAP on 60/80 AW, DMCS Chromosorb G, with temperature programming between 100 and 225°C at 7.5°/min and using FID.

4. Vinegar

There are only a few GC methods for the analysis of vinegars, and selection of methods is therefore academic. Reference may be made to Chapter 14, p. 251, for details of these methods.

H. Recommended methods for the analysis of sugar products and syrups

The analysis of cane sugar, beet sugar, glucose syrup and maple syrup for their sugar patterns can be achieved using GC. The determination of amino acids and aconitic acid in cane sugar products and the determination of lactic acid in sugar beet liquors can be made by GC and are important to the food industry, as also is the flavour composition of maple syrup.

The introduction of TMS derivatives by Sweeley *et al.*[55] to facilitate sugar determinations by GC has largely replaced the conventional methyl and acetyl derivatives, and many methods have been put forward for the determination of sugars in sugar products and syrups, based on the separation of their TMS derivatives. As a general method for the determination of pentoses, hexoses and sugar alcohols in food, the method of Mueller and Goeke[56] is recommended. Appropriate methods for the preparation of samples is given before TMS derivative formation and GLC is performed using a column of 3% OV-1 on 100/120

Gas Chrom Q with temperature programming between 180 and 300°C at 4°/min, and using FID.

For the specific analysis of glucose syrup for glucose, maltose, maltotriose and maltotetraose, the GC conditions of Brobst and Lott[57] are recommended. TMS derivatives are prepared and chromatographed using a column of 0.25% SE-52 on 60/80 glass microbeads. FID is used and temperature programming is carried out between 75 and 245°C at 6.4°/min.

Some workers have had difficulty in separating sucrose and lactose by GC of their TMS derivatives, and the TFA derivatives are more suitable. In the recommended method, due to Luke[58], the TFA derivatives are prepared and chromatographed using a column of 20% SE-30 on 80/100 AW Chromosorb W, operated at 200°C and using FID.

The amino acids of cane syrup are best determined by the method of Johnson, Corliss and Fernandez-Flores[59]. The amino acids are concentrated on an ion exchange column and, after their elution, the N-TFA butyl esters are prepared. These derivatives are chromatographed using a column of 0.65% EGA on 60/80 Chromosorb W, with temperature programming between 80 and 205°C at 4°/min, and using FID.

The predominant organic acid in sugar cane and sorghum syrups is aconitic acid, and its concentration is important since a high level can cause clarification problems. The method recommended for its determination is that of Mehltretter and Otten[60]. Aconitic acid is completely removed from aqueous solution of table syrup by its precipitation as lead aconitate, which is then converted into the TMS derivative. GLC of this and the TMS derivative of glucose, added as internal standard, is performed using a column of 10% SE-30 on 80/100 S810, operated at 140°C and using FID. If S810 is unavailable, it is suggested that a Chromosorb or equivalent support material be substituted.

In the processing of sugar beet the concentration of lactic acid in the molasses and in raw and refined juices is a measure of the microbiological action which has taken place. The recommended method for the determination of lactic acid in these liquors is that of Oldfield, Parslow and Shore[61]. The liquors are diluted so that the sucrose content is approximately 5% and periodic acid is added, and the mixture is transferred to the injection port of the chromatograph, where the temperature is set high enough to evaporate water, thereby concentrating the periodic acid. This causes oxidation of lactic acid to acetaldehyde, which is swept on to the column with the carrier gas. GLC is performed using a column of 15% PEG 20M on 100/120 HMDS Chromosorb W, operated at 75°C and using FID. This method is applicable to sugar beet molasses and refined juice, but in the examination of raw juice, an ion exchange clean-up step must be incorporated.

The main GC interest in maple syrup concerns its flavour constituents, and the concentrations of such compounds as vanillin, syringaldehyde, dihydroconiferyl alcohol and 1-hydroxypropan-2-one are important from the quality control standpoint. The method recommended for the determination of these flavour constituents in maple syrup is that of Underwood, Filipic and Bell[62]. The sample is extracted with chloroform and the extract concentrated for GLC analysis using a column of 20% Carbowax 20M on 60/80 AW Chromosorb W. TC detection is used and also temperature programming from 50 to 240°C at 3.5°/min.

I. Recommended methods for the analysis of cereals and their products

The composition of the free and bound lipids present in wheat, its flour, dough and bread has been deduced, using GC, by an examination of the glyceride fatty acids after separation of glycerides and their hydrolysis. This composition is important to the flour confectionery industry because of its influence on the ageing of flour and on the mixing behaviour of dough in breadmaking. The recommended method for this analysis is that of Burkwall and Glass[63]. The sample is extracted with diethyl ether, which removes the free lipids. The residue is further extracted with water-saturated butanol, which removes the bound lipids. Both lipid fractions are hydrolysed to yield fatty acids, which are then methylated using diazomethane. The methyl esters are chromatographed using a column of 25% DEGS plus 2% H_3PO_4 on 80/100 AW Chromosorb W, operated at 200°C and using FID.

Graveland's[64] method for the examination of the lipid compositions of wheat flour and dough is also recommended. Both free and bound lipids are extracted by percolating the sample with a solvent mixture of 10:10:1 benzene: ethanol:water. Silica gel TLC is used to separate polar and non-polar lipids, which are subsequently determined by GLC using a column of 12% EGA on Chromosorb W at 195°C and using FID.

TMS derivatives are utilised for GC purposes in the recommended method of Mason and Slover[65] for the determination of sugars in wheat, flour, bread and also wheat-flakes. The reducing sugars are converted into their oximes, and these derivatives plus the free non-reducing sugars are trimethylsilylated to yield derivatives for GC analysis. This is carried out using a column of 1% SE-30 on 100/120 Gas Chrom Q. FID is used and, after keeping the column temperature at 170°C for 38 min, a programme is carried out between 170 and 270°C at 4°/min.

Although bread and flour are not highly flavoured, there has been some interest in the flavour constituents, particularly carbonyl compounds, in bread. The recommended method for the determination of aldehydes in bread is that of Hunter and Walden[66]. The semicarbazones of the aldehydes are formed and extracted with dichloromethane, and the aldehydes are subsequently regenerated with phosphoric acid. The cleaned-up aldehydes are chromatographed using FID and a column of 1% TCEP on 60/80 AW brick-dust at a temperature of either 50, 100 or 150°C, depending on the range of aldehydes being determined. It is suggested that a Chromosorb or equivalent support material be used in place of brick-dust.

The amino acid composition of the proteins in maize grain and soyabean meal has been the subject of study by GC methods, and it is considered that the method used by Zumwalt, Kuo and Gehrke[67] would be generally applicable to the analysis of amino acids from the protein of any cereal product. In this method, the protein is hydrolysed using 6N hydrochloric acid and the hydrolysate is cleaned up on a cation exchange resin. The N-TFA derivatives of the butyl esters of the amino acids are used for GLC which is performed on two different columns, using FID. One column contains 0.65% EGA on 80/100 AW Chromosorb W and is temperature programmed from 70 to 230°C at 6°/min, and the second column comprises 2% OV-17 plus 1% OV-210 on 100/120 Supelcoport, and is similarly temperature programmed.

J. Recommended methods for the analysis of preservatives and antioxidants

The widespread use of preservatives and antioxidants in food has necessitated the development of methods for their analysis, from both a commercial and a legislative point of view. Compared with most others, GC methods offer advantages of sensitivity, specificity and the ability to separate, as well as determine, several compounds in one analysis.

1. Preservatives

Benzoic acid and esters of its 4-hydroxy derivative, propionic acid and sorbic acid are the preservatives which are used generally in food. Two methods are recommended for their determination which, between them, cover all these preservative groups.

Benzoic acid, the methyl, ethyl and propyl esters of 4-hydroxybenzoic acid and sorbic acid are determined by the method of Fogden, Fryer and Urry [68]. The solvent extract of the food sample is evaporated to dryness at 60–70°C in the presence of tris(hydroxymethyl) methylamine in order to minimise loss of benzoic and sorbic acids. GSC is performed on 100/120 Porapak Q at 240°C, using FID and 2-phenoxyethanol in aqueous methanol as internal standard.

The second recommended method is that of Graveland[69], which uses a simple procedure for the determination of propionic and sorbic acids in bread and for the determination of benzoic acid in margarine. The bread sample is blended with diethyl ether containing 3% H_3PO_4 at 0°C for 5 min, using valeric acid as internal standard. The extract is chromatographed using a column of 5% Carbowax 20M–terephthalic acid on 60/80 AW, DMCS Chromosorb W, and using FID. Temperature programming between 100 and 210°C at 5°/min is used.

The margarine sample is extracted similarly, centrifuging the extract and taking the supernatant for GC analysis using the same conditions.

Diethyl pyrocarbonate is an additive used to preserve wine, beer and non-alcoholic beverages. The method recommended for its determination in these beverages is that of Wunderlich[70], which has become an official first action method of the FDA. This method, like most others for the determination of diethyl pyrocarbonate, depends on its hydrolysis to ethanol and carbon dioxide with subsequent reaction between the ethanol and diethyl pyrocarbonate to form diethyl carbonate, the concentration of the latter being a measure of the original preservative concentration. The diethyl carbonate is extracted from the sample with carbon disulphide, which is separated by centrifugation. The concentrated extract is chromatographed using a column of 15% trimethylol propantripelargonate on 60/80 Celite 545, operated at 80°C, and using FID.

Diphenyl and *o*-phenyl phenol are preservatives used to prevent mould growth, particularly on citrus fruit in storage, and residues may therefore be found on citrus fruit or in the juice. The recommended method for their determination is that of Thomas[71], as applied to orange juice. The sample is steam-distilled and the distillate is extracted with chloroform. This extract is used for GLC, which is performed using a column of 20% silicone oil on 100/120

Celite at 160°C and using an AI detector. This method was published in 1960, and it is recommended that it be updated by using FID in place of the AI detector and a Chromosorb or equivalent support material in place of Celite.

2. Antioxidants

The antioxidants which are used mainly in foods are BHA, BHT and gallate esters. Because they are used to prevent oxidation of fats and oils, the foods which contain fatty materials are the ones most likely to contain these antioxidants, and GC methods of analysis have been applied to such foods.

The simple method of Hartman and Rose[72] is recommended for the determination of BHA and BHT in vegetable oils. Methyl undecanoate in carbon disulphide is added as internal standard to the sample of oil, which is dissolved in carbon disulphide, the resulting solution being chromatographed on a column of 10% DC-200 on 80/100 Gas Chrom Q at 160°C, and using FID. There is no clean-up procedure in this method, but a 0.25 in plug of siliconised glass wool is inserted in a sleeve at the injection end of the column, which collects most of the fatty substances which would otherwise foul the column. This plug is replaced daily. The DC-200 column elutes BHA before BHT, but the elution of these antioxidants is reversed on a column of 10% Carbowax 20M on 80/100 Gas Chrom Q, operated at 190°C.

In the recommended method of Stoddard[73] the antioxidants determined, i.e. BHA, BHT, propyl gallate, NDGA and TBHQ, are made into TMS derivatives for GC. In this method, as applied to the determination of these antioxidants in vegetable oils, the sample is dissolved in hexane and extracted several times with acetonitrile and then with 80% aqueous ethanol. After evaporating the combined extracts to dryness, the TMS derivatives are formed and GLC is performed using a column of 3% JXR on 100/120 Gas Chrom Q. FID is used and a temperature programme between 105 and 250°C at 7.5°/min, using methyl pentadecanoate as internal standard.

K. Recommended methods for the analysis of emulsifiers and stabilisers

Many foods contain emulsifiers or stabilisers, natural or synthetic, and the complexity of some of them, e.g. polyglycerides and sorbitan esters of fatty acids, has made absolute analysis extremely difficult.

GC has found a place as a technique to be used with considerable advantage in the analysis of emulsifiers and stabilisers, particularly in conjunction with other separation techniques, such as column chromatography and TLC, which usually precedes GC. This approach is used in the recommended methods for the determination of both glycerides and polyglycerides.

Glycerides may be present naturally in foods or they may be added, and in the latter case the types comprise glyceryl mono- and diesters of mixed, long-chain fatty acids, which include myristic, palmitic, stearic and oleic. In the recommended method of Sahasrabudhe and Legari[74] mono- and diglycerides are separated by silicic acid column chromatography, and each fraction is trimethylsilylated and the derivatives are chromatographed using a column of

3% JXR on 100/120 Gas Chrom Q. FID is used and a temperature programme between 125 and 325°C at 10°/min.

In the recommended method for the determination of polyglycerides, due to Sahasrabudhe[75], the free polyglycerol, cyclic, tri-, di- and monoglycerides which make up commerical polyglycerides are separated by silica gel TLC, prior to the preparation of TMS derivatives for GC, using the same conditions as in the method for mono- and diglycerides[74].

Sorbitan esters of fatty acids are synthetic, oil-soluble emulsifiers and thickeners used widely in the confectionery industry. The recommended method for their determination is that of Sahasrabudhe and Chadha[76]. The emulsifiers are separated, initially, on a silicic acid column and also by silica gel TLC, before the preparation of the TMS derivatives. These derivatives of the polylol moiety are chromatographed as described in the method for mono- and diglycerides[74].

Brominated vegetable oils are used as dispersing agents and stabilisers for the flavouring oils added to soft drinks. The recommended method for the determination of these additives in soft drinks is that of Conacher[77]. The sample is decarbonated and saturated with salt; methyl pentadecanoate is added as internal standard; and the mixture is extracted with diethyl ether. GLC is carried out on a column containing 3% JXR on 80/90 Anakrom ABS. FID is used and temperature programming between 150 and 270°C at 10°/min.

In some countries sucrose diacetate hexa-isobutyrate has replaced brominated vegetable oils as a dispersing agent and stabiliser for the flavouring essential oils incorporated in soft drinks. The recommended method for the determination of this additive in soft drinks is that of Conacher, Chadha and Iyengar[78]. The sample is extracted in the same manner as for brominated vegetable oils[77], but with hexanoic acid replacing methyl pentadecanoate as internal standard. After evaporation of the ether, acid-catalysed transesterification is carried out using sulphuric acid in decanol, and the resulting esters, decyl acetate and decyl isobutyrate, are chromatographed using a column of 3% SE-30 on 100/120 Varaport 30. FID is used and a temperature programme between 100 and 290°C at 10°/min.

Triethyl citrate and triacetin are both used in liquid and dried egg-whites to improve the whipping properties. The recommended method for the detection of these two additives in egg-whites is that of Kogan and Strezleck[79]. Diethyl ether is used to extract triethyl citrate at a pH of 3 and triacetin from neutral solution, and the extracts are concentrated for analysis. GLC is performed using a column of 20% SE-30 on 30/60 Chromosorb W at 200°C, and using TC detection.

L. Recommended methods for the analysis of artificial sweeteners and flavourings

Although GC has been used for the determination of such artificial sweeteners as saccharin and dulcin, the necessity to analyse soft drinks for cyclamate has brought the technique into greater use than hitherto for the analysis of artificial sweeteners in general.

It is sometimes necessary to examine food flavouring substances for the presence of certain components which might be toxicologically dangerous, and GC methods have been used in this context.

1. Artificial sweeteners and cyclohexylamine

The recommended method for the determination of cyclamates in soft drinks is that of Richardson and Luton[80]. Water is added to the sample plus 10N sulphuric acid, so that the solution has a pH between 0.95 and 1.05. Protein materials are precipitated with zinc ferrocyanide and, after filtration, the filtrate is extracted with chloroform and then with petrol. The aqueous solution containing the cyclamate is caused to react with sodium nitrite solution after the addition of petrol containing benzene as internal standard. The mixture is shaken for 3 min at intervals and the organic layer is sampled for GLC analysis. The GLC column consists of 10% PEG 400 on 100/120 Celite, and is temperature programmed between 50 and 120°C at 16°/min, using FID.

If the determination of saccharin is desired, with or without that of cyclamates, the method of Conacher and O'Brien[81] is recommended. The soft drink sample is made alkaline and extracted with diethyl ether. After acidification of the aqueous layer, and the addition of stearic acid as internal standard, the mixture is extracted with ethyl acetate. The ester extract is evaporated to dryness in the presence of diazomethane, and the resulting methyl esters are chromatographed using a column of 3% SE-30 on 100/120 Varaport 30. FID is used and a temperature programme from 125 to 275°C at 15°/min.

There is evidence to suggest that cyclohexylamine present in, or produced by, cyclamates may have some adverse toxicological effects, and its analysis in conjunction with that of cyclamate has, consequently, received attention. The recommended method for its determination in artificially sweetened foods is that of Howard et al.[82] Liquid samples are made alkaline and steam-distilled into 6N hydrochloric acid, and the distillates are made alkaline and distilled into dichloromethane. The extracts are concentrated and determined by GLC using a column of 10% Carbowax 20M plus 2.5% sodium hydroxide on 90/100 Anakrom SD, operated at 100°C and using FID.

Although sorbitol is not an artificial sweetener, it is especially sought in diabetic foods and has been included in this section. The method recommended for its determination in such foods is that of Fernandez-Flores and Blomquist[83]. The food sample is extracted with 80% aqueous methanol and the TMS derivative is made. GLC is carried out using α-D-glucoheptose as internal standard, on a column of 4.2% SE-30 on 60/80 silanised Diaport S. FID is used and a temperature programme between 160 and 280°C at 4°/min.

2. Flavourings

GC is frequently used for the analysis of flavourings, which are made up of large numbers of organic compounds, some at low concentration, even though they may be important flavour and aroma constituents. There is little or no choice from which to recommend GC methods, and Chapter 19 cites methods for the determination of the flavour constituents benzaldehyde, safrole, β-asarone, monosodium glutamate, ammonium glycyrrhizinate and maltol.

M. Recommended methods for the determination of pesticides

Recommended methods for the determination of pesticide residues are mainly confined to multiresidue ones. Chapter 20 includes methods for the analysis of lesser-known and less important pesticides.

1. OC pesticides, including fumigants

Two different types of method are recommended for the determination of OC pesticide residues in crops and foods, being based on different extraction and clean-up procedures. For either method the choice of detector lies between EC and MCD.

The first method, based on the work of Burke and Giuffrida[84], is the official AOAC method[85]. This method involves acetonitrile maceration of the sample, addition of water to the filtered macerate and extraction with petrol. Further clean-up is performed on a column of Florisil, using petrol—diethyl ether mixtures as eluting solvents.

The second method, based on the work of Goodwin, Goulden and Reynolds[86], involves the maceration of the sample with acetone and the separation of the pesticides into hexane or petrol after addition of 2% aqueous sodium sulphate to the filtered macerate. The first-stage clean-up procedure, in which the pesticides are partitioned into acetonitrile followed by their re-extraction into petrol or hexane after the addition of 2% sodium sulphate solution, is based on that of Jones and Riddick[87]. A second-stage clean-up may be necessary with certain foods, and Hamence, Hall and Caverly[88] incorporated a clean-up stage using weakly active alumina, eluting the OC pesticides with hexane or petrol. The GC column of the AOAC-recommended method can be used, viz. 10% DC-200 on 80/100 Chromosorb W-HP, although lower loadings down to 3% are equally as effective.

It is essential in OC pesticide residue analysis to be certain of the identities of the pesticides, and this may be achieved by using two or three different column materials. The stationary phases QF-1 and XE-60 are recommended as alternative ones to DC-200.

The methyl silicone-type phase is a general-purpose one in OC pesticide analysis, but if the separation of the isomers of BHC or the separation of hexachlorobenzene from these isomers is required, a 2:2:1 mixture of 3% OV-61 : 7.5% QF-1:3% XE-60, a phase used by Di Muccio, Boniforti and Monacelli[89], is recommended. With this mixed phase coated on 80/100 Chromosorb W-HP or equivalent, GLC is carried out at 190°C, using EC detection.

The authors[90] recommend 2% XE-60 as the stationary phase for separating the fungicide, PCNB, from other OC compounds of short retention time, including the isomers of BHC. This phase is coated on 60/80 AW, DMCS Chromosorb G, and GLC is carried out at 155°C, using EC detection.

In the determination of the common insecticidal fumigants, including chloroform, carbon tetrachloride, trichloroethylene and carbon disulphide, Bielorai and

Alumot's method[91] is recommended. Water is added to the sample, e.g. cereal, which is steam-distilled into toluene, which traps the fumigants. Normally no clean-up is necessary, and the toluene solution is chromatographed on a column consisting of 10% DC-710 on 80/100 HMDS Chromosorb W operated at 60°C, and using EC detection.

The fumigants ethylene dibromide and methyl bromide are readily determined in cereals and flour by the method of Heuser and Scudamore[92]. The sample is shaken with a mixture of 5:1 acetone:water, and the supernatant is taken for GLC analysis using a column of 15% UCON LB-550-X on 60/80 Chromosorb W, operated at 85°C and using FID.

Two methods are recommended for the determination of OC herbicides of the type 2,4-D-2,4,5-T-MCPA and dicamba in foods and crops, both depending on derivative formation. In Yip's method[93] the sample is acidified, and a mixture of diethyl ether and petrol is used to extract these acidic herbicides. The clean-up procedure is simply a sodium bicarbonate extraction of the herbicides followed by acidification and re-extraction into chloroform. Diazomethane is used to form the methyl esters, which are best separated and determined according to the GLC conditions described by Munro[94], viz. a column of 2% NPGS on 60/80 Anakrom AD operated at 200°C, and using MCD.

In the other recommended method, due to Garbrecht[95], the TMS derivatives of these acidic OC herbicides are prepared, and GLC is carried out using a column of 5% DC-200 on 80/100 Chromosorb W-HP, operated at 190°C and using TC detection.

2. OP pesticides

The AOAC method[85] is recommended for the multiresidue determination of OP pesticides. This method, based on the work of Storherr and Watts[96], used an ethyl acetate extract of the crop or food sample followed by the clean-up procedure known as sweep codistillation (see Chapter 20, p. 326). No other clean-up is necessary, and GLC is performed using a column of 10% DC-200 on 60/100 Chromosorb W-HP, operated at 205°C and using AFID. This detector or FPD is recommended in all the methods of OP pesticide residue analysis.

The second method is based on that of Laws and Webley[97]. The sample is macerated with dichloromethane and the filtered macerate is partitioned between petrol and aqueous methanol. The petrol layer contains the relatively non-polar OP insecticides, which are cleaned up on alumina, and the aqueous methanol layer contains the relatively polar OP insecticides, which are cleaned up on activated carbon after transfer to chloroform. Either the 10% DC-200 phase can be used[85] or the 5% OV-101 phase suggested by Bowman, Beroza and Hill[98]. This phase is coated on 80/100 Gas Chrom Q and the column is temperature programmed between 150 and 300°C at 10°/min. The 5% Dexsil stationary phase of Bowman and Beroza[99] is a good alternative, since it can be temperature programmed to higher temperatures than most phases and is particularly valuable for the determination of high-boiling OP compounds. This phase is used on 80/100 AW Chromosorb W, and the column is temperature programmed between 150 and 300°C at 10°/min. FPD is used with both columns[98, 99].

3. ON pesticides

The recommended detector system for all non-derivatised ON pesticides is the Coulson conductivity detector as described by Laski and Watts[100], who used a column consisting of 10% DC-200 on 80/100 Chromosorb W-HP, operated at 180°C.

The recommended extraction, clean-up and GC determination of carbamate multiresidues is that of Holden, Jones and Beroza[101]. The crop sample is macerated with dichloromethane, and some of the deleterious material is precipitated with a solution of ammonium chloride in phosphoric acid. Activated charcoal is used as a clean-up material, and 1-fluoro-2,4-dinitrobenzene is used to form 2,4-dinitroaniline derivatives. These derivatives are chromatographed using a column of 2% XE-60 on 50/60 Anakrom ABS, operated at 190°C and using EC detection.

If *N*-methylcarbamates are to be determined, the recommended procedure is that due to Moye[102]. The sample is extracted with methanol, and the supernatant is made 5mM to sodium hydroxide for on-column transesterification. GSC is performed using a 80/100 Porapak P column, operated at 180°C and using AFID.

The recommended method for the general analysis of thiocarbamates, based on the determination of their prime degradation product, ethylene thiourea, is that described by Haines and Adler[103]. Methanol is used to extract these pesticides, and the extract is cleaned up on alumina, using an eluting mixture of 2.5 ml methanol and 97.5 ml 10% acetone in benzene. A solution of 1-bromobutane and sodium borohydride in dimethylformamide is added to form 2-butylthio-2-imidazoline, which is chromatographed on a column of 20% SE-30 on 80/100 Gas Chrom Q, operated at 200°C and using FPD.

Triazine residues have been determined, mainly on cereals, such as maize. In using a column comprising 2% Reoplex 400 on 80/100 Gas Chrom Q operated at 195°C, and using CCD, Westlake, Westlake and Gunther[104] found that clean-up was unnecessary for maize samples extracted with methanol. Although this method was used only for the determination of ACD-15M, any methanol-soluble triazine could be determined similarly by this recommended method.

A second method recommended for general triazine analysis is that of Delley *et al.*[105] Methanol is the extractant, and water is added to the extract, which is made alkaline with sodium carbonate before extraction with diethyl ether. The ether extract is cleaned up on deactivated alumina, the eluate being evaporated and the residue dissolved in carbon tetrachloride. This solution is cooled to −20°C and passed through a column of sodium bisulphate, the triazines being eluted with cooled chloroform. GLC is performed on a column of 8% Reoplex 400 on HMDS Chromosorb W, operated at 180°C and using FID.

The common substituted phthalimides, i.e. folpet, captan and captafol, may be determined together in fruits by the method of Pomerantz, Miller and Kava [106]. Acetonitrile is used to extract the samples, and a column of alumina is used as a clean-up and separation material. An eluting solvent of 20% dichloromethane in petrol removes folpet and also any DDT which might occur in a residue analysis of fruit. A second eluting solvent of 50% dichloromethane in petrol removes captan and captafol and any folpet left on the alumina from the previous

128

elution. The eluates are chromatographed separately on a column of 3% XE-60 on 80/100 Chromosorb W-HP, operated at 178°C and using EC detection.

The recommended method for the determination of residues of the nitrophenolic herbicides DNOC, dinoseb, dinosam and dinex is that of Yip and Howard[107]. The crop sample is extracted with chloroform, and, after evaporation of the extract, the methyl esters are formed using diazomethane. Clean-up is effected on an acidic Celite column and on an alumina column, the methyl esters being eluted from the latter with mixtures of diethyl ether and petrol. GLC is carried out on a column of 10% DC-200 on 80/90 Anakrom ABS, operated at 185°C and using EC detection.

4. Pyrethrins and their synergists

Although no GC method has been cited specifically in Chapter 20 for the determination of pyrethrins and their synergists in food, it is recommended that a method based on that of Kawano and Bevenue[108] be used. The sample is extracted with petrol or hexane, and clean-up is effected on either Florisil or alumina. A column of 5% SE-30 on 60/80 AW, DMCS Chromosorb W, operated at 190°C and using FID, will separate and detect pyrethrins, piperonyl butoxide and N-octylbicycloheptene dicarboximide.

N. Recommended methods for the analysis of PCBs, methyl mercury, N-nitrosamines and solvents

These four groups comprise possible food contaminants and GC methods have been used to detect or determine them in certain foods.

1. PCBs

PCBs are used industrially and can find their way into waterways and the sea and can be transferred to fish. Their determination has presented a problem because of their similar GC behaviour to that of members of the DDT family. The recommended extraction procedure and analysis is that of Hannan, Bills and Herring[109], as applied to fish oils. The sample is extracted with petrol and partitioned with acetonitrile. Florisil column chromatography is used to clean up the extract, eluting with petrol so that DDT and TDE remain on the column while PCBs and DDE are eluted. The eluates are irradiated in a UV source, and, on GC examination, a degradation pattern chromatogram is produced which is distinctive for PCBs. GLC is performed using a column of 4% SE-30 plus 4% QF-1 on 70/80 Anakrom ABS, operated at 180°C and using EC detection.

After this extraction of PCBs, another recommended method for their GC evaluation is that of Asai et al.[110] A column containing a neutral palladium catalyst on 80/100 DMCS Gas Chrom Q is attached to the injection port of the GC and samples are led through this column prior to analysis. This catalytic precolumn can produce reactions of hydrogenation, dehydrogenation or hydrogenolysis, PCBs giving a carbon skeleton chromatogram due to biphenyl and cyclohexylbenzene. Members of the DDT family give a different carbon skeleton chromatogram and are therefore distinguished from PCBs. After the catalytic reaction, GLC of PCB degradation products is performed using a column of Carbowax 400 on 80/100 Porasil S, operated at a temperature of 178°C and using FID.

2. Methyl mercury

Mercury-containing industrial effluents, discharged into waterways, have caused the accumulation of mercury in the form of its methyl derivative in certain fish species. GC methods have been used extensively in the determination of methyl mercury, usually as one of its halides.

The first recommended method for the determination of methyl mercury in fish is that of Newsome[111], based on a modification of the procedure first published by Westoo[112]. Newsome obtained emulsion problems in the extraction of methyl mercury from samples of high lipid content; hence the modification of the Westoo method. The fish sample is homogenised in 1N hydrobromic acid plus potassium bromide, and, after filtration, the filtrate is extracted with benzene. The benzene solution is extracted with cysteine acetate solution, and this extract is acidified with hydrobromic acid before re-extracting with benzene. GLC is carried out on the concentrated benzene extract, using a column of 2% BDS on 100/120 AW, DMCS Chromosorb W, operated at 120°C and using EC detection.

The second recommended method is that of Uthe, Solomon and Grift[113]. The fish sample is homogenised in acidic sodium bromide solution and the resulting methyl mercuric bromide is extracted with toluene. The methyl mercuric bromide is removed from the toluene via its thiosulphate complex with aqueous ethanol, and, after the addition of potassium iodide, the aqueous ethanolic solution is extracted with benzene, which removes methyl mercuric iodide. GLC is carried out using a column of 7% Carbowax 20M on AW, DMCS Chromosorb W, operated at 170°C and using EC detection.

3. N-Nitrosamines

The possibility of the formation of *N*-nitrosamines from secondary amines and nitrite has led to their determination in foods which contain these precursors. GC is an important technique in the separation and determination of nitrosamines, but MS is needed to confirm, or otherwise, their identities. MS is used in conjunction with GC in the recommended method of Crosby *et al.*[114], as applied to meat, bacon, fish and cheese. The minced sample is mixed with salt and steam-distilled, the distillate being treated with dilute sulphuric acid and sodium sulphate and then redistilled. After making the distillate alkaline, it is extracted with dichloromethane and the extract concentrated for analysis using two sets of GC conditions. In the first, the column is composed of 15% FFAP on 80/100 Chromosorb W, operated at 140°C and using CCD set in the reductive mode for nitrogen. In the second, the column is composed of 15% Carbowax 20M on 80/100 Chromosorb W, operated at 120°C and using AFID, although these workers have found the CCD to be more satisfactory.

In another recommended method, due to Alliston, Cox and Kirk[115], salt is added to the minced sample, which is steam-distilled. Two further distillations are made from alkaline and acidic solutions, and the final distillate is equally divided. One portion is made alkaline and reduced in an electrolysis cell, which produces secondary amines. Both halves are then caused to react with hepta-fluorobutanoyl chloride in cyclohexane, and the resulting heptafluorobutana-mides are chromatographed, the unreduced portion yielding adventitious secondary amines. GLC is carried out with EC detection and using either a column of 15% FFAP on 80/100 Chromosorb W at 60°C or a column of 20% FFAP on the same support at 110°C, depending on the type of nitrosamines in the sample.

4. Solvents

Solvent residues sometimes remain in foods which have been treated with them to remove fatty or oily substances. GC methods are used to determine solvent residues, e.g. aliphatic hydrocarbons and chlorinated hydrocarbons, lower aliphatic alcohols and acetone.

The recommended method for the determination of hexane in oilseed meals and flours is that of Fore and Dupuy[116]. A small sample is placed into a glass injection port liner between two small retaining plugs of glass wool, and the liner is inserted into the injection port. A small volume of water is injected into the region above the sample, and volatiles, including hexane, are swept on to the column of 80/100 Porapak P. FID is used and a temperature programme between 70 and 180°C, using various rates.

Chlorinated hydrocarbons, such as methylene chloride, ethylene dichloride and trichloroethylene, are best separated and analysed by Robert's method [117], as applied to oleoresins. The samples are extracted with ethanol and the extract is chromatographed on a 150/200 Porapak Q column, operated at 160°C and using MCD.

For the determination of acetone residues in oilseed meals and flours, the method of Dupuy, Rayner and Fore[118] is recommended. The sample plus water and methanol is placed in a sealed bottle, which is incubated at 70°C for 5 h, and the head space is sampled. GSC is carried out on a column of 80/100 Porapak P with a temperature programme between 70 and 180°C at various rates, and using FID.

Residual methanol is best determined by the method of Litchmann and Upton[119]. The method depends on the formation of methyl nitrite, after the addition of sodium nitrite solution and oxalic acid solution to the liquid sample. The head space is sampled after a short interval at room temperature, and the methyl nitrite is chromatographed using a column of 120/150 Porapak R, op-erated at 150°C and using FID.

The method of Fore, Rayner and Dupuy[120] is recommended for the deter-mination of residues of isopropanol in oilseed meals and flour. The sample is extracted with ethanol, which also serves as internal standard. The concen-trated extract is chromatographed using FID on a 80/100 Porapak P column, using a temperature programme between 80 and 100°C.

131

References

1. Glass, R. L., Lohse, L. W., and Jenness, R., *J. Dairy Sci.,* **51**, 1847 (1968)
2. Sampugna, J., Pitas, R. E. and Jensen, R. G., *J. Dairy Sci.,* **49**, 1462 (1966)
3. Arnold, R. G., Libbey, L. M. and Day, E. A., *J. Food. Sci.,* **31**, 566 (1966)
4. Morgan, M. E. and Day, E. A., *J. Dairy Sci.,* **48**, 1382 (1965)
5. Scanlan, R. A. and Lindsay, R. C., *J. Food Sci.,* **33**, 440 (1968)
6. Reineccius, G. A., Kavanagh, T. E. and Keeney, P. G., *J. Dairy Sci.,* **53**, 1018 (1970)
7. Withington, D. F., *Analyst,* **92**, 705 (1967)
8. Phillips, A. R. and Sanders, B. J., *J. Ass. Publ. Anal.,* **6**, 89 (1968)
9. Eisner, J., Wong, N. P., Firestone, D. and Bond, J., *J. Ass. Off. Agric. Chem.,* **45**, 337 (1962)
10. Langler, J. E., Libbey, L. M. and Day, E. A., *J. Agric. Food Chem.,* **15**, 386 (1967)
11. Liebich, H. M., Douglas, D. R., Bayer, E. and Zlatkis, A., *J. Chromat. Sci.,* **8**, 351 (1970)
12. Castledine, S. A. and Davies, D. R. A., *J. Ass. Publ. Anal.,* **6**, 39 (1968)
13. Hubbard, A. W. and Pocklington, W. D., *J. Sci. Food Agric.,* **19**, 571 (1968)
14. Hoffmann, G. and Meijboom, P. W., *J. Amer. Oil Chem. Soc.,* **45**, 468 (1968)
15. Grieco, D. and Piepoli, G., *Riv. Ital. Sostanze Grasse,* **41**, 283 (1964)
16. Fedeli, E., Lanzani, A., Capella, P. and Jacini, G., *J. Amer. Oil Chem. Soc.,* **43**, 254 (1966)
17. Amati, A., Carraro Zanirato, F. and Ferri, G., *Riv. Ital. Sostanze Grasse,* **48**, 39 (1971)
18. Losi, G. and Piretti, M. V., *Riv. Ital. Sostanze Grasse,* **47**, 493 (1970)
19. Ackman, R. G., *J. Gas Chromat.,* **4**, 256 (1966)
20. Nonaka, M., Black, D. R. and Pippen, E. L., *J. Agric. Food Chem.,* **15**, 713 (1967)
21. Cross, C. K. and Ziegler, P., *J. Food Sci.,* **30**, 610 (1965)
22. Wong, N. P., Damico, J. N. and Salwin, H., *J. Ass. Off. Anal. Chem.,* **50**, 8 (1967)
23. Keay, J. N. and Hardy, R., *J. Sci. Food Agric.,* **23**, 9 (1972)
24. Salwin, H. and Bond, J. F., *J. Ass. Off. Anal. Chem.,* **52**, 41 (1969)
25. Johansen, R. G. and Voris, S. S., *Cereal Chem.,* **48**, 576 (1971)
26. Hunter, G. L. K. and Brogden, W. B., *J. Food Sci.,* **30**, 1 (1965)
27. Hunter, G. L. K. and Moshonas, M. G., *Anal. Chem.,* **37**, 378 (1965)
28. Moshonas, M. G., *J. Agric. Food Chem.,* **19**, 769 (1971)
29. Stanley, W. L., Ikeda, R. M., Vannier, S. H. and Rolle, L. A., *J. Food Sci.,* **26**, 43 (1961)
30. Cicplinski, E. W. and Averill, W., *Gas Chromatography Applications,* No. GC-AP-002, Perkin-Elmer Corporation (1962)
31. Fernandez-Flores, E., Kline, D. A. and Johnson, A. R., *J. Ass. Off. Anal. Chem.,* **53**, 17 (1970)
32. Fernandez-Flores, E., Kline, D. A., Johnson, A. R. and Leber, B. L., *J. Ass. Off Anal. Chem.,* **53**, 1203 (1970)
33. Kline, D. A., Fernandez-Flores, E. and Johnson, A. R., *J. Ass. Off. Anal. Chem.,* **53**, 1198 (1970)
34. Romani, R. J. and Ku, L., *J. Food Sci.,* **31**, 558 (1966)
35. Baerwald, G. and Miglio, G., *Mschr. Brau.,* **23**, 288 (1970)
36. Mueller-Stock, A., Joshi, R. K. and Buechi, J., *J. Chromatog.,* **63**, 281 (1971)
37. Brodnitz, M. H., Pollock, C. L. and Vallon, P. P., *J. Agric. Food Chem.,* **17**, 760 (1969)
38. van Praag, M., Stein, H. S. and Tibbetts, M. S., *J. Agric. Food Chem.,* **16**, 1005 (1968)
39. Pierce, A. R., Graham, H. N., Glassner, S., Madlin, H. and Gonzalez, J. G., *Anal. Chem.,* **41**, 298 (1969)
40. Newton, J. M., *J. Ass. Off. Anal. Chem.,* **52**, 653 (1969)
41. Trachman, H., *Communs. Wallerstein Labs.,* **32**, 111 (1969)
42. Clapperton, J. F. and Holliday, A. G., *J. Inst. Brew.,* **74**, 164 (1968)
43. Feil, M. F. and Marinelli, L., *Proc. Amer. Soc. Brew. Chem.,* 29 (1969)
44. Kurosky, A. and Bars, A., *J. Inst. Brew.,* **74**, 200 (1968)

45. Kieser, M. E., Pollard, A., Stevens, P. M. and Tucknott, O. G., *Nature*, **204**, 887 (1964)
46. Powell, A. D. G. and Brown, I. H., *Proc. 9th Conv. Inst. Brewing, Australian Section*, 257 (1966)
47. Richardson, P. J. and Mocek, M., *Proc. Amer. Soc. Brew. Chem.*, 128 (1971)
48. Capella, P. and Losi, G., *Ind. Agrar.*, **6**, 7 (1968)
49. Brunelle, R. L., Schoeneman, R. L. and Martin, G. E., *J. Ass. Off. Anal. Chem.*, **50**, 329 (1967)
50. Martin, G. E. and Caress, E. A., *J. Sci. Food Agric.*, **22**, 587 (1971)
51. Hardy, P. J., and Ramshaw, E. H., *J. Sci. Food Agric.*, **21**, 39 (1970)
52. Ashmead, H. L., Martin, G. E. and Schmit, J. A., *J. Ass. Off. Agric. Chem.*, **47**, 730 (1964)
53. Research Committee on the Analysis of Potable Spirits, *J. Ass. Publ. Anal.*, **10**, 49 (1972)
54. Guymon, J. F., and Crowell, E. A., *Amer. J. Enol. Viticult.*, **20**, 76 (1969)
55. Sweeley, C. C., Bentley, R., Makita, M. and Wells, W. W., *J. Amer. Chem. Soc.*, **85**, 2497 (1963)
56. Mueller, B. and Goeke, G., *Dt. LebensmittRdsch.*, **68**, 222 (1972)
57. Brobst, K. M. and Lott, C. E., *Proc. Amer. Soc. Brew. Chem.*, 71 (1966)
58. Luke, M. A., *J. Ass. Off. Anal. Chem.*, **54**, 937 (1971)
59. Johnson, A. R., Corliss, R. L. and Fernandez-Flores, E., *J. Ass. Off. Anal. Chem.*, **54**, 61 (1971)
60. Mehltretter, C. L. and Otten, J. G., *Int. Sug. J.*, **73**, 235 (1971)
61. Oldfield, J. F. T., Parslow, R. and Shore, M., *Int. Sug. J.*, **72**, 35 (1970)
62. Underwood, J. C., Filipic, V. J. and Bell, R. A., *J. Ass. Off. Anal. Chem.*, **52**, 717 (1969)
63. Burkwall, M. P. and Glass, R. L., *Cereal Chem.*, **42**, 236 (1965)
64. Graveland, A., *J. Amer. Oil Chem. Soc.*, **45**, 834 (1968)
65. Mason, B. S. and Slover, H. T., *J. Agric. Food Chem.*, **19**, 551 (1971)
66. Hunter, I. R. and Walden, M. K., *J. Gas Chromat.*, **4**, 246 (1966)
67. Zumwalt, R. W., Kuo, K. and Gehrke, C. W., *J. Chromatog.*, **55**, 267 (1971)
68. Fogden, E., Fryer, M. and Urry, S., *J. Ass. Publ. Anal.*, **12**, 93 (1974)
69. Graveland, A., *J. Ass. Off. Anal. Chem.*, **55**, 1024 (1972)
70. Wunderlich, H., *J. Ass. Off. Anal. Chem.*, **55**, 557 (1972)
71. Thomas, R., *Analyst*, **85**, 551 (1960)
72. Hartman, K. T. and Rose, L. C., *J. Amer. Oil Chem. Soc.*, **47**, 7 (1970)
73. Stoddard, E. E., *J. Ass. Off. Anal. Chem.*, **55**, 1081 (1972)
74. Sahasrabudhe, M. R. and Legari, J. J., *J. Amer. Oil Chem. Soc.*, **44**, 379 (1967)
75. Sahasrabudhe, M. R., *J. Amer. Oil Chem. Soc.*, **44**, 376 (1967)
76. Sahasrabudhe, M. R. and Chadha, R. K., *J. Amer. Oil Chem. Soc.*, **46**, 8 (1969)
77. Conacher, H. B. S., *J. Ass. Off. Anal. Chem.*, **56**, 602 (1973)
78. Conacher, H. B. S., Chadha, R. K. and Lyengar, J. R., *J. Ass. Off. Anal. Chem.*, **56**, 1264 (1973)
79. Kogan, L. and Strezleck, S., *Cereal Chem.*, **43**, 470 (1966)
80. Richardson, M. L. and Luton, P. E., *Analyst*, **91**, 520 (1966)
81. Conacher, H. B. S. and O'Brien, R. C., *J. Ass. Off. Anal. Chem.*, **54**, 1135 (1971)
82. Howard, J. W., Fazio, T., Klimeck, B. A. and White, R. H., *J. Ass. Off. Analyt. Chem.*, **52**, 492 (1969)
83. Fernandez-Flores, E. and Blomquist, V. H., *J. Ass. Off. Anal. Chem.*, **56**, 1267 (1973)
84. Burke, J. and Giuffrida, L., *J. Ass. Off. Agric. Chem.*, **47**, 326 (1964)
85. *Official Methods of Analysis of the Association of Official Analytical Chemists*, 11th edn, AOAC, Washington, 475 (1970)
86. Goodwin E. S., Goulden, R. and Reynolds, J. G., *Analyst*, **86**, 697 (1961)
87. Jones, L. R. and Riddick, J. A., *Anal. Chem.*, **24**, 569 (1952)
88. Hamence, J. H., Hall, P. S. and Caverly, D. J., *Analyst*, **90**, 649 (1965)
89. Di Muccio, A., Boniforti, L. and Monacelli, R., *J. Chromatog.*, **71**, 340 (1972)
90. Dickes, G. J. and Nicholas, P. V., *J. Ass. Publ. Anal.*, **7**, 14 (1969)
91. Bielorai, R. and Alumot, E., *J. Agric. Food Chem.*, **14**, 622 (1966)
92. Heuser, S. G. and Scudamore, K. A., *Analyst*, **93**, 252 (1968)

93. Yip, G., *J. Ass. Off. Agric. Chem.,* **45,** 367 (1962)
94. Munro, H. E., *Pesticide Sci.,* **3,** 371 (1972)
95. Garbrecht, T. P., *J. Ass. Off. Anal. Chem.,* **53,** 70 (1970)
96. Storherr, R. W. and Watts, R. R., *J. Ass. Off. Agric. Chem.,* **48,** 1154 (1965)
97. Laws, E. Q. and Webley, D. J., *Analyst,* **86,** 249 (1961)
98. Bowman, M. C., Beroza, M. and Hill, K. R., *J. Ass. Off. Anal. Chem.,* **54,** 346 (1971)
99. Bowman, M. C. and Beroza, M., *J. Ass. Off. Anal. Chem.,* **54,** 1086 (1971)
100. Laski, R. R. and Watts, R. R., *J. Ass. Off. Anal. Chem.,* **56,** 328 (1973)
101. Holden, E. R., Jones, W. M. and Beroza, M., *J. Agric. Food Chem.,* **17,** 56 (1969)
102. Moye, H. A., *J. Agric. Food Chem.,* **19,** 452 (1971)
103. Haines, L. D. and Adler, I. L., *J. Ass. Off. Anal. Chem.,* **56,** 333 (1973)
104. Westlake, W. E., Westlake, A. and Gunther, F. A., *J. Agric. Food Chem.,* **18,** 685 (1970)
105. Delley, R. G., Friedrich, K., Karlhuber, B., Szekely, G. and Stammbach, K., *Z. Anal. Chem.,* **228,** 23 (1967)
106. Pomerantz, I. H., Miller, L. J. and Kava, G., *J. Ass. Off. Anal. Chem.,* **53,** 154 (1970)
107. Yip, G. and Howard, S. F., *J. Ass. Off. Anal. Chem.,* **51,** 24 (1968)
108. Kawano, Y. and Bevenue, A., *J. Chromatog.,* **72,** 51 (1972)
109. Hannan, E. J., Bills, D. D. and Herring, J. L., *J. Agric. Food Chem.,* **21,** 87 (1973)
110. Asai, R. I., Gunther, F. A., Westlake, W. E. and Iwata, Y., *J. Agric. Food Chem.,* **19,** 396 (1971)
111. Newsome, W. H., *J. Agric. Food Chem.,* **19,** 567 (1971)
112. Westoo, G., *Acta Chem. Scand.,* **20,** 2131 (1966)
113. Uthe, J. F., Solomon, J. and Grift, B., *J. Ass. Off. Anal. Chem.,* **55,** 583 (1972)
114. Crosby, N. T., Foreman, J. K., Palframan, J. F. and Sawyer, R., *Nature,* **238,** 342 (1972)
115. Alliston, T. G., Cox, G. B. and Kirk, R. S., *Analyst,* **97,** 915 (1972)
116. Fore, S. P. and Dupuy, H. P., *J. Amer. Oil Chem. Soc.,* **49,** 129 (1972)
117. Roberts, L. A., *J. Ass. Off. Anal. Chem.,* **51,** 825 (1968)
118. Dupuy, H. P., Rayner, E. T. and Fore, S. P., *J. Amer. Oil Chem. Soc.,* **48,** 155 (1971)
119. Litchmann, M. A. and Upton, R. P., *Anal. Chem.,* **44,** 1495 (1972)
120. Fore, S. P., Rayner, E. T. and Dupuy, H. P., *J. Amer. Oil Chem. Soc.,* **48,** 140 (1971)

Part 3

Food Composition

9
Dairy Products

A. Introduction

The quality of dairy products is the concern of many food analysts. Interest in the analysis of such products is shared between those who manufacture and sell them to the public and those various Authorities which enforce the appropriate legal standards.

In the first instance, milk, cream, butter and cheese must be wholly those substances, and assessment of their quality largely relies on the qualitative and quantitative examination of the fat. Dairy fat is composed of a mixture of glycerides, which are compounds made up of glycerol and various fatty acids. After hydrolysis of the glycerides, the common method employed to differentiate between fatty acids is to split them into three fractions: (1) water-soluble volatile acids, (2) water-insoluble volatile acids, (3) water-soluble volatile acids which also form water-soluble silver salts. Placed on an empirical basis, these three fractions give Reichert, Polenske and Kirschner values[1]. Another common method, which gives the Hydroxamic Acid Index[2], is based on the reaction of the glycerides of the fat with hydroxylamine under alkaline conditions to form hydroxamic acid derivatives.

The main reason for using GC in the quality control of dairy fat is to divide further the Reichert, Polenske and Kirschner fractions into their component fatty acids, thereby increasing specificity, leading to an increase in sensitivity. It follows that the increase in sensitivity afforded by GC can be best utilised in the micro-examination of dairy fat, e.g. in butter confectionery, cream soups, etc., where the classical methods already mentioned are under a sensitivity strain. Similarly, GC makes possible the differentiation of glycerides, which are grouped together in the determination of the Hydroxamic Acid Index.

Most vegetable oils and fats can be manufactured more cheaply than butter, and therefore a small addition of such an oil or fat to butter becomes a lucrative proposition. Most countries have experienced this type of adulteration, particularly Italy and France. The Reichert, Polenske and Kirschner values can indicate an impure butter but not the absolute identity and quantity of the adulterant.

The presence of vegetable oils and fats in butter can also be detected by an examination of the sterols in the unsaponifiable fraction. Pure butter contains cholesterol but no sitosterol. Cholesterol can be isolated from butter and

Table 9.1 GC conditions used in the analysis of dairy products

Class of compound	Ref.	Column dimensions	Stationary phase	Support material	Carrier gas and conditions	Temperature or temperature programme/°C	Detector
Fatty acids	7	2m x 4mm	10% PPGA plus 1% $H_3 PO_4$	Celite 545	N_2	210	FI
Fatty acids	7	2.4m x 4mm	2.5% PEGS plus 2.5% PEG plus 1% $H_3 PO_4$	Chromosorb W	N_2	80–210 at 4.5/min	FI
Butyric acid	50	5ft x 0.25in O.D.	5% Carbowax 20M plus 0.5% terephthalic acid	100/120 AW Supasorb	N_2, 50ml/min	125	FI
Methyl esters of fatty acids	9	6ft x 0.25in O.D.	20% DEGS	35/80 Chromosorb	He, 60ml/min	196	TC
Methyl esters of fatty acids	15	1.5m x 4mm I.D.	10% DEGS	80/90 Anakrom F6	He, 120ml/min	90–210 at 6/min after 3 min at 90	FI
Methyl esters of fatty acids	33	2m x	PEGA	firebrick	He, 7–67ml/min	220	–
Methyl esters of fatty acids	38	8ft x 0.125in O.D.	12% DEGS	42/60 Chromosorb W	–	100 until C_{10} peak, then 4/min, up to 176	FI
Ethyl esters of fatty acids	42	1m x 6mm	23.75% silicone oil plus 1.25% stearic acid	Celite C22	He, 100ml/min	150	–

Compound	No.	Column	Stationary phase	Support	Carrier gas	Temperature	Detector
Ethyl esters of fatty acids	42	2m x 4mm	25% PEGS	Celite C22	He, 42ml/min	200	—
Ethyl butyrate	48	5ft x	5% SE-30	60/80 DMCS Chromosorb W	—	70–200 at 8/min	FI
Propyl esters of fatty acids	44	5ft x 0.125in	20% DEGS	60/80 HMDS Chromosorb W	—	50–200 at 4/min	—
Butyl esters of fatty acids	17	3m x 6.4mm	18% DEGS	Anakrom ABS	—	50–115 at 12/min 115–150 at 4/min 150–215 at 24/min	FI
Biacetyl	24	13.5ft x 0.125in O.D.	20% TCEP	80/100 Celite 545	$Ar:CH_4$, 95:5, 24ml/min	70	EC
Aldehydes	58	100ft x	Apiezon L	Capillary	Ar	—	EC
Lactones	59	0.9m x 4mm	22.6% Apiezon L	120/150 Celite	–, 0.77 atm pressure	190	—
Alcohols, tocopherols	57	6ft x 0.25in I.D.	1.5% SE-52	100/120 silanised Gas Chrom P	Ar, 12 p.s.i. inlet pressure	210	AI
Sterols	51	6ft x 0.25in O.D.	2% SE-52	80/100 DMCS Gas Chrom P	Ar, 52ml/min	212	AI
Sterol acetates	54	6ft x 0.25in I.D.	1% SE-52	130/140 Celite	Ar, 60–80ml/min	245	AI

Class of compound	Ref.	Column dimensions	Stationary phase	Support material	Carrier gas and conditions	Temperature or temperature programme/°C	Detector
TMS derivatives of sterols	56	2m x 0.125in	3% JXR	100/120 Gas Chrom Q	He, 3 atm pressure	210	–
TMS derivatives of sugars	30	1.83m x 3.2mm	5% SE-30	80/100 AW, DMCS Chromosorb G	He, 15ml/min	130–270 at 6/min	FI
TMS derivatives of vitamin D	27	1.8m x 4mm	5% BDS	Gas Chrom Q	–	–	–
Vitamins D_2, D_3	28	6ft x 5mm I.D.	3% OV-1	100/120 Gas Chrom Q	He, 110ml/min	150–270 at 5/min	FI
General flavour compounds	74	12ft x 0.125in O.D.	20% TCEP	60/80 AW Celite 545	He, 30ml/min	50	FI
General flavour compounds	74	12ft x 0.125in O.D.	20% Carbowax 600	60/80 AW Celite 545	He, 30ml/min	80	FI

140

acetylated and the melting point of the acetate determined. Vegetable oil yields sitosterol acetate under the same conditions, and this affects the melting point of the cholesterol acetate. This method of detection has obvious limitations, whereas GC differentiation of these sterols or their acetates has led to the detection of as little as 2% of vegetable oil in butter.

Another important reason for using GC in dairy products analysis is that these should be fresh and show no evidence of rancidity. Indication of oxidative rancidity is shown by the total free fatty acid content, the peroxide content or the colour reactions of the oxidised fat with phloroglucinol or 2-thiobarbituric acid. Alternatively, the examination of the volatile fractions of dairy products by GC indicates some of the compounds responsible for rancidity, and therefore gives a more specific measure of off-flavours. On the other hand, in the manufacture of cheese it is the true flavour of the product which is of paramount importance. Identification of aroma constituents of cheese by GC gives analytical 'backing' to organoleptic testing, which has been the main way of assessing cheese flavour and also of differentiating one variety of cheese from another.

In 1952 James and Martin[3] separated the volatile fatty acids from formic acid to dodecanoic acid by GC. This classical study not only heralded the new analytical technique, but also gave particular impetus to investigations of the fatty acid constitutions of dairy products.

Most of the literature concerning fatty acids includes references to either their common names or their systematic names or their numbers of carbon atoms together with indications of the positions of any double bonds. A table of such data is given in Appendix 3.

Much has been written about the GC of dairy products but few attempts have been made to collect together the analytical information. A general review has been given by the authors[4] and a review of the fatty acid analysis of milk has been given by Jensen *et al.*[5]

The essential GC details, where known, of some of the more important references in this chapter are listed in *Table 9.1*.

B. Milk and cream

Milk is essentially an emulsion of fatty acid glycerides, water, proteins and sugars, and it is logical that the quality of milk is determined by analysis of these constituents. The total fat can be extracted by the established methods of Rose–Gottlieb, Werner–Schmidt, Gerber or Mojonnier, and some workers have achieved separation with silica gel or ion exchange column chromatography.

The fatty acid content of the glycerides of milk fat can be obtained by the GC examination of the free fatty acids, the simple alkyl esters of the fatty acids or the glycerides themselves. Although the free fatty acids are more difficult to volatilise than their esters, Hankinson, Harper and Mikolajcik[6] achieved separation of seven milk fat acids on a very early GC apparatus. More recently, Palo, Hrivnak and Goerner[7] separated the milk fatty acids from C_2 to C_{18} using two sets of GC conditions to ensure good recoveries from the complete range.

By far the most popular esterification technique is methylation. The three basic methods of methylation are (1) acid-catalysed methanolysis;(2) methanol—boron trifluoride reaction, and (3) methanol—sodium methoxide reaction under sealed conditions. These methods have been critically assessed by de Man[8], who concluded that serious losses of lower fatty acids occurred with methods (1) and (2). Method (3) not only produced satisfactory recoveries of butyric, caproic, caprylic and capric acids, but also gave better recoveries of linoleic and linolenic acids.

In 1961 Smith[9] separated the methyl esters of the C_4 to C_{18} fatty acids of milk fat using a small sample injection. For minor constituents such as methyl arachidate (0.5% in the fat) and methyl arachidonate (0.3% in the fat) he used a large sample injection and also tentatively identified in trace amounts the acids of C_{13}, C_{15} and C_{17}. A typical chromatogram of methyl esters of fatty acids of milk as found by Smith is shown in *Figure 9.1*. Bills, Khatri and Day[10], Christopherson and Glass[11] and Leemann and Stahel[12] have achieved similar

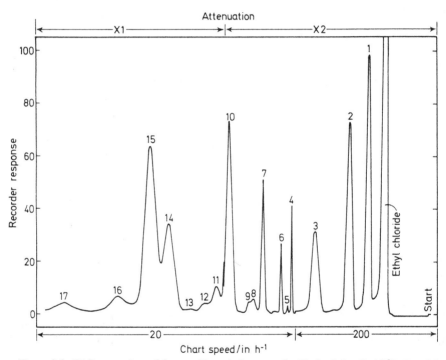

Figure 9.1. GLC separation of fatty acid methyl esters of milk fat (after Smith[9]). For GC details, see Table 9.1. Peaks in order of appearance: (1) C–4, (2) C–6, (3) C–8, (4) C–10, (5) C–10$^{1=}$, (6) C–12, (7) C–14, (8) C–14$^{1=}$, (9) C–15, (10) C–16, (11) C–16$^{1=}$, (12) C–17, (13) probably C–18^{iso}, (14) C–18, (15) C–18$^{1=}$, (16) C–18$^{2=}$, (17) C–18$^{3=}$

separations of the methyl esters of the fatty acids of milk fat. Whereas Hadorn and Zuercher[13] in a similar application used the butyric acid concentration to calculate the milk fat content, Jensen, Gander and Sampugna[14] extracted milk fat by silica gel column chromatography prior to GC. They examined 106 samples of raw milk, and discovered that milk from pasture-fed cattle contained a higher than average content of $C_{18:0}$ and $C_{18:1}$ acids and a lower than average content of C_{12}, C_{14} and C_{16} acids. The average milk was taken as that produced by cattle given a balanced diet, including hay as well as fresh pasture.

Glass, Lohse and Jenness[15] used a transesterification technique prior to the GC of the methyl esters of the fatty acids in order to determine the fat content of milk. This technique has particular value when the available sample is too small for conventional methods to be used, i.e. 30–60 mg of milk. The fat is extracted in a mixture of methanol, dimethyl carbonate and benzene, and methyl tridecanoate is added to this mixture as internal standard. The areas under the GC peaks are related directly to the area given by the internal standard, and the fat content is calculated. A correction is added for methyl butyrate which is obscured by the solvent peak. The method is quick and is in good agreement with the Rose–Gottlieb method.

Butyl esters as well as methyl esters of fatty acids have been used by Gander, Jensen and Sampugna[16], and led Sampugna, Pitas and Jensen[17] to explore a superior method of preparing butyl esters. The esters are prepared in a few minutes by heating the fat with di-n-butyl carbonate and sodium butoxide, and are then chromatographed using a three-speed temperature programme. In this way only one chromatogram is necessary to include the whole range of fatty acids in milk fat, whereas with methyl esters it is so often necessary to make two chromatograms.

In order to find the compounds which give rise to the change in milk flavour when undergoing various processings, Patton[18] examined evaporated and dried whole milk as well as raw milk. He found the presence of δ-deca- and δ-dodeca-lactones in the processed milks, and suggested that these lactones were formed by heat processing. This opinion was supported by his findings of small quantities of δ-lactones in pasteurised milk. This work was extended by Wong and Patton[19] to include the flavour constituents of cream, where methyl sulphide, acetone, 2-butanone and ethanol were identified.

As flavour research has developed, the methods of extraction have become more elaborate. The main flavour interest in milk and cream is associated with those compounds which constitute rancid flavour or those which are the precursors of rancidity and sourness, whereas in yoghurt production there is an interest in desirable flavour constituents. *Table 9.2* summarises the volatile constituents of milk and cream. Hennings, Viljhalmsson and Dunkley[20] and Bassette, Ozeris and Whitnah[21] used a head space technique to study the volatiles of rancid milk, whereas Mabbitt and McKinnon[22] examined volatiles by a cold trapping technique. These authors showed that 2-butanone increased as souring progressed.

Normal sterilised milk will keep for long periods, and attempts have been made commercially to concentrate such milk for ease of storage. The major obstacle to the storage of concentrated sterilised milk is the development of a stale flavour. This flavour has been investigated by Arnold, Libbey and Day[23], and the components are listed in *Table 9.2*. Scanlan and Lindsay[24] determined the biacetyl content of heated milk and found concentrations of $29.0–31.5$ p.p. 10^9.

Acetic acid was steam-distilled from commercial sour cream by Hempenius and Liska[25] in an assessment of quality, a method which would be suitable for acetic acid in other cultured dairy products.

Table 9.2 Milk and cream volatile constituents

Product	Isolation technique	Volatile constituents	Ref.
Rancid milk	head space	methyl sulphide, acetone	20
Rancid milk	head space	acetaldehyde, propionaldehyde, acetone, 2-butanone	21
Sour milk	cold trapping	ethanol, biacetyl, 3-hydroxy-2-butanone, 2-butanone, 2,3-butanediol	22
Stored, sterilised milk	on-column trapping	2-heptanone, 2-nonanone, 2-tridecanone, acetophenone, benzothiazole, benzaldehyde, naphthalene, dichlorobenzene, δ-decalactone, o-amino-aceto-phenone	23
Milk	on-column trapping	biacetyl	24
Sour cream	steam distillation	acetic acid	25
Milk	none	methanol, ethanol, 2-propanol, 1-propanol, acetaldehyde, acetone, biacetyl	26

Palo and Ilkova[26] examined milk directly and separated volatiles on a 1:1 Porapak P:Q column. By this technique they also analysed kefir, cream and yoghurt cultures.

Some interest has been shown in the vitamin D content of milk and non-fat dried milk. Janecke and Brendel[27] separated the unsaponifiable material from the milk, and followed clean-up and precipitation of the sterol fraction with GC of the TMS derivative of vitamin D. The method is particularly applicable to milk which has been exposed to UV radiation in order to increase vitamin D up to 1000 I.U. Panalaks[28] isolated vitamins D_2 and D_3 from non-fat dried milk by extraction and column chromatographic clean-up, and subjected them to GC after modification with antimony trichloride.

Bell and Christie[29] determined vitamin D_2 in full-cream dried milk using a two-stage column chromatographic clean-up stage after fat saponification, followed by GC of the TMS derivatives.

Little attempt has been made to analyse milk for sugars by GC, which must be partly attributed to their non-volatility. However, Reineccius, Kavanagh and Keeney[30] have overcome this problem by using the more volatile TMS derivatives, and have obtained glucose, galactose, lactose and their isomers from whole and skimmed milk. These authors had previously tried methylated and acetylated carbohydrate derivatives without much success. The free monosaccharide content of whole milk was found to be 13.8 and 11.7 mg per 100 ml milk of glucose and galactose, respectively, which agrees well with reported figures by other techniques.

Jaynes and Asan[31] have determined the ratios of lactose anomers in whole and skimmed milk, utilising TMS derivatisation.

C. Butter

As already mentioned, most GC investigations of butter have concerned studies of its fatty acid composition. Butter fat is isolated by conventional extraction procedures, and butter oil is normally obtained by centrifugation. The GC of the various fatty acids and their alkyl esters is obviously very similar to that of milk and cream, but the data produced are not only used to detect extraneous fatty materials, but also to quantify such adulteration.

A very early attempt was made by Hawke[32] to use methyl esters of the butter fatty acids in a study of oxidation of butter. In 1960 Wolff[33] applied a similar technique in order to assess butter quality. This work now ranks as a classical piece of research in that it introduced the conception of using fatty acid ratios as a means of determining the true nature of an oil or fat. He carried out methylation in one of two ways: (1) 2 g of fatty acids in methanol was treated with sulphuric acid for 30 min, water was added and the methyl esters were extracted into hexane; (2) between 2 and 5 g of neutral fat in methanol was saponified with potassium hydroxide for 30 min and the methyl esters were extracted into hexane.

Wolff calculated the proportions of acids found in butter and also in palm kernel oil and coconut oil, both of which are used in the manufacture of margarine. *Table 9.3* shows the large difference between butter and these vegetable oils

Table 9.3 Acids in weight % in various oils and fats (after Wolff[33] and after Luddy *et al.*[38])

Oil or fat	C_4	C_6	C_8	C_{10}	C_{12}	C_{14}	C_{16}	C_{18}	C_{20}	Others
Palm kernel oil	–	–	3.5	3.3	48.9	15.9	8.5	19.5	–	0.4
Coconut oil	–	1.1	8.2	5.8	47.3	18.0	9.0	10.3	–	0.3
Butter	–	1.6	1.4	2.8	3.2	15.1	33.8	40.5	1.5	0.1
Butter	4.2	1.6	1.3	2.7	3.2	12.8	29.9	41.4	–	2.9

with respect to the proportion of the C_{12} acid. This led Wolff to calculate the C_{12}/C_{10} acid ratio for pure butter and for butter with various additions of vegetable oils. *Table 9.4* gives values of this ratio for butter and vegetable oils and mixtures of the two. An examination of eight French butters showed that the C_{12}/C_{10} acid ratio was between 1.07 and 1.13 and that a 5% addition of margarine increased this ratio to 1.6.

Following this classical work, many workers, particularly in Italy and France, examined ratios of the butter fatty acids. As a means of assessing the purity of butter, Francesco and Avancini[34] showed that genuine butter had a C_{12}/C_{10} acid ratio never exceeding 1.6 and a $C_4/C_6 + C_8$ acid ratio never exceeding 1.8. These workers examined the butter with added suet, lard and coconut oil, and showed that the ratios given above were greatly exceeded. Doro and Gabucci[35] examined 42 samples of butter and obtained a C_{12}/C_{10} acid ratio

145

Table 9.4 Effect on the C_{12}/C_{10} acid ratio of the addition of vegetable oils to butter (after Wolff[33])

Oil or fat	C_{12}/C_{10} ratio
8 French butters	1.07–1.13
95% Butter, 5% palm kernel oil	1.96
97% Butter, 3% palm kernel oil	1.64
98.5% Butter, 1.5% palm kernel oil	1.44
90.4% Butter, 9.6% margarine	2.06
95% Butter, 5% margarine	1.60
Commercial margarine	8.30, 8.20
Coconut oil	8.20, 7.90
Palm kernel oil	17.1, 15.1, 13.5

of between 1.1 and 1.2. They also studied the C_{14}/C_{12} acid ratio and obtained a range of 3.2–3.8. Olivari and Benassi[36] examined 25 samples of butter, and agreed with Wolff's findings that the C_{12}/C_{10} acid ratio was between 1.0 and 1.2. Valussi and Cofleri[37] used a GC system to study the methyl esters of the fatty acids from butter incorporating two different helium carrier gas flow rates. The flow rate was increased from 2 1/h after emergence of the C_{12} peak to 6.5 1/h. Besides a study of the C_{12}/C_{10} acid ratio and the $C_4/C_6 + C_8$ acid ratio, these workers looked at the differences between the C_{18} saturated and unsaturated acids. In genuine butter the ratio of the C_{18} unsaturated/C_{18} saturated acids is approximately 3. If adulteration by vegetable oils has taken place, the ratio decreases towards 2.

In 1968 Luddy *et al.*[38] used a methylation technique directly on the glycerides of butter oil, using potassium methylate as catalyst and heating the mixture in a closed vial for 2 min at $65°C$. The fatty acid composition of a typical butter oil analysed in this way is given in *Table 9.3* and compares favourably with that given by Wolff for butter fat. Wolff's GC conditions 8 years earlier did not include butyric acid. Acid ratios calculated from figures given by Luddy and co-workers for butter oil show the C_{12}/C_{10} figure to be 1.18, the $C_4/C_6 + C_8$ figure to be 1.45, the C_{14}/C_{12} figure to be 4.00 and the C_{18} unsaturated/C_{18} saturated figure to be 2.39.

Various workers have determined the fatty acid composition of butter glycerides by chromatographing either the free acids after saponification[39–41] or their ethyl esters[40–42] or butyl esters[43] or propyl esters[44].

The most satisfactory method of obtaining the lower free fatty acids is by transesterification. The fat is treated with the appropriate alcohol in the presence of a suitable catalyst and the glycerides are converted directly into the esters. Thus Lozano, Arias and Pera[45] prepared methyl esters for the analysis of Spanish butters.

The fact that most vegetable oils contain very little butyric acid, whereas butter fat contains between 3 and 5%, has been used successfully to determine the butter content of margarine and other vegetable oils. Hadorn and Zuercher[46] used GC of the fatty acids after saponification and acidification, and used valeric acid as internal standard. Karleskind, Valmalle and Wolff[47] saponified the glycerides and distilled the acidified extract according to the Reichert–Meissl method. The methyl esters were made and GC was applied using methyl valerate as internal standard. These authors showed that French butter contained between 3.2 and 4.0% of butyric acid in the fat, whereas lard, tallow, copra, palm kernel and palm oils contained 0.02, 0.04, 0.01, 0.01 and 0.03% butyric acid, respectively. These results showed that the amount of butyric acid in these oils is equivalent to that in approximately 1% of butter, and it is claimed, therefore, that 2% of butter in an oil can be detected by a butyric acid analysis.

Withington[48] found that the determination of butyric acid is the most suitable method for the analysis of butter fat where it is present in foods less than 10%, e.g. in margarine containing butter and in butter confectionery. He used a transesterification technique incorporating sodium hydroxide in ethanol and measured the ethyl butyrate against the methyl hexanoate internal standard. The ethyl butyrate content of 19 butters of various countries of origin showed an average of 4.79% for pure butter. This value was then used as the standard for calculation of butter content in various samples. Withington compared the individual ethyl butyrate contents of the 19 butters with Kirschner values. Kirschner values are essentially a measure of butyric acid as found in the classical Reichert–Polenske–Kirschner method. *Figure 9.2* demonstrates the good agreement between Kirschner values and ethyl butyrate concentration by GC analysis for the 19 butters.

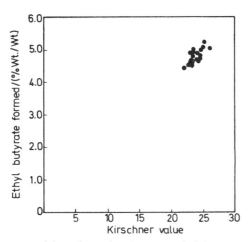

Figure 9.2. Relationship between ethyl butyrate formed by transesterification of butter fats and Kirschner values (after Withington[48])

Hadorn and Zuercher[49] transmethylated the fatty acids obtained from butter confectionery items, and calculated the butter content from the resulting methyl butyrate concentration in a similar fashion to that of Withington.

A semimicro method of determining butter fat in fat mixtures has been devised by Phillips and Sanders[50], using the butyric acid content. These authors found it simpler to deal with butyric acid after saponification of the fat followed by acidification, rather than use transesterification, since the esters of the higher fatty acids have to be eluted from the GC column prior to the next analysis. They found an average butyric acid content of 3.6% for butter fat, and have recommended this be used to calculate the butter fat content of such items as milk chocolate coatings, cream soups, fish or meat pastes with butter, butter or milk biscuits and butter sugar confectionery. This figure of 3.6% butyric acid in butter fat is in excellent agreement with Withington's 4.79% ethyl butyrate, which is equivalent to 3.63% butyric acid.

The fact that butter contains no sitosterol, which constitutes the main sterol fraction of vegetable oils, is another means of detecting the presence of margarine in butter. In 1962 Eisner et al.[51] extracted sterols from the unsaponifiable matter of margarine–butter mixtures by using Florisil column separation. β-sitosterol, γ-sitosterol and δ-sitosterol were separated by GC, and 1% of margarine in butter was detected. The peak for β-sitosterol, the major sterol constituent of margarine, can be seen with as little as 0.2% addition of margarine to butter. Cumont and Richou-Bac[52] chromatographed the unsaponifiable matter either directly or after TLC clean-up, and detected the presence in butter of 0.03% and 0.002% β-sitosterol by these respective procedures.

A comparison of four separation procedures prior to the examination of butter for the presence of vegetable oils has been conducted by Thorpe[53]. He concluded that preparative TLC clean-up was superior to digitonin precipitation methods. In 1964 Cannon[54] used digitonin precipitation followed by acetylation, and separated the cholesterol and sitosterol acetates derived from butter and margarine, respectively. From his findings he concluded that most analysts could detect the presence of 2% margarine in butter by this method, In a collaborative study concerning seven laboratories La Croix[55] concluded that 2% addition of either cottonseed, soyabean, safflower or peanut oils to butter could be detected by the appearance of a β-sitosterol peak, although some difficulty might be experienced at this level with safflower oil. After preparation of sterol digitonides, Cerutti, Volonterio and Resmini[56] prepared TMS derivatives and examined these by GC. These authors claimed that as little as 1% of vegetable oil could be detected using the appearance of the β-sitosterol peak as evidence.

The unsaponifiable matter of butter has also been the focus of attention by Eisner, Iverson and Firestone[57]. These authors fractionated the unsaponifiable matter of fats and oils on Florisil columns and separated tocopherols, high molecular weight aliphatic alcohols and triterpenoid alcohols, these fractions being subjected to GC. The patterns obtained can be used to characterise the oils or fats.

There has been only a small interest in butter volatiles relative to the fatty acids. Parks, Keeney and Schwartz[58] examined butter oil for bound aldehydes by capillary column GC and found saturated C_9 to C_{18} aldehydes and C_{11} to C_{18} branched-chain aldehydes.

There is strong evidence to suggest that fresh butter contains γ- and δ-keto acids. These were found in steam-deodorised butter fat at a total concentration in the region of 100 p.p.m. by van der Ven[59]. He deduced this after heating the butter fat after reduction and observing the presence of δ-octa-, δ-deca- and δ-dodeca-lactones with smaller amounts of γ-deca-, γ-undeca- and γ-dodeca-lactones. Boldingh et al.[60] also investigated the δ-lactone content of butter fat, and, in finding the same compounds as van der Ven, showed them to be the principal contributors of butter flavour. Jurriens and Oele[61] found that the total lactone content of butter fat lay between 1 and 40 p.p.m. and was at the upper end of this range if the butter was preheated at 140°C.

The flavour components of butter culture were isolated by Lindsay, Day and Sandine[62] by a steam-distillation technique followed by capillary column GC. Aldehydes, methyl ketones, primary and secondary alcohols, alkyl esters of aliphatic acids and sulphur-containing compounds were among the 50 compounds tentatively identified.

D. Cheese

Most GC investigations on cheese have dealt with flavour and aroma. The importance of flavour is obvious since very often only subtle changes in flavour are evident between one variety of cheese and another. It is therefore necessary for the cheese manufacturer to keep constant the flavour of a particular variety of cheese, and GC methods have been developed with that end in view. Where a GC analysis of cheese has been used to find the fatty acid content of the glycerides, the reason is usually to assess the influence of such fatty acids on flavour content.

Iyer *et al.*[43] studied the fatty acid composition of the glycerides of Provolone cheese fat using butyl esters, and concluded that the short-chain volatile fatty acids were important contributors to flavour. Langler and Day[63] suggested that 2-methyl-butyric acid was an essential flavour component of Swiss cheese but that 3-methyl-butyric acid probably was not.

The influence of fatty acids on flavour quality can be a deleterious one and can lead to rancidity. In contrast to the results obtained following the lipolysis of milk, de Man[64] found that Cheddar cheese monoglycerides contained short-chain fatty acids. Rancid cheese was found to have a high content of monoglycerides, particularly monobutyrin, and de Man has suggested that it could be a major contributor to rancid defect.

As early as 1958, Patton, Wong and Forss[65] examined Cheddar cheese for volatile components and found that dimethyl sulphide was an important aroma character in good-quality cheese. They also considered that biacetyl and 2-heptanone were major flavour contributors. Other compounds found were ethanol, acetone, 2-butanone and 3-hydroxy 2-butanone. McGugan and Howsam[66] obtained good resolution of neutral volatiles of Cheddar cheese by the use of capillary columns. They found 2-butanol, 1-propanol, 2-pentanone, 3-pentanone, ethyl butyrate and valeraldehyde in the volatile fraction. In a similar study, ten commercial Cheddar cheeses were used by Bills, Willits and Day[67]. They used the gas-entrainment and on-column technique of Morgan and Day (see Chapter 4, p.64) to trap the volatiles and found, in descending order of magnitude, ethanol, 2-butanol, 2-butanone, 1-propanol, ethyl butyrate, ethyl acetate, acetone and acetaldehyde.

Liebich *et al.*[68] used three methods of extracting cheese oil for examination of flavour components: (1) centrifugation, (2) low-temperature vacuum distillation, (3) methanol extraction. These combined techniques yielded more than 150 components in Cheddar cheese and showed that the main flavour agents were methyl ketones.

In another method of entrainment of volatile Leibich *et al.*[69] placed cheese in a liner attached to the injector. The liner was held for 10 min at 190°C to allow the volatiles to pass directly on to the GC column, leaving the oil behind. These authors examined Cheddar, Blue, Roquefort, Romano, Swiss and Limburger cheeses in this way, and found that Blue and Roquefort varieties contained the highest concentration of volatiles. Roquefort cheese contained 110 p.p.m. of 2-heptanone and also the C_3, C_5, C_7 and C_9 methyl ketones together with butanoate, hexanoate and octanoate. Blue cheese contained all these methyl

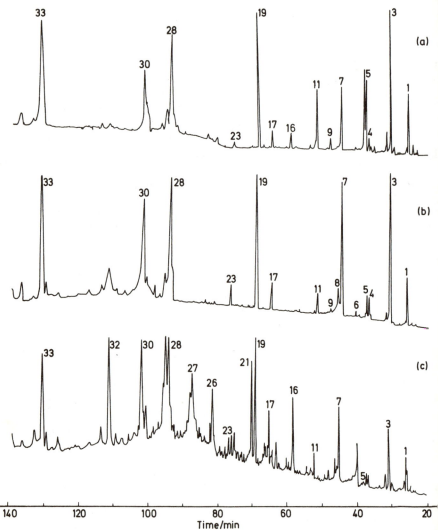

Figure 9.3. Chromatograms of (a) Cheddar cheese aged 3 months; (b) Cheddar cheese aged 24 months; (c) Cheddar cheese powder from a 24 month old cheese. Attenuation 4 in all three chromatograms. Conditions: 600 ft x 0.02 in I.D.; Dowfax 9 N 15; temperature programmed from room temperature to 150°C at 2°/min; inlet pressure 15 p.s.i.; sample size 10 μl of cheese oil. (After Liebich et al.[69])

ketones but not the esters. Although Cheddar and Romano cheeses contained less volatile constituents than Blue or Roquefort, Cheddar contained volatiles in the region 5 to 10 p.p.m. Three chromatograms of various Cheddar cheeses, showing these volatiles, are given in *Figure 9.3*. The individual components are not identified, but it is possible to compare the different overall pictures of (a) a 3 month old Cheddar cheese, (b) a 24 month old Cheddar cheese, (c) a 24 month old Cheddar cheese powder. Some changes have taken place after ageing of the Cheddar cheese from 3 to 24 months: some components have increased as others have decreased. The cheese powder shows many more volatile components than does the ordinary cheese.

Blue cheese is a popular variety for GC study. Coffman, Smith and Andrews[70] identified 29 neutral substances in this cheese, including 2-heptanone, 2-nonanone, 2-undecanone, 2-heptanol and 2-nonanol; and Jackson and Hussong[71] suggested that 2-pentanol, 2-heptanol and 2-nonanol were formed during ripening from the corresponding methyl ketones.

In a series of ripening experiments Scarpellino and Kosikowski[72] showed that acetic acid and ethanol increased during the ripening of Cheddar cheese, whereas there was a decrease of the trace of acetyl methyl carbinol originally present. Butyric acid, 2-butanol and 2-butanone appeared as ripening progressed.

Day and Anderson[73] identified 32 aroma constituents of Blue cheese using combined GC and MS. These constituents included methyl ketones, secondary alcohols, and methyl and ethyl esters of short-chain fatty acids. GC coupled with MS was used also by Langler, Libbey and Day[74] to separate the major neutral volatiles of Swiss cheese, including the varieties Emmental and Dagano. *Table 9.5* gives a quantitative breakdown of the volatiles of a typical Swiss cheese. As Patton, Wong and Forss[65] had found earlier, dimethyl sulphide can be a very important flavour contributor in cheese, and at 0.11 p.p.m. in this typical Swiss cheese, it is above its flavour threshold.

Ethanol is invariably the neutral volatile compound which is found in the greatest concentration in cheese. This is not surprising since it should be formed

Table 9.5 Typical analysis of neutral volatiles of Swiss cheese (after Langler *et al.*[74])

Compound	Concentration/p.p.m.
Dimethyl sulphide	0.11
Biacetyl	0.8
Acetaldehyde	1.4
Acetone	1.6
2-Butanone	0.3
2-Methyl butyraldehyde	0.42
2-Pentanone	0.98
2-Heptanone	0.45
Ethanol	16.3
2-Butanol	0.3
1-Propanol	2.9
1-Butanol	0.7
Methyl hexanoate	1.5
Ethyl butyrate	0.6

in reasonable quantity in any process of fermentation. Although a volatile, it is not a flavour constituent and it is used as a guide to the extent of fermentation. The 'nose' volatiles of cheese have been separated by Kroger and Patton[75]. They heated grated cheese at 90°C for 8 min in the barrel of a syringe and plunged the ensuing vapour directly on to the chromatographic column. Under standard conditions each cheese gives its own vapour 'profile', and changes can be seen in this profile as a given cheese ripens. As Cheddar cheese ripens and ethanol decreases, so 1-propanol and 2-butanol increase and also 2-butanone is formed.

In the past few years cheese technology has advanced in the area of the production of cheese powders. It is very desirable in this process that as little as possible flavour is lost of the original cheese, and in comparing methods of spray-drying, Bradley and Stine[76] concluded that foam spray-drying gave a higher yield of volatiles than the more conventional spray-drying processes.

Although aliphatic amines and piperidine have been associated with cheese flavour in a GC study of matured Russian cheeses by Golovnya, Mironov and Abdullina[77], it is clear that the important flavour components of cheese are methyl ketones, secondary alcohols, fatty acids and their esters, methyl sulphide and biacetyl.

References

1. Report of the Analytical Methods Committee of the Society of Public Analysts, *Analyst*, **61**, 404 (1936)
2. Bassette, R. and Keeney, M., *J. Ass. Off. Agric. Chem.*, **39**, 469 (1956)
3. James, A.T. and Martin, A.J.P., *Biochem. J.*, **50**, 679 (1952)
4. Dickes, G.J. and Nicholas, P.V., *J. Ass. Publ. Anal.*, **10**, 87 (1972)
5. Jensen, R.G., Quinn, J.G., Carpenter, D.L. and Sampugna, J., *J. Dairy Sci.*, **50**, 119 (1967)
6. Hankinson, C.L., Harper, W.J. and Mikolajcik, E., *J. Dairy Sci.*, **41**, 1502 (1958)
7. Palo, V., Hrivnak, J. and Goerner, F., *Nahrung*, **12**, 225 (1968)
8. de Man, J.M., *Lab. Pract.*, **16**, 150 (1967)
9. Smith, L.M., *J. Dairy Sci.*, **44**, 607 (1961)
10. Bills, D.D., Khatri, L.L. and Day, E.A., *J. Dairy Sci.*, **46**, 1342 (1963)
11. Christopherson, S.W. and Glass, R.L., *J. Dairy Sci.*, **52**, 1289 (1969)
12. Leemann, W. and Stahel, O.F., *Beckmann Rep.*, **1**, 11 (1970)
13. Hadorn, H. and Zuercher, K., *Dt. LebensmittRdsch.*, **66**, 77 (1970)
14. Jensen, R.G., Gander, G.W. and Sampugna, J., *J. Dairy Sci.*, **45**, 329 (1962)
15. Glass, R.L., Lohse, L.W. and Jenness, R., *J. Dairy Sci.*, **51**, 1847 (1968)
16. Gander, G.W., Jensen, R.G. and Sampugna, J., *J. Dairy Sci.*, **45**, 323 (1962)
17. Sampugna, J., Pitas, R.E. and Jensen, R.G., *J. Dairy Sci.*, **49**, 1462 (1966)
18. Patton, S., *J. Dairy Sci.*, **44**, 207 (1961)
19. Wong, N.P. and Patton, S., *J. Dairy Sci.*, **45**, 724 (1962)
20. Jennings, W.G., Viljhalmsson, S. and Dunkley, W.L., *J. Food Sci.*, **27**, 306 (1962)
21. Bassette, R., Ozeris, S. and Whitnah, C.H., *J. Food Sci.*, **28**, 84 (1963)
22. Mabbitt, L.A. and McKinnon, G., *J. Dairy Res.*, **30**, 359 (1963)
23. Arnold, R.G., Libbey, L.M. and Day, E.A., *J. Food Sci.*, **31**, 566 (1966)
24. Scanlan, R.A. and Lindsay, R.C., *J. Food Sci.*, **33**, 440 (1968)
25. Hempenius, W.L. and Liska, B.J., *J. Dairy Sci.*, **51**, 221 (1968)
26. Palo, V. and Ilkova, H., *J. Chromatog.*, **53**, 363 (1970)
27. Janecke, H. and Brendel, R., *Naturwissenschaften*, **58**, 54 (1971)
28. Panalaks, T., *Analyst*, **95**, 862 (1970)
29. Bell, J.G. and Christie, A.A., *Analyst*, **99**, 385 (1974)
30. Reineccius, G.A., Kavanagh, T.E. and Keeney, P.G., *J. Dairy Sci.*, **53**, 1018 (1970)

31. Jaynes, H.O. and Asan, T., *J. Milk Food Technol.*, **36**, 333 (1973)
32. Hawke, J.C., *J. Dairy Res.*, **24**, 366 (1957)
33. Wolff, J-P., *Ann. Falsif. Exp. Chim.*, **53**, 318 (1960)
34. Francesco, F. De and Avancini, D., *Boll. Lab. Chim. Prov.*, **12**, 422 (1961)
35. Doro, B. and Gabucci, G., *Boll. Lab. Chim. Prov.*, **14**, 3 (1963)
36. Olivari, L. and Benassi, R., *Boll. Lab. Chim. Prov.*, **14**, 304 (1963)
37. Valussi, S. and Cofleri, G., *Boll. Lab. Chim. Prov.*, **13**, 3 (1962)
38. Luddy, F.E., Barford, R.A., Herb, S.F. and Magidman, P., *J. Amer. Oil Chem. Soc.*, **45**, 549 (1968)
39. Sulser, H. and Buechi, W., *Z. Lebensmitt. Forsch.*, **141**, 145 (1969)
40. Haenni, H. and Ritter, W., *Milchwissenschaften*, **19**, 1 (1964)
41. Anselmi, S., Boniforti, L. and Monacelli, R., *Boll. Lab. Chim. Prov.*, **11**, 317 (1960)
42. Boniforti, L., *Ann. Falsif. Exp. Chim.*, **55**, 255 (1962)
43. Iyer, M., Richardson, T., Amundson, C.H. and Boudreau, A., *J. Dairy Sci.*, **50**, 285 (1967)
44. Grosjean, M.H. and Fouassin, A., *Rev. Ferment. Ind. Aliment.*, **23**, 57 (1968)
45. Lozano, J.S., Arias, A.C. and Pera, F.G., *An. Bromat.*, **18**, 401 (1966)
46. Hadorn, H. and Zuercher, K., *Mitt. Geb. Lebensmitt. Hyg.*, **59**, 369 (1968)
47. Karleskind, A., Valmalle, G. and Wolff, J-P., *J. Ass. Off. Anal. Chem.*, **53**, 1082 (1970)
48. Withington, D.F., *Analyst*, **92**, 705 (1967)
49. Hadorn, H. and Zuercher, K., *Mitt. Geb. Lebensmitt. Hyg.*, **62**, 123 (1971)
50. Phillips, A.R. and Sanders, B.J., *J. Ass. Publ. Anal.*, **6**, 89 (1968)
51. Eisner, J., Wong, N.P., Firestone, D. and Bond, J., *J. Ass. Off. Agric. Chem.*, **45**, 337 (1962)
52. Cumont, G. and Richou-Bac, L., *Recl. Med. Vet.*, **146**, 735 (1970)
53. Thorpe, C.W., *J. Ass. Off. Anal. Chem.*, **53**, 623 (1970)
54. Cannon, J.H., *J. Ass. Off. Agric. Chem.*, **47**, 577 (1964)
55. La Croix, D.E., *J. Ass. Off. Anal. Chem.*, **53**, 535 (1970)
56. Cerutti, G., Volonterio, G. and Resmini, P., *Riv. Ital. Sostanze Grasse*, **46**, 356 (1969)
57. Eisner, J., Iverson, J.L. and Firestone, D., *J. Ass. Off. Anal. Chem.*, **49**, 580 (1966)
58. Parks, O.W., Keeney, M. and Schwartz, D.P., *J. Dairy Sci.*, **44**, 1940 (1961)
59. van der Ven, B., *Rec. Trav. Chim.*, **83**, 976 (1964)
60. Boldingh, J., Haverkamp, P., Begeman, A.P. de J. and Taylor, R.J., *Rev. Franc. Corps. Gras.*, **13**, 235 (1966)
61. Jurriens, G. and Oele, J.M., *J. Amer. Oil. Chem. Soc.*, **42**, 857 (1965)
62. Lindsay, R.C., Day, E.A. and Sandine, W.E., *J. Dairy Sci.*, **48**, 1566 (1965)
63. Langler, J.E. and Day, E.A., *J. Dairy Sci.*, **49**, 91 (1966)
64. de Man, J.M., *J. Dairy Sci.*, **49**, 343 (1966)
65. Patton, S., Wong, N.P. and Forss, D.A., *J. Dairy Sci.*, **41**, 857 (1958)
66. McGugan, W.A. and Howsam, S.G., *J. Dairy Sci.*, **45**, 495 (1962)
67. Bills, D.D., Willits, R.E. and Day, E.A., *J. Dairy Sci.*, **49**, 681 (1966)
68. Liebich, H.M., Douglas, D.R., Bayer, E. and Zlatkis, A., *J. Chromat. Sci.*, **8**, 355 (1970)
69. Liebich, H.M., Douglas, D.R., Bayer, E. and Zlatkis, A., *J. Chromat. Sci.*, **8**, 351 (1970)
70. Coffman, J.R., Smith, D.E. and Andrews, J.S., *J. Food Sci.*, **25**, 663 (1960)
71. Jackson, H.W. and Hussong, R.V., *J. Dairy Sci.*, **41**, 920 (1958)
72. Scarpellino, R. and Kosikowski, F.V., *J. Dairy Sci.*, **45**, 343 (1962)
73. Day, E.A. and Anderson, D.F., *J. Agric. Food Chem.*, **13**, 2 (1965)
74. Langler, J.E., Libbey, L.M. and Day, E.A., *J. Agric. Food Chem.*, **15**, 386 (1967)
75. Kroger, M. and Patton, S., *J. Dairy Sci.*, **47**, 296 (1964)
76. Bradley, R.L. and Stine, C.M., *J. Gas Chromat.*, **6**, 344 (1968)
77. Golovnya, R.V., Mironov, G.A. and Abdullina, R.M., *Zh. Anal. Khim.*, **23**, 766 (1968)

10
Fats and Oils

A. Introduction

Fats and oils are used extensively in cooking and are the essential media for frying. Animal fats are solid at room temperature, and this quality is directly proportional to the degree of saturation of the fatty acids which are part of the glycerides. In contrast. the liquid nature of oils is proportional to the unsaturation of the glyceride fatty acids, this unsaturation being most pronounced in fish oils.

The fats and oils discussed in this chapter are depot fats and oils as distinct from milk fat, which is covered in Chapter 9. Depot fat is that fat which is the adipose tissue in animals and the fruity substance in nuts and fruits. The chemical constitution of fats and oils has been extensively reviewed by Hilditch and Williams[1].

In this chapter the fatty acids are frequently designated in the mnemonic shorthand devised by Stoffel, Insull and Ahrens[2]. The carbon symbol is followed by two numbers separated by a colon, the first being the number of carbon atoms in the chain and the second being the number of double bonds. Appendix 3 gives this shorthand numbering system of fatty acids with their common and systematic names.

The analysis of fats and oils is necessary for three reasons: (1) confirmation of identity, (2) presence or absence of adulterants and (3) assessment of edibility.

Confirmation of identity of a fat or oil depends on its physical and chemical constitution. Classical physical characteristics, such as density, melting point and refractivity, go some way in classifying them but have their limitations. Similarly, chemical analyses, such as the determinations of iodine, thiocyanogen, hydroxyl, acetyl and saponification values and also the examination of the constituents of the unsaponifiable matter, are of great importance. Nevertheless, these data are based on whole fats or oils, whereas the actual analysis, if possible, must give more specific information. It is in this way that GC applied to the analysis of glycerides, fatty acids, sterols, triterpene alcohols, tocopherols, etc., can be most useful.

Ideally, GC is used in conjunction with the classical physical and chemical tests in qualitative examinations. La Croix, Prosser and Sheppard[3] compared GC fatty acid analysis of oils and fats with that of thiocyanogen number and with the lead salt—ether method, and concluded that the GC results were the most accurate; they recommended the technique particularly for evaluating the ratios of saturated to unsaturated acids and of saturated to polyunsaturated acids. In 1966 the Canadian delegation to the Fats and Oils Committee of the FAO/WHO Codex Alimentarius Commission proposed that the classical methods for standardisation of identification of fats and oils, viz. iodine value, relative density, refractive index and saponification value, be replaced by the GC fatty acid analysis. The USA delegation counter-proposed that the classical and GC methods be used jointly, and this was accepted. This counter-proposal is a sensible approach to the general analytical problems confronting the fat and oil chemist. Very often a cursory examination is all that is required and, in such cases, it would not only be time-consuming but also costly to employ GC when it was not necessary to do so. On the other hand, where the chemist is faced with analytical problems which require investigations of possible adulteration by traces of contaminants, GC is invaluable.

One of the biggest problems confronting the oil manufacturer is how to store the commodity without the onset of rancidity or deterioration in some other manner. Frequently a fat or oil will develop an unpleasant odour or taste on storage which renders it unsuitable for the purpose originally intended. A general indication of rancidity is given by the production of free fatty acids produced by enzymic action on the glycerides. Such free acidity can be titrated with alkali and calculated as oleic acid. When this rancidity occurs, chemical tests can be made to confirm it, e.g. peroxide value, Kreis test and thiobarbituric acid number[4]. Another type of deterioration, known as ketone rancidity, is the result of microbiological attack on the fat or oil to produce carbonyls, which often lead to strong off-flavours. The ability of GC to differentiate such volatiles in head space analysis is utilised in the examination of rancid fats and oils. An examination of flavour components is useful not only with storage problems but also with possible deterioration of an oil which has been used perpetually in cooking. There has also been considerable interest shown in the specific natural flavours of certain oils, e.g. soyabean and cottonseed.

In summary, the oil technologist uses GC to assess the quality of the product in relation to its state of refinement, its storage potential and its variety and country of origin.

The essential GC details, where known, of some of the more important references in this chapter concerning animal fats, vegetable oils and fish oils are listed in *Table 10.1*

B. Animal fats

Animal fats are solids at room temperature and melt over a range of temperature. Attempts have been made to quantify these ranges in order to detect the presence of adulterants, but these have met with limited success since most fats melt within the same range. Specific gravities of different fats are also similar, and therefore are of little value. The determination of refractive indices and iodine values give some information regarding the purity of a fat, particularly when the results are used in conjunction with more specific chemical tests. However, the examination of the fatty acids and glycerides by GC yields the most comprehensive information.

Table 10.1 GC conditions used in studies of fats and oils

Class of compound	Ref.	Column dimensions	Stationary phase	Support material	Carrier gas and conditions	Temperature or temperature programme/°C	Detector
Methyl esters of fatty acids	15	5ft x -	10% PEGA	100/120 Celite	N_2, 9 p.s.i. inlet pressure	190	FI
Methyl esters of fatty acids	19	1.2m x -	5% Apiezon L	Acid/alkali-washed 100/120 Celite	Ar, 50–60 ml/min	185–200	AI
Methyl esters of fatty acids	41	2m x 6mm I.D.	20% PEGS	60/80 Chromosorb W	He, 80ml/min	213	TC
Methyl esters of fatty acids	72	6ft x 0.25in	15% DEGS	60/70 Anakrom AK	Ar, 12 p.s.i. inlet pressure	178	AI
Methyl esters of fatty acids	103	6ft x 3mm I.D.	3% EGSP-Z	100/120 Gas Chrom Q	Ar, 16 p.s.i. inlet pressure	180	FI
Methyl esters of fatty acids	118	150ft x 0.01in I.D.	BDS	Capillary	He, 40 p.s.i. inlet pressure	170	FI
Methyl azelaoglycerides	43	4ft x 0.375in	2% SE-30	60/70 Anakrom ABS	He, 100 ml/min	250–325	FI
Lactones	95	6ft x 0.125in	3% QF-1	100/120 Gas Chrom Q	He, 75ml/min	120	FI
Alcohols, tocopherols	58	6ft x 0.25in I.D.	1% SE-52	100/120 silanised Gas Chrom P	Ar, 12 p.s.i. inlet pressure	210	AI
Triterpene alcohols, sterols	47	2m x 2mm	1% SE-30	100/120 silanised Gas Chrom P	N_2, 20ml/min	230	FI

156

Sample	Ref	Column	Stationary phase	Support	Carrier gas	Temp (°C)	Detector
TMS derivatives of tocopherols	75	2m x 4mm	1% SE-30	Gas Chrom P	N$_2$, 30ml/min	240	FI
Sterols	45	1.5m x -	10% SE-30	60/80 AW Chromosorb W	–	265	–
Sterol acetates	46	2.6ft x 0.25in	4% SE-30	AW silanised 80/100 Gas Chrom P	N$_2$, 60ml/min	220	FI
TMS derivatives of sterols	74	–	3% JXR	100/120 Gas Chrom P	N$_2$, 25ml/min	260	FI
General	25	5m x 2mm	5% PEGA	Celite	N$_2$, 0.5 atm inlet pressure	104	FI
General	88	0.9m x 4mm	30% DC silicone oil	Celite 545	N$_2$, 29ml/min	93	Gas density balance
Pentane	93	10ft x -	15% Carbowax 20M	Gas Chrom P	He, 75ml/min	150	FI

1. Lard

Lard is used as a cooking and frying medium, and may be defined as the rendered fresh fat from pigs. It should be clear on melting and free from rancidity. Lard has been the subject of adulteration, particularly by fats and oils which are less expensive to manufacture. Beef and mutton fats in the form of tallow, horse fat and many hardened vegetable and fish oils have been used for this adulteration. The Bomer value, which depends on the differences between the melting points of the glycerides of lard and beef fat, gives an indication of adulteration of lard with beef tallow. The GC examination of the fatty acids of these animal fats is one of the best ways of assessing adulteration, and several workers have made such a study. In 1963 Heyes[5] reviewed the use of GC in the analysis of oils and fats, and Jaarma[6], Armandola[7] and also Luddy *et al.*[8] used methyl esters of the glyceride fatty acids to characterise lard. Magidman *et al.*[9] used the GC of the methyl esters after their separation into six fractions by silicic acid chromatography. This type of column chromatographic separation is useful for obtaining the minor acid constituents which are often obscured by the larger peaks when the mixture is subjected to direct GC. Herb *et al.*[10] made a prior separation of the methyl esters according to chain length on a non-polar stationary phase before re-chromatography on a polar column. The fatty acid composition of 61 samples of lard was determined by Lotito and Cucurachi[11]. Among the minor constituents were the $C_{8:0}$, $C_{15:0}$, $C_{17:0}$, $C_{17:1}$, $C_{20:2}$ and $C_{20:4}$ acids.

Doro and Remoli[12] made use of the ratios of certain fatty acids of lard in order to ascertain the presence of adulterants. This approach, made successful by Wolff in the determination of the purity of butter (see Chapter 9), has not proved quite so successful in detecting adulteration of animal fats. The maximum $C_{14:0}/C_{16:0}$ acid ratio for lard is given as 0.0677, and a 20% addition of tallow gives a higher ratio. The $C_{18:3}/C_{18:1}$ and the $C_{14:0 + 16:0 + 18:0}/C_{18:2}$ acid ratios are not trustworthy since 20% additions of tallow can give ratios within the normal range for pure lard. Doro[13] followed up this approach in the detection of the addition of tallow or horse fat to lard. Pascussi and Paolini[14] were more successful in finding that the $C_{14:0 + 16:0 + 18:0}/C_{18:2}$ acid ratio for pure lard lies between 4 and 5. A ratio greater than 5.5 could mean the presence of an adulterant. Similarly, the $C_{18:3}/C_{18:1}$ acid ratio is normally 100–200 for pure lard, and values in excess of 200 could indicate adulteration.

Castledine and Davies[15] used a transesterification technique to obtain methyl esters of the fatty acids of a number of animal fats. These authors calculated the $C_{16:0}/C_{14:0}$, $C_{18:1}/C_{18:0}$, $C_{18:1}/C_{14:0}$ and $C_{18:1}/C_{16:0}$ acid ratios for these fats, and these values are given in *Table 10.2*. The $C_{16:0}/C_{14:0}$ and the $C_{18:1}/C_{14:0}$ acid ratios are normally greater than 6 for pure lard, where beef and lamb tallow give ratios below 4. In addition, lard has a greater C_{18} unsaturated acid concentration than does beef or mutton tallow.

The most satisfactory way of determining the presence of beef and lamb tallow in lard is to analyse the fat for specific branched-chain and minor fatty acids. Grieco[16] found C_{14} and C_{16} branched-chain acids in small concentration in beef tallow but virtually absent from lard. Using this approach, it is possible to detect 5–10% tallow in lard. Bastijns[17] also used the presence of C_{14} and C_{16} branched-chain acids in a similar study. He found that these acids were of

Table 10.2 Fatty acid ratios of some animal fats (after Castledine and Davies[15])

Type of fat	$C_{16:0}/C_{14:0}$	$C_{18:1}/C_{18:0}$	$C_{18:1}/C_{14:0}$	$C_{18:1}/C_{16:0}$
Lard	7.5	2.8	6.9	0.9
Lamb	3.4	0.8	3.1	0.9
Beef	3.5	0.8	3.2	0.9
Dog	2.8	7.1	3.9	1.4
Cat	2.2	2.5	1.9	0.9
Horse	2.3	4.4	1.2	0.5
Duck	14.8	6.9	16.7	1.1

a concentration of less than 0.01% for C_{14} and less than 0.05% for C_{16} in lard, whereas in beef fat there could be 5–10 times greater concentrations. Bastijns also discovered that the total C_{15} acids, including the iso- and anteiso-acids, was less than 0.03% in lard but 0.2–1.0% in beef fat. He concluded that a combination of GC and the determination of Bomer value increased the validity of analyses concerning lard–beef fat mixtures. A false conclusion by this method could arise if the pigs yielding the lard had been fed on beef offal as part of their diet. However, since modern lard production would invariably include fat of pigs from numerous sources, such a possibility is deemed to be rare. In a similar study Armandola[18] examined lards of various origins, and found that $C_{14:1}$ and $C_{15:0}$ acid were seldom present at a concentration greater than 0.1%.

In a comprehensive survey of a number of animal fats from various origins, Hubbard and Pocklington[19] found that lard contained less $C_{14:0}$, $C_{17:0}$ and $C_{17:1}$ acids than either beef or lamb fat, but that there was more $C_{18:2}$ acid in lard. These authors also found significant levels of branched-chain acids in ruminant fat, but not in lard, and their figures for the C_{15} acids support the findings of Bastijns. As little as 5% tallow in lard can be detected by using these relative acid concentrations. *Figure 10.1* shows four chromatograms, each of the fatty acids of beef and pork fat using two different stationary phases. The differences in the acid composition of the two fats are clearly demonstrated. These workers have tabulated typical fatty acid compositions of the glycerides of lard, beef, lamb, chicken, turkey, duck, game birds and rabbit.

The triglyceride composition of lard has been determined by Kuksis and McCarthy[20] but is of little value in determining lard quality. A similar study, which included beef tallow, was made by Sato, Matsui and Ikekawa[21].

The unsaponifiable matter of lard, beef fat and horse fat has been examined by Cook and Sturgeon[22] with a view to using some of the information to identify horse meat.

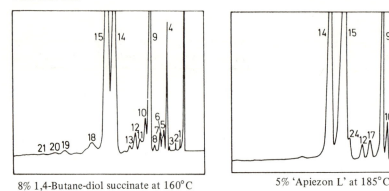

8% 1,4-Butane-diol succinate at 160°C   5% 'Apiezon L' at 185°C

Fatty acids of beef fat

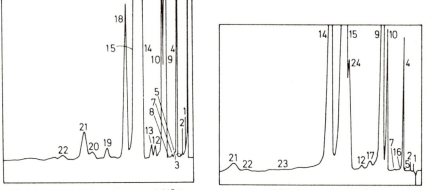

8% 1,4-Butane-diol succinate at 160°C   5% 'Apiezon L' at 185°C

Fatty acids of pork fat

Figure 10.1. Chromatogram peak identification (after Hubbard and Pocklington[19]). 1,10:0; 2,12:0; 3,14:Obr; 4,14:0; 5,14:1+15:0 br; 6,15:0 br; 7,15:0; 8,16:0 br; 9,16:0; 10,16:1; 11,17:0 br; 12,17:0; 13,17:1; 14,18:0; 15,18:1; 16,15:0 br; 17,17:1+17:0 br; 18,18:2; 19,18:3; 20,20:0; 21,20:1; 22,20:2; 23,20:4+20:5; 24,18:2+28:3. For GLC conditions, see Table 10.1

2. Beef and lamb fats

Beef and lamb fats are obtained either as dripping or as suet. Dripping comes from the fat or bones of cattle or sheep and suet from the fatty tissues associated with the loins or kidneys of those animals.

Terrell *et al.*[23] examined nine different sites of bovine depot fat in order to discover if there were significant differences between the fatty acid compositions. They deduced that outer layers of fat contained a higher proportion of unsaturated acids than inner layers.

It is virtually impossible to differentiate beef and lamb fats by chemical tests, and *Table 10.2* shows the closeness of the ratios of the fatty acids of these two fats.

From time to time horse fat has been used to adulterate beef fat. Castledine and Davies[15] found more total C_{18} unsaturated acids in horse fat than in beef or lamb fats and that the $C_{18:1}/C_{18:0}$ acid ratios for horse, beef and lamb were 4.4, 0.8 and 0.8, respectively. Langner[24] examined the fatty acid compositions of beef fats containing different concentrations of horse fat, and by various

160

combinations of ratios of $C_{16:0}, C_{18:0}, C_{18:1}, C_{18:2}$ and $C_{18:3}$ acids claimed to be able to detect as little as 5% horse fat.

Although it is generally acknowledged that beef fat has superior organoleptic qualities to those of lamb fat, little is known concerning the volatile constituents responsible for their flavours. Hoffmann and Meijboom[25] deduced that the major volatiles in beef and mutton tallow were 4-heptenal and 2,6-nonadienal.

3. Other animal fats

Hubbard and Pocklington[19] have determined the fatty acid compositions of the fats of chicken, turkey, grouse, partridge, duck, pheasant and rabbit.

Castledine and Davies[15] have calculated various fatty acid ratios for duck, dog and cat fats, and these are included in *Table 10.2*.

Goose fat has been the subject of adulteration by lard, and Pupin and Vuillaume[26] used the $C_{18:0}/C_{18:1}$ and $C_{10:0}/C_{12:0}$ acid ratios as a means of detecting the presence of lard. Goose fat does not contain the $C_{10:0}$ acid, whereas lard contains small amounts. In addition, lard has a greater concentration of $C_{18:0}$ and $C_{14:0}$ acids than goose fat.

There has been local interest in Australia regarding the composition of kangaroo lipids. Payne[27] found that kangaroo fat contained less $C_{18:2}$ and $C_{18:3}$ acids than horse fat, but more than ruminant fat. Since horse fat has been used in admixture with kangaroo fat, Payne used the relative concentrations of these acids to determine 15% or more horse fat in that of kangaroo.

C. Vegetable oils and fats

Vegetable oils are used mainly for the dressing of salads, in margarine manufacture and as cooking oils. Salad dressings must be fluid at room temperature and vegetable oils with their delicate flavours are most suitable for this purpose. Olive oil is the major ingredient of the majority of salad oils, for which there is a particularly large market in Mediterranean countries.

The suspicions surrounding the possible harmful effects of cholesterol which is present in animal fats has acted as a stimulus to the use of vegetable oils in cooking, particularly in frying. Common oils used in frying are corn, cotton seed, soyabean, peanut and sunflower. For similar reasons, margarine is used as a substitute for butter and may be made from such vegetable oils as palm kernel, coconut, soyabean, sunflower and rape seed. Some margarine compositions have been listed in the publication *Which*[28], the authors making the recommendation that margarine composition should appear on its label.

Like animal fats, vegetable oils are composed of glycerides which, in turn, are made up of fatty acids. Both animal fats and vegetable oils contain more or less the same acids, and oleic is the most common to both. The relative concentrations of the acids differ between fats and oils, the latter usually containing a higher percentage of unsaturated acids. Hydrogenation of oils produces a more saturated product and, hence, a more solid one. In this way so-called hardened vegetable or fish oils are used as substitutes for lard and are termed cooking fats. Although hydrogenation confers solid properties on the resulting

161

fat, some unsaturated acids remain and the detection of one of these, namely iso-oleic acid, has been used for determining whether or not hardened oils are present in lard.

Specific chemical tests are used to detect the presence of certain vegetable oils in other fats and oils, but these have limited value when small concentrations of adulterants are present. In elucidating the fatty acid structure, GC is used in support of these tests and is also utilised in the examination of the unsaponifiable matter in order to determine the presence of sterols, triterpene alcohols and tocopherols. The identification of phytosterol is indisputable evidence of the presence of a vegetable oil, and this type of GC analysis is often more rewarding than the fatty acid analysis. In an attempt to supply another distinction between animal fats and vegetable oils, Fedeli, Camurati and Jacini[29] have separated the tocopherols of vegetable oils by using silica gel TLC prior to making the TMS derivatives for GC analysis.

Very little of the fatty acid and unsaponifiable fraction territories have remained unexplored, but the determination of flavour components of vegetable oils and fats is probably only just beginning. Much attention has been paid to the role of lactones as flavour constituents in various foods, particularly in dairy products. Martin and Berner[30] extracted the lactones from various edible fats and were able to determine C_8 to C_{12} δ-lactones and C_8 and C_9 γ-lactones at the 1 p.p.m. level. Carbonyls, too, have received attention in studies of volatile constituents of fats and oils. Johnson and Hammond[31] have produced a method whereby alkan-2-ones, alkanals, alk-2-enals and alka-2,4-dienals are converted to their 2,4,6-trichlorophenylhydrazones for GC purposes. If these derivatives are used in conjunction with a preliminary separation by TLC, this affords a sensitive way of analysing carbonyls in general foods.

GC has been used in the analysis of over 30 vegetable oils, and these can be classified in various ways. They can be classified within their range of iodine number, which is directly proportional to the degree of saturation. Thus non-drying oils, such as olive, peanut and almond, have iodine values in the range 80–110; semidrying oils, such as cotton seed, sesame and soyabean, have iodine values in the range 80–140; and the drying oils of linseed and sunflower[4] a range of 125–200.

In general works concerning the study of oils it is convenient to classify them either botanically or according to their usage. For the purposes of this chapter, the vegetable oils have been divided according to the classification given by Ullmann[32]. The oils are separated into six groups, each group being designated by the predominant acid present in the oil. These groups are: (1) lauric/myristic acid, (2) palmitic acid, (3) stearic acid, (4/5) oleic/linoleic acid, (6) linolenic acid, (7) erucic acid.

The oleic/linoleic acid group contains a large number of vegetable oils, and these have been further divided into oleic-predominant and linoleic-predominant subgroups.

1. Lauric/myristic acid group

This group comprises coconut and palm kernel oils and also nutmeg butter.

Coconut and palm kernel oils are used in the production of margarine and also in many foods where a solid vegetable fat is ultimately required, this being

Table 10.3 Triterpene alcohols of some vegetable oils (after Fedeli et al.⁴⁷)

Vegetable oil	24-Methylene cycloartanol	Cyclolaudenol	Cycloartenol	Euphorbol	α-Amyrin	Butyrospermol	Cycloartanol	β-Amyrin
Coconut	✓	x	✓	x	✓	x	x	✓
Palm kernel	✓	x	✓	x	✓	x	x	x
Palm	✓	x	✓	x	✓	x	✓	x
Cocoa butter	✓	x	✓	x	x	x	x	x
Olive	✓	x	✓	x	✓	✓	x	✓
Peanut	✓	x	✓	x	✓	x	x	✓
Tea seed	✓	x	✓	x	x	✓	x	✓
Castor	✓	x	✓	x	✓	x	x	✓
Sesame	✓	x	✓	x	✓	x	x	x
Rice-bran	✓	x	✓	x	x	x	✓	✓
Soyabean	x	✓	✓	x	✓	x	x	✓
Corn	✓	x	✓	x	✓	x	✓	✓
Sunflower	✓	x	✓	✓	x	x	x	x
Grape seed	✓	x	✓	x	✓	x	x	x
Poppy seed	✓	x	✓	✓	x	x	x	✓
Linseed	✓	x	✓	x	x	x	x	x
Rape seed	✓	x	✓	x	✓	x	✓	✓

✓ , present; x, absent.

163

particularly the case in the confectionery industry, where the refined products 'stearine' and 'oleine' are used. These oils have also been found as additions to cocoa butter. Both oils are characterised by high Reichert values and very high Polenske and Saponification values[33]. Palm kernel oil, being one containing a high concentration of lauric acid, has been used to replace cocoa butter. The glycerides and fatty acids of these oils have been studied by several workers using GC[34-43]. About half of both coconut and palm kernel oil glycerides contain lauric acid, and nutmeg butter glycerides contain mostly myristic acid. Analysis of the unsaponifiable matter of the oils in this group has yielded useful information in respect of their triterpene alcohol and sterol compositions[44-48].

Table 10.3 lists 17 vegetable oils indicating the presence or absence of the eight identified alcohols. There are 22 other unidentified alcohols which contribute to the fingerprint chromatogram. Coconut oil contains β-amyrin and palm kernel oil does not. This fact could be exploited if attempts were made to analyse mixtures of these two oils.

The volatile flavour constituents of coconut oil have been isolated using a vacuum distillation technique by Allen[49]. The methyl ketones C_7, C_9, C_{11}, C_{13} and C_{15} were identified as anticipated and also the corresponding δ-lactones. Allen considers that δ-octa-lactone is largely responsible for coconut flavour and aroma.

2. Palmitic acid group

The sole member of this group is palm oil. This oil is made from the outer pulp of the fruit of the oil palm, whereas palm kernel oil is made from the nut. Palm oil is used as a food particularly by the populace of the countries of origin, e.g. West Africa.

The glyceride content of palm oil has been investigated by Sato *et al.*[37], Hadorn and Zuercher[40] and Grieco and Piepoli[41]. The major fatty acids are palmitic and oleic.

The sterol and triterpene alcohol composition of the unsaponifiable matter of palm oil has been studied[45-47] and *Table 10.3* shows the triterpene alcohol analysis.

3. Stearic acid group

This group comprises cocoa butter and illipe butter. Cocoa butter is used in chocolate manufacture and has been substituted in this context by illipe butter.

The glyceride composition of cocoa butter has been investigated by several workers[50-55], the major acids being stearic and oleic.

There have been some examples of the adulteration of cocoa butter with palm kernel or coconut oils. Iverson[56] found that these oils could be used as 'hard butters' in vegetable fat coatings and that the difference in lauric acid content between these oils and cocoa butter could be exploited analytically. Whereas palm kernel and coconut oils contain approximately 50% of lauric acid, cocoa butter contains only about 0.01%. Since it is possible by GC to detect 0.25% lauric acid, it should be possible to detect 0.5% addition of either palm kernel or coconut oil to cocoa butter.

Illipe butter is very similar in fatty acid composition to cocoa butter. Iverson and Harrill[42] found 45% stearic acid, 31% oleic acid and 16% palmitic acid, results which are in close agreement with those of Grieco and Piepoli[41].

Fincke[57] examined the sterol content in the unsaponifiable matter of cocoa butter and also the possible substitutes illipe butter, calvetta and coberine. He found that calvetta and coberine fats could be distinguished from the two butters by their much lower ratio of stigmasterol to campesterol.

Aliphatic alcohols, tocopherols and triterpene alcohols of some vegetable oils, including cocoa butter, were determined by Eisner, Iverson and Firestone[58]. These gave fingerprint GC patterns which could prove useful in identification of the fat or oil. *Table 10.3* includes the major triterpene alcohol composition of cocoa butter.

4. Oleic acid group

This group comprises olive oil, peanut oil, almond kernel oil, peach kernel oil, shea nut oil, pecan nut oil, hazel nut oil, avocado oil, cashew nut oil, coriander seed oil, tea seed oil and castor oil.

Olive oil Olive oil is used universally as an edible oil but finds particular favour in the Mediterranean countries. The mass of literature concerning its quality which emanates from Italy indicates its economic importance to that country. It is widely used as a salad oil and as a packing medium for fish of the sardine type.

Many workers have examined olive oil for the fatty acid composition of its glycerides, and oleic acid has been found between 66 and 88% of the total acids[50,59-69].

The detection of adulteration of olive oil with animal fats and vegetable oils is not a very difficult task using classical chemical tests and by GC. The high price of olive oil relative to that of other oils encourages such adulteration, but it has been more difficult to detect the presence of low-quality olive oil in the top-quality product. In the production of the low-quality oil an esterification procedure is frequently employed and in this process some measure of heat treatment is necessary. Under these conditions some of the oleic acid is isomerised to give elaidic acid and, hence, the presence of this *trans* form of oleic acid indicates the presence of the low-quality oil. Averill[70] gives GC conditions for the separation of methyl oleate and methyl elaidate. Pallotta and Losi[71] used a preparative TLC technique to concentrate the elaidic acid prior to GC and IR spectroscopic identification.

A more conventional sophistication of olive oil is by the addition of seed oils. Galanos, Kapoulas and Voudouris[72] used argentation—silica gel TLC to separate the polyunsaturated glycerides of olive oil. This means that a glyceride containing a considerable proportion of unsaturated acids of four or more double bonds can be segregated by this technique and removed from the TLC plate and analysed by GC of the methyl esters. In an examination of 17 pure olive oils this fraction contained between 35 and 37% of linoleic acid, whereas the same fraction of most seed oils contains 65–70% of linoleic acid. These workers claim that a 2% addition of seed oil to olive oil will give a significant increase of linoleic acid above the 35–37%. Although this method depends on the comparatively high level of linoleic acid in seed oil glyceride, there is a commercial safflower oil which has a low linoleic acid content. Addition of this oil to olive oil would, therefore, not be detectable by this means.

The sterol, triterpene alcohol and tocopherol contents of the unsaponifiable fraction of olive oil have been determined[45,73–75], and *Table 10.3* shows the triterpene alcohol composition.

Peanut oil (arachis oil, ground nut oil)   Peanut oil is extensively used as a stable cooking oil, particularly in frying, and as a salad oil ingredient. Adulteration of peanut oil with cotton seed, sesame, tea seed, rape seed, corn and poppy seed oils has occurred and there are specific chemical tests which can be applied to peanut oil in order to detect these adulterants. The tests for the presence of peanut oil in other oils usually depends on the presence in peanut oil of arachidic and lignoceric acids in reasonable concentration, an analysis for which GC is particularly suited.

The fatty acid composition of peanut oil glycerides has been deduced by several workers[51,68,76–78], and oleic acid is normally present between 40 and 60%.

Peanuts have been used to adulterate sesame seeds, and Letan, Turzynski and Raveh[79] used methylation of the glycerides and a urea-complexing method prior to GC in order to concentrate the methyl behenate. Under standard conditions peanut oil contained between 2.3 and 4.3% behenic acid, whereas sesame oil contained a maximum of 0.3%. This difference was used to detect the presence of peanuts in tehina and halva, which are sesame seed products. The unsaponifiable matter of peanut oil has been investigated by several workers[44–46,58,75] and the triterpene alcohol composition is shown in *Table 10.3.*

Almond kernel oil   Almond kernel oil is too expensive to produce to have any value to the food industry. The fatty acid compositions of the glycerides of almond kernel oil were determined by Hadorn and Zuercher[40] and by Bruno[80].

In the examination of the sterols by Karleskind *et al.*[45] almond kernel oil was found to contain 97% of β-sitosterol and 3% of campesterol. No cholesterol, brassicosterol or stigmasterol was present.

Peach kernel oil   Peach kernel oil has been used as a substitute for the more expensive almond kernel oil and is differentiated chemically by Bieber's test. The fatty acid composition has been deduced using GC by Iverson[81] and, more recently, by Lotti and Anelli[82], who examined 60 different oils.

Shea nut oil   Shea nut oil is used as a baking fat and does not keep well unless completely refined. The 'oleine' fraction of the oil has been used for the manufacture of margarine and the 'stearine' fraction as a cocoa butter substitute.

Its proportion of unsaponifiable matter is larger than average for a vegetable oil, and although the analysis of this fraction has received little attention, Copius-Peereboom[46] found that it contained 0.5% cholesterol and no brassicosterol.

Pecan nut oil　Pecan nut oil is not widely used and the range of fatty acid composition of the glycerides of 12 samples has been given by French[76].

Hazel nut oil　Hazel nut oil is a non-drying oil resembling olive oil in composition. For this reason it has been used as an ingredient for salad oil.

The fatty acid composition of the glycerides of hazel nut oil has been given by Hadorn and Zuercher[40] and shows its similarity to olive oil, particularly in the high proportion of oleic acid.

The unsaponifiable matter has been investigated by Karleskind *et al.*[45], who found that the sterol fraction contained 1% of brassicosterol.

Avodaco oil　Avocado oil is of academic interest and the fatty acid composition of the glycerides of two samples has been given by French[76]. It has a notably high concentration of palmitoleic acid of between 7 and 11%.

Cashew nut oil　Cashew nuts are used in the confectionery industry and are not grown in large enough quantity for the oil to be used in margarine or salad oil, for which it would be well suited.

Coriander seed oil　Subbaram and Youngs[51] have listed the fatty acids of the glycerides of coriander seed oil but it is of no commercial importance. The interest in this seed oil lies in its 71.8% of petroselenic açid (octadec-6-enoic acid).

Tea seed oil　Tea seed oil is used as a salad and frying oil. It is almost identical with olive oil in fatty acid composition and has been used as an adulterant of that oil. It is surprising that very little GC study has been made of this oil, although Fedeli *et al.*[47] have included it in their triterpene alcohol analyses (see *Table 10.3*).

Castor oil　Castor oil is a high viscosity, high specific gravity vegetable oil with medical and industrial applications but with few food outlets.

The unsaponifiable matter was examined by Eisner *et al.*[58], and *Table 10.3* includes the triterpene alcohol analysis of Fedeli *et al.*[47]

5. Linoleic acid group

This group comprises two oils with an approximately equal concentration of linoleic and oleic acid — sesame oil and rice bran oil—and 13 oils which have a predominance of linoleic acid—cotton seed oil, soyabean oil, corn oil, sunflower oil, safflower oil, milo maize oil, walnut oil, cherry kernel oil, grape seed oil, poppy seed oil, citrus seed oil, pumpkin seed oil and tomato seed oil.

In the examination of oils with a high proportion of linoleic acid care must be taken with the saponification conditions. Jamieson and Reid[83] found that if the alkaline saponification temperature was raised above 120°C, the linoleic acid could become isomerised. This reaction increased with increase in temperatures above 120°C and also with time.

Sesame oil Sesame oil is used as a salad oil and as a frying oil. It has also been an ingredient of margarine, and in some countries this inclusion was a legal requirement because sesame seed oil can be detected easily, this fact being exploited should the margarine be substituted for butter. Baudouin's test is used to detect sesame oil and depends on the presence of the sterols, sesamol and sesamolin.

Several workers have determined the fatty acid composition of sesame oil glycerides[40,41,81,84]. Fatty acid analysis of the extractives of tehina and halva, which were made from both Ethiopian and Israeli sesame, was used as a means of detecting possible peanut adulteration by Letan *et al.*[79] Sesame and peanut oils are similar in fatty acid composition, except that peanut oil has a greater concentration of $C_{22:0}$ and $C_{24:0}$ acids.

The triterpene alcohol analysis of sesame oil is shown in *Table 10.3*.

Rice bran oil Rice bran oil is used as a food in India and Pakistan. The fatty acid composition of the glycerides of four samples of the oil has been determined by Iverson[81].

The triterpene alcohol analysis of rice bran oil is shown in *Table 10.3*.

Cotton seed oil Cotton seed oil is an important edible oil since it is produced relatively cheaply and is used as a salad oil, a frying oil and a constituent of margarine. When unrefined it can become rancid, but it is often hydrogenated to obtain greater stability. Halphen's test is the chemical method by which cotton seed oil is detected. This test depends on the presence of cyclopropenoid acids in the oil which are absent from most other vegetable oils. The GC analysis of cyclopropenoid acids has been carried out by Schneider, Sook and Hopkins[85]. Besides a conventional analysis of straight-chain acids, they found approximately 0.5% of the cyclopropenoid acids, malvalic and sterculic.

Other workers have also investigated the composition of cotton seed oil glycerides[40,67,84,86].

Synodinos *et al.*[64] have compared the glyceride fatty acid compositions of cotton seed and olive oils. There is in cotton seed oil a small but significant amount of myristic acid which is absent in olive oil and this fact can be used for the detection of the former in the latter. Further comparison of the two oils shows that cotton seed contains a much larger proportion of linoleic acid, and this can be exploited semiquantitatively to detect as little as 10% cotton seed oil in olive oil.

Autoxidation of cotton seed oil and soyabean oil has been studied and is discussed in the next section under soyabean oil.

Soyabean oil Soyabean oil is a relatively cheap oil to produce and is extensively used in margarine manufacture.

The composition of the oil in terms of the glyceride fatty acids and the flavour constituents has received considerable attention, and intact glyceride composition has been investigated by Youngs and Subbaram[50]. Several workers, including Luddy *et al.*[8], Hadorn and Zuercher[40], Grieco and Piepoli[41], Hivon *et al.*[67], Iverson[81], Kato and Yamaura[87] and O'Connor and Herb[84], have determined the fatty acid composition of the glycerides.

The triterpene alcohols, sterols and tocopherols of soyabean oil have received the attention of Eisner *et al.*[58] and Capella *et al.*[44] Four triterpene alcohols, namely cycloartenol, α- and β-amyrin and cycloaudenol, were identified by Fedeli *et al.*[47] Soyabean oil is very unusual in containing cycloaudenol and not containing 24-methylene cycloartanol. This triterpene alcohol analysis could form the basis of a specific method for the detection of soyabean oil in other oils.

One of the major problems in the storage of soyabean oil is its tendency to autoxidation, which leads to an off-flavour frequently described as 'beany'. This oxidation is known as reversion and has led to investigations of the compounds which are responsible for this off-flavour. Hoffmann[88] separated the neutral volatiles of the oil into six fractions by GC, and described them in odiferous terms. The fraction called 'green beans' was examined by GC and IR spectroscopy and was found to contain 3-*cis*-hexenal, which was thought to be the compound responsible for the off-flavour. *Figure 10.2* is a chromatogram with six peaks corresponding to the six named odiferous fractions. Chang *et al.*[89] described the reversion flavour as 'beany and grassy'; they examined soyabean oil along similar lines to those of Hoffmann and recorded the presence of ethyl formate, ethyl acetate, ethyl alcohol, n-butyraldehyde, 2-heptanone and 2-heptenal. In a later publication Chang *et al.*[90] attributed the reversion

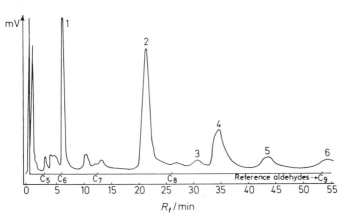

Figure 10.2. GLC of soyabean oil stripping products. Column:length, 900 mm; diameter, 4mm; carrier, Celite 545; immobile phase, silicone oil (30% Dow-Corning); mobile phase, nitrogen. Temperature 93°C; gas pressure, 24.5 cmHg; gas rate, 29 ml/min. For reference the retention times (R_t values) of a homologous series of saturated aliphatic aldehydes are given on a second abscissa. (After Hoffmann[88])

R_t values and odours of the GLC fractions of the neutral volatile matter from oxidised soyabean oil

Fraction number	R_t at 93°C (min)	Odour description
1	6[a]	Green beans
2	20	Brown beans
3	29	Rotten apples
4	33	Rancid hazelnut
5	42	Brown beans
6	53	Citrus

[a]R_t equal to that of n-hexanal.

flavour to 2-pentyl-furan which was present in the oil at a level between 1 and 10 p.p.m. The flavour threshold of 2-pentyl-furan is 1 p.p.m.; however, it does not have a beany odour on its own but only when present in the oil. These authors broadly separated the volatiles into five fractions, and each fraction was re-chromatographed twice. Krishnamurthy *et al.*[91] proposed that 2-pentyl-furan was formed by autoxidation of the linoleic acid of soyabean oil and also of cotton seed oil. It is known that hydrogenation of soyabean oil helps prevent autoxidation and thereby prevents reversion flavour, which supports the theory that the reversion flavour is formed from unsaturated components. Selke, Moser and Rohwedder[92] used combined GC–MS in the examination of the neutral volatiles of soyabean oil. They found that the predominance of C_5 hydrocarbons in the volatile fraction of fresh oil diminished on oxidation from 91% to 22% in 6 weeks. Simultaneously, the C_3 aldehyde increased from 9% to 42%, accompanied by small increases in the C_5 and C_6 aldehydes. Contrary to these findings, Scholz and Ptak[93] found that pentane is a direct indicator of rancidity. These workers assert that pentane is present in low concentration in fresh soyabean, cotton seed and peanut oils and increases with autoxidation of methyl linoleate, and that monitoring of pentane could be used as a measure of rancidity.

The volatile fractions of slightly autoxidised soyabean and cotton seed oils contain dec-1-yne, according to Smouse, Mookherjee and Chang[94], and the flavour threshold of this compound is 0.1 p.p.m.

In highly peroxidised soyabean and cotton seed oils γ- and δ-lactones have been found by Fioriti, Krampl and Sims[95]. Four different GC systems were used to cover the whole range of lactones and the γ-C_9 lactone was found in largest concentration.

Corn oil (maize oil) Corn oil is widely used as a salad oil and frying oil. The theory that oils containing polyunsaturated acids are less likely to contribute to thrombosis has prompted greater usage of oils containing these acids, and corn oil is a good example.

The composition of corn oil glycerides has been investigated by several workers[39–41, 67, 76, 81, 84, 96], and the triterpene alcohol content is included in *Table 10.3*.

Sunflower oil Sunflower oil is similar in composition to both corn and safflower oils. It is used extensively as a salad oil and as an ingredient of margarine, particularly the 'easy-to-spread' variety.

The composition of sunflower oil glycerides has been investigated by several workers[40, 55, 84, 97].

Triterpene alcohol analysis carried out by Fedeli *et al.*[47] showed sunflower oil to contain only 24-methylene cycloartanol and euphorbol. However, in his survey of vegetable oils, given in *Table 10.3*, only one other oil, namely poppy seed, contained euphorbol, and this fact would be an analytical aid in the detection of sunflower oil in most other oils.

Safflower oil Safflower oil, being very similar to sunflower oil, is also used in margarine manufacture and as an ingredient in salad oil.

Several workers have investigated the composition of safflower oil glycerides[40, 81, 84, 98].

Milo maize oil Iverson[81] has determined the fatty acid composition of the glycerides of milo maize oil and Eisner *et al.*[58] have examined the unsaponifiable matter for aliphatic alcohols, tocopherols and triterpene alcohols.

Walnut oil Although walnut oil is uncommon owing to the popularity of the whole nut as a food, it has some value as a salad oil ingredient. It has also been used to adulterate olive oil.
   The fatty acid composition of its glycerides has been investigated by Hadorn and Zuercher[40] and Iverson[81].

Cherry kernel oil Cherry kernel oil appears to have no commercial value, but Pifferi[99] has examined the glyceride fatty acids and found that it was very similar to soyabean oil, except that it contained about 14% of elaeostearic acid (octadeca-9,11,13-trienoic acid) and only 2% of linolenic acid, which is the more usual $C_{18:3}$ acid. It also contained 45% linoleic acid and 21% oleic acid.

Grape seed oil Refined grape seed oil can be used as a cooking oil, and its fatty acid glyceride composition has been investigated by Hadorn and Zuercher[40] and Grieco and Piepoli[41].
   Fedeli *et al.*[47] include grape seed oil in their triterpene alcohol results in *Table 10.3*.

Poppy seed oil Poppy seed oil is expensive and can be used as a salad oil ingredient. It has been adulterated with sesame oil, and although Hadorn and Zuercher's[40] figures show that poppy seed oil has less oleic and more linoleic acid than sesame oil, it would be difficult to detect small amounts of the latter in the former. However, in the triterpene alcohol analysis by Fedeli *et al.*[47] sesame oil contains α-amyrin and poppy seed oil does not (see *Table 10.3*).
   The presence of poppy seed oil in other oils (except sunflower) could also be detected by its constituent triterpene alcohol, euphorbol.

Citrus seed oil French[76] has listed the glyceride fatty acid content of citrus seed oil, although it has no commercial importance.

Pumpkin seed oil For academic purposes only, Hadorn and Zuercher[40] have included the glyceride fatty acid analysis of pumpkin seed oil.

Tomato seed oil The glyceride fatty acid composition of tomato seed oil has been elucidated by Tsatsaronis and Boskou[100].

6. Linolenic acid group

The only member of this group which can be used for edible purposes is linseed oil, and it can be used only when completely refined. Several workers have analysed the fatty acids of the glycerides of linseed oil, including Jamieson and Reid[39], Hadorn and Zuercher[40] and Grieco and Piepoli[41]. The oil is characterised by the high linolenic acid content, which is normally found in the range 45–60% of the total acid concentration.

7. Erucic acid group

This group comprises rape seed and mustard seed oils. These oils, when refined and hydrogenated, are used as substitutes for ghee. Rape seed oil is used in margarines which are high in polyunsaturated acids. They are characterised by the high concentration of the $C_{22:1}$ acid (erucic acid). Several workers have analysed the glycerides for fatty acids, and erucic acid is found in the range 25–60%.

Wachs and May[101] have described a GC method where the presence of small amounts of marine oils can be detected in rape seed oil. The method depends on the premise that marine oils contain certain saturated multibranched chain acids which are absent in rape seed oil.

Rape seed oil is also characterised by its relatively high concentration of brassicosterol. Copius-Peereboom[46] found 8% brassicosterol plus 1.4% cholesterol, and Karleskind *et al.*[45] found 10% brassicosterol and no cholesterol.

D. Fish oils

Fish oils contain relatively high concentrations of unsaturated fatty acids, and this unsaturated character means that these oils are more unstable than most vegetable oils. To be of value to the food industry, fish oils are hydrogenated to give fats melting in the range 30–45°C. Hydrogenation also eliminates the fishy odour and flavour associated with fish lipids and the hydrogenated oils can then be used in margarine and shortening manufacture. Margarines can be made from hardened oils of herring, mackerel, capelin and small members of the cod family.

Fish liver oils have also been included in this section of this chapter, although they are more noteworthy for their vitamin A and D content, e.g. cod liver oil.

Most of the GC examinations of fish and fish liver oils have centred around their fatty acid compositions, but there has been some investigation of the unsaponifiable fractions, usually the analysis of hydrocarbons and glycerol ethers.

Marine oils generally contain polyenoic acids with as many as six double bonds in some cases. Lambertsen, Myklestad and Braekkan[102] exploited the difference in the number of double bonds of the fatty acids to make a preliminary separation of the methyl esters by TLC into six fractions. Each fraction was analysed by GC, and these workers found that the major polyenoic acids of marine oils were $C_{18:4}$, $C_{20:5}$, $C_{22:5}$ and $C_{22:6}$. Many marine oils also contain branched and unbranched C_{17} acids which are uncommon in vegetable oils.

There is such a broad range of unsaturated acids together with branched-chain acids in fish oils that it has often been necessary to use two sets of GC conditions to complete the analysis. Normally, either two polar stationary phases are used or one polar and one non-polar stationary phase. Ackman[103] has found that the organosilicone polyester, EGSP-Z, is capable of separating almost all acids encountered in a fish oil in a single chromatogram.

Fish oils which have received GC analytical attention are herring, capelin, salmon, mullet, tuna, menhaden, pilchard, shrimp and sardine. The essential

GC details, where known, of some of the more important references to fish oil analyses are included in *Table 10.1*.

The fatty acid composition of herring oil glycerides has been studied by several workers[104-107], and in an examination of the unsaponifiable fraction, Gershbein and Singh[108] found that a quarter of that portion was made up of hydrocarbons, predominantly pristane and squalene.

The chemical testing carried out to determine the rancidity of oils usually involves the peroxide number, carbonyl number and thiobarbituric acid number.

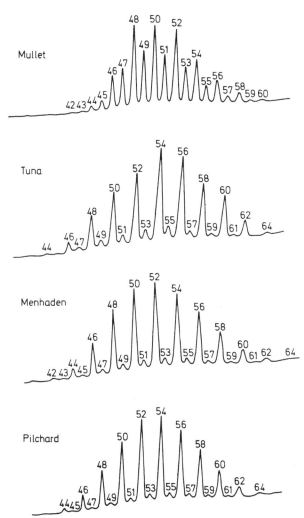

Figure 10.3. Chromatograms of fully hydrogenated mullet, tuna, menhaden and pilchard oil triglycerides. GC conditions are: 1.83 m × 2.5 mm O.D. column containing 3% JXR on 100/120 Gas Chrom Q and temperature programmed from 210 to 375°C. The carbon numbers of the triglycerides are indicated. (After Litchfield et al.[115])

Fats and Oils

These numbers increase as rancidity increases but only up to a point in time. Thereafter these values can decrease, and therefore they have limited use where an oil is kept for periods of months or years. Kiesvaara, Nikkila and Westergen[109] have evaluated a GC method which depends on the increase in heptanal as rancidity takes place in herring oil. The method also measures the formation of hexane as rancidity becomes absolute, the volatiles being flushed from the oil into cold traps and then analysed by GC.

Capelin oil is similar in fatty acid composition to herring oil and has been analysed using GC by Ackman *et al.*[110]

Fatty acid composition of the glycerides of Coho salmon oil has been determined[111, 112] and also that of pilchard oil[113] and Alaskan shrimp oil[114].

Litchfield, Harlow and Reiser[115] found that fish oils contain up to 5% of odd carbon number fatty acids and that mullet is unusual in that it contains more than 10% of these acids. These workers studied the hydrogenated glyceride GC patterns of various fish oils, and *Figure 10.3* shows these for mullet, tuna, menhaden and pilchard.

The most important fish liver oils are those of cod and shark. Fatty acid compositions have been determined for the glycerides of cod liver oil[116-118] and for the glycerides of shark liver oil[119].

References

1. Hilditch, T.P. and Williams, P.N., *The Chemical Constitution of Natural Fats*, 4th edn, Chapman and Hall, London (1964)
2. Stoffel, W., Insull, W. and Ahrens, E.H., *Proc. Soc. Exp. Biol.*, **99**, 238 (1958)
3. La Croix, D.E., Prosser, A.R. and Sheppard, A.J., *J. Ass. Off. Anal. Chem.*, **51**, 20 (1968)
4. Pearson, D., *The Chemical Analysis of Foods*, 6th edn, Churchill, London, 514 (1970)
5. Heyes, T.D., *Chem. & Ind.*, 660 (1963)
6. Jaarma, M., *Acta Chem. Scand.*, **18**, 300 (1964)
7. Armandola, P., *Riv. Ital. Sostanze Grasse*, **41**, 587 (1964)
8. Luddy, F.E., Barford, R.A., Herb, S.F. and Magidman, P., *J. Amer. Oil Chem. Soc.*, **45**, 549 (1968)
9. Magidman, P., Herb, S.F., Luddy, F.E. and Riemenschneider, R.W., *J. Amer. Oil Chem. Soc.*, **40**, 86 (1963)
10. Herb, S.F., Magidman, P., Barford, R.A. and Riemenschneider, R.W., *J. Amer. Oil Chem. Soc.*, **40**, 83 (1963)
11. Lotito, A. and Cucurachi, A., *Riv. Ital. Sostanze Grasse*, **43**, 499 (1966)
12. Doro, B. and Remoli, S., *Riv. Ital. Sostanze Grasse*, **42**, 138 (1965)
13. Doro, B., *Riv. Ital. Sostanze Grasse*, **44**, 349 (1967)
14. Pascussi, E. and Paolini, F., *Riv. Ital. Sostanze Grasse*, **41**, 315 (1964)
15. Castledine, S.A. and Davies, D.R.A., *J. Ass. Publ. Anal.*, **6**, 39 (1968)
16. Grieco, D. *Riv. Ital. Sostanze Grasse*, **41**, 261 (1964)
17. Bastijns, L.J., *Chem. & Ind.*, 721 (1968)
18. Armandola, P., *Riv. Ital. Sostanze Grasse*, **44**, 123 (1967)
19. Hubbard, A.W. and Pocklington, W.D., *Sci. Food Agric.*, **19**, 571 (1968)
20. Kuksis, A. and McCarthy, M.J., *J. Amer. Oil Chem. Soc.*, **41**, 17 (1964)
21. Sato, K., Matsui, M. and Ikekawa, N., *Bunseki Kagaku*, **16**, 1160 (1967)
22. Cook, H.R. and Sturgeon, J.D., *J. Ass. Off. Anal. Chem.*, **49**, 877 (1966)
23. Terrell, R.N., Lewis, R.W., Cassens, R.G. and Bray, R.W., *J. Food Sci.*, **32**, 516 (1967)
24. Langner, H.J., *Fette Seifen Anstrmitt.*, **71**, 893 (1969)
25. Hoffmann, G. and Meijboom, P.W., *J. Amer. Oil Chem. Soc.*, **45**, 468 (1968)

26. Pupin, F. and Vuillaume, R., *Ann. Falsif. Exp. Chim.*, **61**, 13 (1968)
27. Payne, E., *J. Sci. Food Agric.*, **22**, 520 (1971)
28. 'Butter and Margarine', *Which*, 49 (1973)
29. Fedeli, E., Camurati, F. and Jacini, G., *Riv. Ital. Sostanze Grasse*, **48**, 565 (1971)
30. Martin, E., and Berner, C., *Mitt. Geb. Lebensmitt. Hyg.*, **63**, 122 (1972)
31. Johnson, D.C. and Hammond, E.G., *J. Amer. Oil Chem. Soc.*, **48**, 653 (1971)
32. Ullmann, F., in *Encyclopadie der Technischen Chemie*, Ch. 7, Urban and Schwartz, München-Berlin, 477 (1956)
33. Williams, K.A., in *Oils, Fats and Fatty Foods*, 4th edn, Churchill, London, 272 (1966)
34. Leegwater, D.C. and van Gend, H.W., *Fette Seifen Anstrmitt.*, **67**, 1 (1965)
35. Kuksis, A. and McCarthy, M.J., *Can. J. Biochem. Physiol.*, **40**, 679 (1962)
36. Bezard, J. and Bugaut, M., *J. Chromat. Sci.*, **7**, 639 (1969)
37. Sato, K., Matsui, M. and Ikekawa, N., *Bunseki Kagaku*, **15**, 954 (1966)
38. Carracedo, C.F. and Prieto, A., *Grasas Aceit.*, **20**, 289 (1969)
39. Jamieson, G.R. and Reid, E.H., *J. Chromatog.*, **17**, 230 (1965)
40. Hadorn, H. and Zuercher, K., *Mitt. Geb. Lebensmitt. Hyg.*, **58**, 351 (1967)
41. Grieco, D. and Piepoli, G., *Riv, Ital. Sostanze Grasse*, **41**, 283 (1964)
42. Iverson, J.L. and Harrill, P.G., *J. Ass. Off. Anal. Chem.*, **50**, 1335 (1967)
43. Subbaram, M.R. and Youngs, C.G., *J. Amer. Oil Chem. Soc.*, **44**, 425 (1967)
44. Capella, P., Fedeli, E., Cirimele, M., Lanzani, A. and Jacini, G., *Riv. Ital. Sostanze Grasse*, **40**, 660 (1963)
45. Karleskind, A., Audiau, F. and Wolff, J-P., *Rev. Franc. Corps Gras*, **13**, 165 (1966)
46. Copius-Peereboom, J.W., *J. Gas Chromat.*, **3**, 325 (1965)
47. Fedeli, E., Lanzani, A., Capella, P. and Jacini, G., *J. Amer. Oil Chem. Soc.*, **43**, 254 (1966)
48. Rao, M.K.G. and Perkins, E.G., *J. Agric. Food Chem.*, **20**, 240 (1972)
49. Allen, R.R., *Chem. & Ind.*, 1560 (1965)
50. Youngs, C.G. and Subbaram, M.R., *J. Amer. Oil Chem. Soc.*, **41**, 218 (1964)
51. Subbaram, M.R. and Youngs, C.G., *J. Amer. Oil Chem. Soc.*, **41**, 445 (1964)
52. Iverson, J.L., *J. Ass. Off. Anal. Chem.*, **50**, 1118 (1967)
53. Provvedi, F. and Ciallella, G., *Riv. Ital. Sostanze Grasse*, **38**, 361 (1961)
54. Hadorn, H. and Zuercher, K., *Mitt. Geb. Lebensmitt. Hyg.*, **60**, 109 (1969)
55. Shoeb, Z.E., *Grasas Aceit.*, **20**, 4 (1969)
56. Iverson, J. L., *J. Ass. Off. Anal. Chem.*, **55**, 1319 (1972)
57. Fincke, A., *Fette Seifen Anstrmitt.*, **73**, 534 (1971)
58. Eisner, J., Iverson, J.L. and Firestone, D., *J. Ass. Off. Anal. Chem.*, **49**, 580 (1966)
59. Munari, S., Biagini, E., Grieco, D. and Ratto, G., *Riv. Ital. Sostanze Grasse*, **45**, 745 (1968)
60. Grieco, D., *Riv. Ital. Sostanze Grasse*, **39**, 432 (1962)
61. Montefredine, A. and Laporta, L., *Riv. Ital. Sostanze Grasse*, **40**, 379 (1963)
62. Cucurachi, A., *Riv. Ital. Sostanze Grasse*, **41**, 234 (1964)
63. Cucurachi, A., *Riv. Ital. Sostanze Grasse*, **44**, 260 (1967)
64. Synodinos, E., Kotakis, G.A. and Kokkoti-Kotakis, E., *Rev. Franc. Corps Gras*, **10**, 285 (1962)
65. Capella, P. and Jacini, G., *Riv. Ital. Sostanze Grasse*, **40**, 317 (1963)
66. Losi, G. and Pallotta, U., *Riv. Ital. Sostanze Grasse*, **43**, 425 (1966)
67. Hivon, K.J., Hagan, S.N. and Wile, E.B., *J. Amer. Oil Chem. Soc.*, **41**, 362 (1964)
68. Kuemmel, D.F., *J. Amer. Oil Chem. Soc.*, **41**, 667 (1964)
69. Riva, M., Poy, F. and Gagliardi, P., *Riv. Ital. Sostanze Grasse*, **41**, 267 (1964)
70. Averill, W., *Instrument News*, **17**, 13 (1967)
71. Pallotta, U. and Losi, G., *Riv. Ital. Sostanze Grasse*, **42**, 538 (1965)
72. Galanos, D.S., Kapoulas, V.M. and Voudouris, E.C., *J. Amer. Oil Chem. Soc.*, **45**, 825 (1968)
73. Eisner, J., Iverson, J.L., Mozingo, A.K. and Firestone, D., *J. Ass. Off. Agric. Chem.*, **48**, 417 (1965)
74. Amati, A., Carraro Zanirato, F. and Ferri, G., *Riv. Ital. Sostanze Grasse*, **48**, 39 (1971)
75. Losi, G. and Piretti, M.V., *Riv. Ital. Sostanze Grasse*, **47**, 493 (1970)

76. French, R.B., *J. Amer. Oil Chem. Soc.*, **39**, 176 (1962)
77. Iverson, J.L., Firestone, D. and Horwitz, W., *J. Ass. Off. Agric. Chem.*, **46**, 718 (1963)
78. Worthington, R.E. and Holley, K.T., *J. Amer. Oil Chem. Soc.*, **44**, 515 (1967)
79. Letan, A., Turzynski, B. and Raveh, A., *J. Ass. Off. Agric. Chem.*, **48**, 897 (1965)
80. Bruno, S., *Farmaco Ed. Prat.*, **20**, 85 (1965)
81. Iverson, J.L., *J. Ass. Off. Agric. Chem.*, **48**, 902 (1965)
82. Lotti, G. and Anelli, G., *Riv. Ital. Sostanze Grasse*, **46**, 110 (1969)
83. Jamieson, G.R. and Reid, E.H., *J. Chromatog.*, **20**, 232 (1965)
84. O'Connor, R.T. and Herb, S.F., *J. Amer. Oil Chem. Soc.*, **47**, 186A (1970)
85. Schneider, E.L., Sook, P.L. and Hopkins, D.T., *J. Amer. Oil Chem. Soc.*, **45**, 585 (1968)
86. Iverson, J.L., *J. Ass. Off. Anal. Chem.*, **52**, 1146 (1969)
87. Kato, A. and Yamaura, Y., *Chem. & Ind.*, 1260 (1970)
88. Hoffmann, G., *J. Amer. Oil Chem. Soc.*, **38**, 1 (1961)
89. Chang, S.S., Brobst, K.M., Tai, H. and Ireland, C.E., *J. Amer. Oil Chem. Soc.*, **38**, 671 (1961)
90. Chang, S.S., Smouse, T.H., Krishnamurthy, R.G., Mookherjee, B.D. and Reddy, R.B., *Chem. & Ind.*, 1926 (1966)
91. Krishnamurthy, R.G., Smouse, T.H., Mookherjee, B.D., Reddy, B.R. and Chang, S.S., *J. Food Sci.*, **32**, 372 (1967)
92. Selke, E., Moser, H.A. and Rohwedder, W.K., *J. Amer. Oil Chem. Soc.*, **47**, 393 (1970)
93. Scholz, R.G. and Ptak, L.R., *J. Amer. Oil Chem. Soc.*, **43**, 596 (1966)
94. Smouse, T.H., Mookherjee, B.D. and Chang, S.S., *Chem. & Ind.*, 1301 (1965)
95. Fioriti, J.A., Krampl, V. and Sims, R.J., *J. Amer. Oil Chem. Soc.*, **44**, 534 (1967)
96. Kaderavek, G. and Gay, G., *Olearia*, **17**, 145 (1963)
97. Earle, F.R., Vanetten, C.H., Clark, T.F. and Wolff, I.A., *J. Amer. Oil Chem. Soc.*, **45**, 876 (1968)
98. Ibrahim, W., Iverson, J.L. and Firestone, D., *J. Ass. Off. Agric. Chem.*, **47**, 776 (1964)
99. Pifferi, P.G., *Riv. Ital. Sostanze Grasse*, **43**, 505 (1966)
100. Tsatsaronis, G.C. and Boskou, D.G., *J. Ass. Off. Anal. Chem.*, **55**, 645 (1972)
101. Wachs, W. and May, A., *Dt. Lebensmitt-Rdsch.*, **64**, 412 (1968)
102. Lambertsen, G., Myklestad, H. and Braekkan, O.R., *J. Food Sci.*, **31**, 48 (1966)
103. Ackman, R.G., *J. Gas Chromat.*, **4**, 256 (1966)
104. Ackman, R.G. and Castell, J.D., *J. Gas Chromat.*, **5**, 489 (1967)
105. Ackman, R.G., Hooper, S.N. and Hingley, J., *J. Amer. Oil Chem. Soc.*, **48**, 804 (1971)
106. Linko, R.R. and Karinkanta, H., *J. Amer. Oil Chem. Soc.*, **47**, 42 (1970)
107. Jangaard, P.M., *J. Amer. Oil Chem. Soc.*, **42**, 845 (1965)
108. Gershbein, L.L. and Singh, E.J., *J. Amer. Oil Chem. Soc.*, **46**, 554 (1969)
109. Kiesvaara, M., Nikkila, O.E. and Westergren, K., *Acta Chem. Scand.*, **21**, 2887 (1967)
110. Ackman, R.G., Jangaard, P.M., Burgher, R.D., Hughes, M.L. and MacCallum, W.A., *J. Fish. Res. Bd Can.*, **20**, 591 (1963)
111. Saddler, J.B., Lowry, R.R., Krueger, H.M. and Tinsley, I.J., *J. Amer. Oil Chem. Soc.*, **43**, 321 (1966)
112. Ackman, R.G., *J. Amer. Oil Chem. Soc.*, **44**, 372 (1967)
113. Ackman, R.G. and Sipos, J.C., *J. Fish. Res. Bd Can.*, **21**, 841 (1964)
114. Krzeczkowski, R.A., *J. Amer. Oil Chem. Soc.*, **47**, 451 (1970)
115. Litchfield, C., Harlow, R.D. and Reiser, R., *Lipids*, **2**, 363 (1967)
116. Reed, S.A. and de Witt, K.W., *Chem. & Ind.*, 393 (1963)
117. Ackman, R.G. and Burgher, R.D., *J. Fish. Res. Bd Can.*, **21**, 319 (1964)
118. Ackman, R.G., Sipos, J.C. and Jangaard, P.M., *Lipids*, **2**, 251 (1967)
119. Gelpi, E. and Oro, J., *J. Amer. Oil Chem. Soc.*, **45**, 144 (1968)

11
Meat, Fish and Eggs

A. Introduction

In addition to their appeal as appetising foods, meat and fish are major contributors to the protein and fat contents of the diet. The composition of meat and fish protein has been studied by macro-organic analytical techniques, and the relatively small GC application has involved the analysis of fish protein concentrate. On the other hand, meat fats and fish oils have attracted considerable GC study, which has been described fully in Chapter 10.

The difficult field of meat and fish flavour research has involved numerous analytical techniques, and the bulk of this chapter deals with the application of GC to the detection and determination of meat and fish volatiles. A review of the literature regarding the use of GC in the investigation of meat flavour shows that most study has been concerned with beef and has been carried out in the United States. That country has also shown considerable interest in the detection of degradation products in stored eggs by GC.

The essential GC details, where known, of some of the more important references in this chapter, concerning meat, fish and eggs, are listed in *Table 11.1*.

B. Meat

Apart from its nutritional value, meat is an appetising food. Most of this appeal is found in its aroma and flavour after cooking, and therefore the analysis of the volatiles responsible has been the subject of investigation by many flavour chemists. Although many microanalytical techniques have been used for such investigation, GC alone and GC coupled with MS or IR spectroscopy have been the most popular. All the commonly eaten meats have received attention, particularly beef and chicken, and a review of general meat flavour has been given by Hornstein and Crowe[1].

As early as 1964, there was considerable evidence to suggest that volatiles from the lean meat of beef, pork and lamb have similar compositions and that the true flavour, and, hence, the difference in meat variety, lay in the fatty tissues. Hornstein, Crowe and Sulzbacher[2] had already suggested that the volatiles in the lean portions of beef and whale meat were identical except

177

Table 11.1. GC conditions used in the analysis of meat, fish and eggs

Class of compound	Ref.	Column dimensions	Stationary phase	Support material	Carrier gas and conditions	Temperature or temperature programme/°C	Detector
General	7	13ft x 0.25in O.D.	15% Apiezon L	60/80 Chromosorb W	He, 100ml/min.	100	FI
General	19	14ft x 0.25in O.D. 6ft x 0.25in O.D.	1% SE-30 -UCON LB-550X	microbeads diatomaceous earth	He, 40ml/min He, 35ml/min	40 60–80	FI FI
General	25	300ft x 0.01in I.D.	Apiezon L	capillary	He, 8ml/min	75–200 at 2/min	FI
General	26	9ft x 0.25in O.D.	9% DEGS	Chromosorb G	N_2, 40ml/min	31–175 at 3/min	–
General	34	6ft x 0.25in	5% β,β'-oxydi-propionitrile	60/80 Gas Chrom P	He, 55ml/min	– 60 to + 50 at 1/min	FI
Amines	33	3m x 3mm I.D.	20% Dowfax 9N9 plus 2.8% KOH	80/100 Silocell C22	N_2, 25ml/min	100	FI
		3m x 3mm I.D.	17% Dowfax 9N9 plus 14% KOH	80/100 Silocell C22	N_2, 25ml/min	100	FI

Compound	No.	Column	Stationary phase	Support	Carrier gas	Temperature	Detector
Amines	42	15ft x 0.125in	5% THEED-15% TEP	80/100 Chromosorb W	–	40	FI
Carbonyls	21	1.2m x 1mm	Porapak S	Porapak S	N_2, 25ml/min	100–180 at 10/min	FI
Carbonyls	38	1.5m x 4mm	10% PEGA	100/120 Celite 545	N_2, 60 ml/min	125	FI
Sulphur-containing compounds	40	1.8m x 4mm I.D.	10% Carbowax 1500	60/80 AW, DMCS Chromosorb W	Ar, 50ml/min	70	AI
Propyl esters of aliphatic acids	46	10ft x 4mm I.D.	10% DEGS	100/120 Gas Chrom Z	He, 80ml/min	130	FI
TMS derivatives of sterols	52	5ft x 0.125in O.D.	5% QF-1	80/100 Gas Chrom Q	He, 32ml/min	230	FI

that the latter contained trimethylamine. Although Stahl[3] identified methyl mercaptan, ethyl mercaptan and hydrogen sulphide in raw meat by GC, it was evident that a great number of volatiles were produced on heating and that only by experimenting with cooking conditions could the true aroma and flavour character of meat be found. Hornstein[4] suggested that two approaches were valid in flavour studies, viz. the examination of raw meat for flavour precursors and the examination of cooked meat for the flavour constituents themselves. He further suggested that the similarity of basic meaty flavour of beef, pork and lamb could be attributed to their similar concentrations of free amino acids and reducing sugars, both classes of compounds acting as flavour precursors. The influence of fat on meat flavour could be attributed to oxidation of unsaturated fatty acids after cooking to yield various carbonyls. Under controlled conditions the carbonyl concentration contributes to the desirable flavour characteristic, but if too much total carbonyl is produced, leads to undesirable off-flavours.

In a study of beef and pork Hornstein and Crowe[5] concluded that the flavour was associated with the fat, and they examined the free fatty acid compositions of these two meats before and after heating in air at 100°C for 4 h. Their findings reveal the predominance of oleic acid in both beef and pork, and also show the increase in all acid concentrations following the heat treatment.

Among workers using a head space technique for analysis of meat volatiles, Sandoval and Salwin[6] examined beef which had been left to decompose for a week at 40°F. This beef was cooked and freeze-dried and the head space volatiles of the homogenised sample were analysed. The GC profiles of the volatiles were very different for fresh and decomposed beef, but no attempt was made to identify the compounds. Some of the volatiles present in the decomposed beef disappeared on subsequent cooking.

El-Gharbawi and Dugan[7] obtained the volatiles from freeze-dried beef by a steam distillation technique aided by a stream of nitrogen to carry over the carbonyls and sulphur-containing compounds. The GC evidence was supported by chemical tests, and the compounds positively identified were acetaldehyde, propanal, pentanal, hexanal, acetone, methyl mercaptan and methyl disulphide. *Figure 11.1* shows the difference between flavour volatiles of freeze-dried raw beef which has been stored for 6 months at − 40°C under nitrogen and those volatiles from the same meat stored at room temperature.

Kramlich and Pearson[8] used a series of cold taps to collect the volatiles from the heating of a beef—water slurry. Supporting their GC findings with PC and with qualitative chemical tests, they found carbon dioxide, acetone and methyl mercaptan in this volatile fraction. When the volatiles were passed through a water trap, the resulting solution had a beef-like odour, indicating a correlation between the volatiles and flavour. In a similar study, Yueh and Strong[9] found that GC was particularly sensitive to sulphur-containing volatiles and found dimethyl sulphide in the volatile fraction of beef cooked in boiling water. They also examined this lean meat broth for fatty acids by GC of the methyl esters and found formic, acetic, propionic, butyric and isobutyric acids. More recently, Liebich *et al.*[10] used GC coupled with MS to identify roast beef volatiles in the lean and fatty portions. They used a vacuum distillation method followed by diethyl ether extraction prior to capillary column GC analysis, finding that the major classes of volatiles were alkanals, alk-2-enals

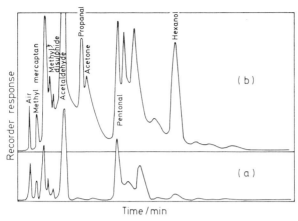

Figure 11.1. Gas–liquid chromatograms of flavour volatiles from freeze-dried raw beef stored 6 months under purified nitrogen at –40° C (a) and room temperature (b) (after El-Gharbawi and Dugan[7]). For GC details, see Table 11.1

and alka-2,4-dienals. High concentrations of 3-hydroxy-2-butanone and γ-butyrolactone were also found, together with many minor components. These authors considered that the prime site of the aldehydes was in the fat portion.

In comparison with cooked meat, raw meat has little flavour. However, a hot water extract of meat yields a product which has the flavour character of the cooked meat itself and this fact has been exploited commercially in the production of meat extracts. Bender and Ballance[11] found that immersion of beef in boiling water gave a mildly flavoured extract, but if prolonged heating was used, the extract became much more highly flavoured. Nitrogen was bubbled through the extract and the carbonyls were trapped in 2,4-dinitrophenylhydrazine, prior to GC analysis. A total of 12 carbonyls and sulphur-containing compounds have been identified as constituents of beef extract.

Irradiation of meat has been introduced as a method of preservation but there have been problems associated with the production of off-flavours by this treatment. Burks *et al.*[12] attributed some off-flavour of irradiated beef to the presence of volatile bases such as methylamine and ethylamine; Wick *et al.*[13] found several volatiles after irradiation treatment of beef, and considered the major contributor to unpleasant odour and flavour to be 3-(methylthio)-propionaldehyde. Merritt *et al.*[14] determined the volatile constituents of the centre cut of uncooked irradiated beef which included acetaldehyde, acetone, methanol, ethanol and some sulphur-containing compounds. All of these except two of the sulphur compounds were also found in non-irradiated beef but it is possible that these were below the detection limit of this early GC analysis. All the volatiles increased in concentration with increase in the amount of radiation.

Most GC analyses of flavour and aroma compounds are carried out at temperatures which are not much above normal room temperature. This means that temperature programming is limited when starting at this temperature, which is the lowest at which the analyst would normally commence the programme. This limitation prompted the use of cryogenic temperature programming, which has a starting temperature well below zero and a finishing temperature at the boiling point of the least volatile component.

Merritt *et al.*[15] have made a study of the cryogenic temperature programming of food volatiles using various rates between $-196°C$ and $+200°C$. Polar phases which have proved to be successful in the GC of volatiles under these conditions are TCEP, β,β'-oxydipropionitrile, DEGS, Carbowax 20M and Carbowax 4000. Among the suitable non-polar phases are squalane, Apiezon M and SF-96. Merritt and his co-workers have used their technique in the examination of volatiles from irradiated beef, identifying the components by MS.

The components of lamb flavour have not received as much attention as those of beef flavour, but Jacobson and Koehler[16] isolated roast lamb volatiles by sweeping them with air into cold traps and analysing the resulting yellow oily concentrate. GC was used to separate the concentrate into eight fractions which were analysed separately by PC, IR spectroscopy and chemical means. It was concluded that carbonyls were important contributors to lamb aroma, and the n-alkanals of 2-10C and the 2-alkanones of 2-5C were identified. In a similar study to that which they made of the flavours of beef and pork, Hornstein and Crowe[17] found that the characteristic aroma of lamb lay in the fat.

The volatiles of dry-cured country style hams were the subject of investigation by Ockerman, Blumer and Craig[18]. They found carbonyls and fatty acids to be major volatile contributors. Cured and uncured ham volatiles were compared by Cross and Ziegler[19], who considered that carbonyls and sulphur-containing compounds were the predominant contributors to the flavour. The hams were either kept at $60°C$ and the volatiles swept with nitrogen on to a 1% DC-550 glass microbead precolumn or alternatively an aqueous extract was made at $90°C$ using nitrogen to sweep the volatiles into traps of 2N hydrochloric acid (to remove bases), 2,4-dinitrophenylhydrazine (to remove carbonyls) and mercuric chloride solution (to remove sulphur-containing compounds). The latter extraction procedure has the advantage of a chemical division of the volatiles so that each fraction may be considered separately either by GC again or by some other suitable technique. In this way Cross and Ziegler showed that hexanal and valeraldehyde were present in much greater concentration in uncured ham than in the cured meat, where they were rarely encountered. It was thought that these carbonyls were derived from oxidative cleavage of unsaturated fatty acids, such as linoleic. This would reaffirm the view that fat plays a major role in the production of flavour and aroma compounds in meat. In addition to hexanal and valeraldehyde, acetaldehyde, propionaldehyde and butyraldehyde were found in slightly higher concentration in uncured ham than in the cured meat. Curing with nitrite appeared to have no effect on the volatile composition of ham except for the addition of nitrogen oxides.

The aroma of raw sausage was investigated by Eser and Niinivaara[20], who specifically looked for carbonyls using a 2,4-dinitrophenylhydrazine trap after making an aqueous extract of the meat. The carbonyls were regenerated from their 2,4-dinitrophenylhydrazones by heating at $200°C$ with 2-oxyglutaric acid and were analysed by GC. There was a marked increase of carbonyls during the first 11 days of storage and ripening, followed by a gradual decrease in the next 25 days. In a similar study of sausage volatiles, Halvarson[21] steam-distilled them in an atmosphere of nitrogen, collecting the carbonyls in a 2,4-dinitrophenylhydrazine trap. After regeneration, the carbonyls were analysed by combined GC and MS, acetaldehyde and propionaldehyde comprising the major portion of the 19 carbonyls found.

The large increase in poultry consumption since World War II has, not surprisingly, led to much research into ways of preserving it so that the essential flavour ingredients remain intact. The effect of temperature, oxygen and the type of tissue used on the volatile composition of chicken have been studied by Pippen and Nonaka[22]. They prepared chicken extracts by making a broth, distilling the volatiles and extracting them into isopentane. Cooking in air as opposed to cooking in nitrogen resulted in a much higher and varied volatile composition. Rancid tissue gave a greater quantity of the same volatiles than fresh tissue, and, as with other meats, these workers found that uncooked chicken produced fewer volatiles and these were less odorous than those of the cooked counterpart. They found small differences in the volatile compositions of cooked chicken and turkey, both containing significant proportions of n-hexanal and n-2,4-decadienal.

Minor *et al.*[23] specifically looked for sulphur-containing compounds in the volatile fraction of cooked chicken. They used four analytical techniques, including GC, in examining the aroma constituents of breast and leg tissues, and tentatively identified nine compounds including sulphides, disulphides and mercaptans. The specificity of their GC findings was increased by trapping the sulphur-containing compounds in mercuric chloride or in mercuric cyanide solution prior to GC.

In a review of poultry flavour, Pippen[24] asserted that GC had shown the presence of more than 200 components in the volatile fraction of cooked chicken, most of these being carbonyls and sulphur-containing compounds. There was some limited evidence to show that the sulphur-containing compounds gave the basic 'meaty' character, whereas the carbonyls gave the more specific 'chicken' flavour. The unpleasant fishy off-flavours in poultry were attributed to the presence of unsaturated fatty acids in the carcass fat. In what appears to be an amplification of this review, Nonaka, Black and Pippen[25] obtained cooked chicken volatiles by isopentane extraction of the aqueous distillate. Of the 227 peaks obtained by capillary column GC, 62 compounds were identified using MS, 49 of these for the first time in chicken. These workers positively identified sulphur-containing compounds, carbonyls, aromatics, furans, esters, hydrocarbons, alcohols and a terpene.

Isopentane extraction of cooked chicken extract was also the method of extraction of volatiles employed by Hobson-Frohock[26]. He increased the specificity of the GC performance by using a precolumn of molecular sieve material 5A, which removes all volatiles except aromatic hydrocarbons, branched cyclic compounds and some halogenated derivatives, and therefore the resulting chromatogram contained peaks only of those classes of compounds. The sieve completely removes n-alkanes, n-alkanals, n-alkanones, n-alkanols and methyl esters of fatty acids. Some n-alkanes can be recovered from the sieve by displacement with hexadecane.

In an improved extraction procedure for chicken volatiles, Lea, Swoboda and Hobson-Frohock[27] powdered the meat extract in liquid nitrogen and then mixed it with Celite and continuously extracted the mixture with chloroethane. The aroma compounds were separated from the chloroethane extract by vacuum distillation and concentrated to a small volume after being taken into 2,2,4-trimethyl pentane. The extract was temperature programmed from $-60°C$ to give the best separation of $C_4 - C_{15}$ volatiles.

The GC analyses of meat fats, such as lard, beef and lamb dripping, are discussed fully in the Animal Fats section of Chapter 10. Cook and Sturgeon[28] investigated the unsaponifiable fractions of the fats from cooked beef, pork and horse, with a view to using the information as a means of identifying the meat. The unsaponifiable matter was fractionated on a Florisil column prior to GC on both polar and non-polar columns using temperature programming to give improved peak separation, as compared with isothermal conditions. The peaks were not identified but different GC patterns emerged for the three meats, and this was considered to be the basis for further study with the objective of a method to differentiate types of meat.

The kind of meat used in liver paste has been deduced by Samuelsson, Borgstroem and Young[29], who utilised information gained by examining the fatty acid composition of the extracted fat. The fat was extracted from the liver paste by a modified Gerber method and hydrolysed, and the methyl esters of the resulting acids were chromatographed.

If deterioration of meat has occurred on storage, it is useful to determine the rate of production of free fatty acids. Hornstein *et al.*[30] examined cured and cooked meat fats in this way, using GC of the methyl esters of the free acids.

In addition to the uses of GC in the quality control of meat and meat products, there has been an increasing demand for the use of microanalytical techniques, such as GC, in the determination of nitrosamines in cured meats. This subject is discussed in Chapter 21, p. 354, which deals with the GC analysis of this food contaminant.

C. Fish

Fish, like meat, is composed of protein, oil and water. Although GC has been applied to the analysis of fish protein, most GC interest has been shown in the composition of the oil and in the identity of the fish flavour volatiles.

The application of GC to the analysis of fish and fish liver oils is discussed fully in Chapter 10, in the section entitled Fish and Fish Liver Oils.

Most fish is transported long distances for sale; even when stored in ice, bacterial and enzymic degradation occurs and off-flavours can develop to such a degree that the fish can become inedible. Aliphatic amines and ammonia are common degradation products and GC has been used for their analysis. In an examination of caviar, Golovnya, Mironov and Zhuravleva[31] isolated the amines via their hydrochlorides and liberated the amines with alkali for analysis on four different stationary phases. A correlation was obtained between the taste of the caviar and the proportions of the 19 amines found.

Gasco, Valverde and Barrera[32] concentrated the volatile compounds of cod after aqueous extraction at room temperature. Nitrogen was passed through the extract under reduced pressure and the vapour collected in cold traps prior to GC analysis. The data for 27 amines are given in this paper, although not all would necessarily be found in fish.

The quantitative determination of tri- and dimethylamine in fish has been achieved by Keay and Hardy[33]. The bases are extracted by perchloric acid and steam-distilled from alkaline solution and then collected in dilute hydrochloric acid solution. After trying various stationary phases, which either gave severe tailing or memory effects, these workers discovered that the best phase was composed of Dowfax 9N9, made alkaline with potassium hydroxide. They give results for amines in cod, calculated as mg nitrogen per 100 g of fish. The general volatile composition of cod has been investigated by Wong, Damico and Salwin[34] using GC coupled to MS following extraction by low-temperature vacuum distillation. Fresh fish has little or no odour and the so-called 'fishy' odour develops on ageing. The object of their investigation was to draw up a guide for their Food and Drug Authority Inspectorate on how to assess fish spoilage according to odour and appearance, backing this sensory evaluation with measurement of those volatiles responsible for off-flavours. Among the volatiles which were used to assess the state of deterioration were formic, acetic and succinic acids, ethanol, trimethylamine and histamine. The full list is included in *Table 11.2*, which shows fish volatiles of cod, herring, oyster and clam and also fish protein concentrate.

Haddock flesh has been used as an experimental medium to discover the best extraction method for fish volatiles by Mendelsohn, Steinberg and Merritt[35]. The methods tried were flushing out followed by head space analysis, head space analysis with no flushing out, vacuum distillation and high-vacuum distillation. The GC was carried out by cooling the column to $-65°C$ before sweeping the volatiles on to it and cryogenic temperature programming to room temperature. Vacuum distillation gave the greatest yield of volatiles but no attempt was made to identify them.

Apart from amines, the important volatiles of fish flesh are, as with meat, carbonyls and sulphur-containing compounds. Jones[36] found that fish invariably deteriorated on storage and that fish cooked within minutes of catching can have a metallic flavour. Plaice was an exception in that its flavour improved with a short time of keeping. Fish also showed similar flavour precursors to meat, e.g. amino acids, sugars, sugar phosphates and mononucleotides.

Hughes[37] examined the volatile fraction of cooked herring flesh and found carbonyls and sulphur-containing compounds, which are included in *Table 11.2*.

Vacuum distillation has been used by Golovnya and Uraletz[38] to isolate the volatiles of fresh salted salmon, sturgeon caviar and salmon roe. Special attention was paid to the carbonyls, which were converted into their 2,4-dinitrophenylhydrazones and regenerated in a purer state for GC analysis. *Figure 11.2* is a chromatogram of the 23 carbonyls of the fresh salted salmon volatile fraction.

Instead of using standards for comparison, these workers used the system of isothermal retention index by studying the retention times of compounds chromatographed isothermally at 50°C and at 25°C intervals above 50°C up to 200°C. The same authors[39] had used this system previously in a study of the carbonyls of salmon.

The qualities of flavour which contribute largely to the organoleptic appeal of certain shellfish has prompted investigation of their flavour volatiles. Ronald and Thomson[40] examined oysters with particular reference to sulphur-containing compounds, trapping them in mercuric chloride solution after sweeping

Table 11.2. Volatile compounds of some types of fish as found by GC

Cod	Cooked herring	Oyster	Clam	Protein concentrate
(after Wong et al.[34])	(after Hughes[37])	(after Ronald and Thomson[40])	(after Gadbois et al.[41])	(after Wick et al.[42])
Hydrogen sulphide	hydrogen sulphide	dimethyl sulphide	acetaldehyde	dimethylamine
Carbon disulphide	methanethoil	methanethiol	propionaldehyde	trimethylamine
Dimethyl sulphide	dimethyl sulphide		butyraldehyde	ammonia
Acetaldehyde	acetaldehyde	1-propanethiol	acetone	ethylamine
Propionaldehyde*	propionaldehyde		isobutyraldehyde	diethylamine
Butyraldehyde*	acetone	1-butanethiol	2-butanone	n-propylamine
Methylene chloride		dimethyl disulphide	2-methyl-2-butanone	butylamine
Chloroform			isovaleraldehyde	
Ethanol			biacetyl	
Acetone			valeraldehyde	
Methyl ethyl ketone*			2-pentanone	
Diethyl ketone*			hexanal	
Methyl propyl ketone*			4-heptanone	
Methyl vinyl ketone*			3-heptanone	
Benzene			2-heptanone	
Toluene			heptaldehyde	
Diethyl ether				
Trimethylamine*				

*Not found in fresh cod.

186

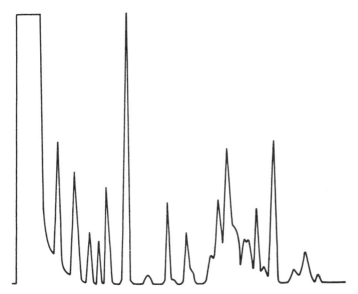

Figure 11.2 Monocarbonyl compounds in volatiles of fresh salted salmon. Chromatogram made on the PEGA column under TPGC conditions. (After Golovnya and Uraletz[38].) For GC details, see Table 11.1

the volatiles with nitrogen from an ice-water extract of the fish. After regeneration of the sulphur-containing compounds with hydrochloric acid, they were chromatographed on three different stationary phases. Listed among the sulphur-containing compounds of oyster given in *Table 11.2* is dimethyl sulphide, the most important contributor to oyster flavour. *Table 11.2* also includes the carbonyls found in clam meat as reported by Gadbois, Mendelsohn and Ronsivalli[41]. They studied the effects of irradiation, heating, storing, packing in air and vacuum packing on the carbonyls, and suggested that these volatiles were not indicators of clam spoilage. Irradiation increased the carbonyl concentration and cooking increased the longer-chain carbonyl concentration at the expense of the shorter-chain carbonyls. Storage also increased carbonyl content, but no link was found between this content and flavour.

Fish protein concentrate is used in animal feeding stuffs and could be used to supplement the protein diet of humans if only the fishy flavour could be removed. Wick, Underriner and Paneras[42] examined the volatile fraction of fish protein concentrate by GC coupled to MS and found six amines and ammonia. These make a considerable contribution to fish flavour and are listed in *Table 11.2*. GC has also been used in the examination of the lipid content of fish protein concentrate. Medwadowski *et al.*[43] examined pout, alewife and menhaden before and after storage for 6 months at both 37 and 50°C. If the original lipid content was less than 0.1%, the concentrate was relatively stable, but if the lipid content was greater than 0.5%, changes in flavour invariably occurred.

In another assessment of the lipid changes associated with degradation of fish tissue, Wood and Hintz[44] stored rockfish fillets at 0°C for 16 days and extracted the lipids in chloroform—methanol. The extract was fractionated by silicic acid column chromatography prior to analysis of the methyl esters of the fatty acids by GC. The experiments showed the expected increase in free fatty acids on storage and that the $C_{20:5}$ and $C_{22:6}$ acids were lost at a faster rate from phospholipids than from neutral lipids.

D. Eggs

Eggs are stored either whole or shelled, homogenised and pasteurised, or shelled, homogenised and dried. If eggs are to be kept any length of time, it is better to shell them so that some form of heat treatment can be applied to minimise the risk of bacterial or other microbial disorder. Such products are liquid egg and dried egg, which are used extensively in the confectionery industry. It is not until the egg in shell is cracked open that any deterioration can be detected (excepting the test where a bad egg floats in water), and The Food and Drug Administration of Washington DC, USA, has taken particular interest in the quality of eggs in shell.

When microbial or enzymic action proceeds in food, aliphatic hydroxy acids are frequently produced, and this has been found to occur in eggs when they commence to decompose. Bethea and Wong[45] examined the volatile acid fraction obtained from Grade A eggs, 'leakers' and 'incubator rejects'. Grade A eggs are top quality, 'leakers' are cracked eggs of acceptable quality, and 'incubator rejects' are eggs which have been incubated but subsequently found to be unhatchable. These workers extracted the acids using a steam distillation method followed by column chromatographic clean-up and followed this with GC analysis supported by MS. In decomposed eggs they found some butyric, isovaleric and n-valeric acids plus three unknown compounds.

An improved method of extraction of the aliphatic acids from eggs was used by Salwin and Bond[46]. This involved liberation of these acids from the homogenised eggs using a sulphuric acid—phosphotungstic acid mixture followed by their extraction with diethyl ether. The propyl esters were made by reaction with BF_3/propanol and analysed by GC. These workers found that fresh eggs contained approximately 4 mg lactic acid per 100 g egg and no succinic acid. After holding the egg for 16 h at room temperature, the lactic acid content had risen to 30.3 mg per 100 g egg and succinic acid appeared at 16.7 mg per 100 g egg. These figures can rise to 100 mg and 60 mg, respectively, in decomposed egg. *Figure 11.3* shows the difference between the chromatogram of a fresh egg extract and one after the egg has been shelled and extracted after 16 h at room temperature.

In a collaborative study conducted by Staruszkiewicz[47] using the Salwin and Bond extraction technique, lactic and succinic acid concentrations of fresh egg were found to be 2.4 mg and 0.8 mg per 100 g egg, respectively. It was suggested that the line of demarcation between passable and decomposed egg should be 7 mg lactic acid per 100 g egg. Bethea[48] also found succinic acid concentration to be a guide to the degree of egg decomposition, and Staruszkiewicz, Bond and Salwin[49] found that 3-hydroxybutyric acid could also be used in this way. Special precautions have to be taken to avoid polymerisation between 3-hydroxybutyric acid and other acids during the analysis.

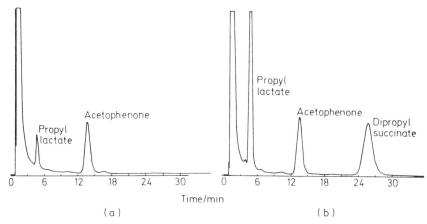

Figure 11.3(a) Gas chromatogram of a sample of frozen fresh whole egg; acetophenone was added as the internal standard. (b) Gas chromatogram of a sample of frozen decomposed whole egg held at room temperature 16 h after being shelled; acetophenone was added as the internal standard. (After Salwin and Bond[46].) For GC details, see Table 11.1

A collaborative study of the GC determination of the propyl esters of lactic, succinic and 3-hydroxybutyric acids in passable and 'incubator reject' eggs was conducted by Staruszkiewicz and Starling[50]. Special attention was given to the 3-hydroxybutyric acid content, and it was agreed that passable eggs contained approximately 0.3 mg per 100 g egg and 'incubator rejects' greater than 5 mg per 100 g egg. This difference ought to be large enough to show the presence of 5–10% of 'incubator rejects' added to passable eggs.

Both the fatty acid and the cholesterol contents of egg have been used to estimate the concentration of egg in other foods. Armandola[51] determined the egg content of noodles by extraction of the lipids and subsequent analysis of the methyl esters of the fatty acids by GC. The oleic and linoleic acid concentrations in egg are very different from those in semolina, and therefore the oleic/linoleic acid ratio can be used to estimate the egg content of semolina products, such as noodles.

Johansen and Voris[52] used egg cholesterol to estimate the egg content of cakes and hamburger buns. The unsaponifiable residue of the confectionery item was dissolved in dimethylformamide and the sterols were converted into their TMS derivatives, which were subsequently chromatographed. Whole dried egg has a cholesterol range of 1.64–1.81%, with an average of 1.72%. This average value is used as for 100% egg and the egg content calculated by direct proportion, although there is a small background level of cholesterol from the confectionery item minus egg. This method can be used to estimate the egg content of egg extenders, egg noodles, pumpkin and custard pies, doughnuts and cake mixes, but the method is obviously invalidated when animal fats or dairy products are used in the confectionery.

In an academic study of the separation of mono- and diglycerides by GC, Watts and Dils[53] analysed the diglyceride content of egg lecithin and found that the diglycerides with carbon numbers 34 and 36 accounted for the majority.

Stabilisers have been used in liquid egg, and the GC analysis of these is dealt with in Chapter 18, p. 304.

References

1. Hornstein, I. and Crowe, P. F., *J. Gas Chromat.*, **2**, 128 (1964)
2. Hornstein, I., Crowe, P. F. and Sulzbacher, W. L., *Nature*, **199**, 1252 (1963)
3. Stahl, W. H., in *Quartermaster Food and Container Institute of the U.S. Armed Forces: a Symposium*, 58 (1957)
4. Hornstein, I., *Chemistry and Physiology of Flavors*, Avi Publishing Co. Inc., Westport, Conn., 228 (1967)
5. Hornstein, I. and Crowe, P. F., *J. Agric. Food Chem.*, **8**, 494 (1960)
6. Sandoval, A. and Salwin, H., *J. Ass. Off. Anal. Chem.*, **51**, 1210 (1968)
7. El-Gharbawi, M. I. and Dugan, L. R., *J. Food Sci.*, **30**, 814 (1965)
8. Kramlich, W. E. and Pearson, A. M., *Food Res.*, **25**, 712 (1960)
9. Yueh, M. H. and Strong, F. M., *J. Agric. Food Chem.*, **8**, 491 (1960)
10. Liebich, H. M., Douglas, D. R., Zlatkis, A., Mueggler-Chavan, F. and Douzel, A., *J. Agric. Food Chem.*, **20**, 96 (1972)
11. Bender, A. E. and Ballance, P. E., *J. Sci. Food Agric.*, **12**, 683 (1961)
12. Burks, R. E., Baker, E. B., Clark, P., Esslinger, J. and Lacey, J. C. *J. Agric. Food Chem.*, **7**, 778 (1959)
13. Wick, E. L., Yamanishi, T., Wertheimer, L. C., Hoff, J. E., Proctor, B. E. and Goldblith, S. A., *J. Agric. Food Chem.*, **9**, 289 (1961)
14. Merritt, C., Bresnick, S. R., Bazinet, M. L., Walsh, J. T. and Angelini, P., *J. Agric. Food Chem.*, **7**, 784 (1959)
15. Merritt, C., Walsh, J. T., Forss, D. A., Angelini, P. and Swift, S. M., *Anal. Chem.*, **36**, 1502 (1964)
16. Jacobson, M. and Koehler, H. H., *J. Agric. Food Chem.*, **11**, 336 (1963)
17. Hornstein, I. and Crowe, P. F., *J. Agric. Food Chem.*, **11**, 147 (1963)
18. Ockerman, H. W., Blumer, T. N. and Craig, H. R., *J. Food Sci.*, **29**, 123 (1964)
19. Cross, C. K. and Ziegler, P., *J. Food Sci.*, **30**, 610 (1965)
20. Eser, H. and Niinivaara, F. P., *Z. Lebensmitt. Forsch.* **124**, 245 (1964)
21. Halvarson, H., *J. Chromatog.*, **66**, 35 (1972)
22. Pippen, E. L. and Nonaka, M., *J. Food Sci.*, **28**, 334 (1963)
23. Minor, L. J., Pearson, A. M., Dawson, L. E. and Schweigert, B. S., *J. Food Sci.*, **30**, 686 (1965)
24. Pippen, E. L., *Chemistry and Physiology of Flavors*, Avi Publishing Co. Inc., Westport, Conn., 251 (1967)
25. Nonaka, M., Black, D. R. and Pippen, E. L., *J. Agric. Food Chem.*, **15**, 713 (1967)
26. Hobson-Frohock, A., *J. Sci. Food Agric.*, **21**, 152 (1970)
27. Lea, C. H., Swoboda, P. A. T. and Hobson-Frohock, A., *J. Sci. Food Agric.*, **18**, 245 (1967)
28. Cook, H. R. and Sturgeon, J. D., *J. Ass. Off. Anal. Chem.*, **49**, 877 (1966)
29. Samuelsson, E. G., Borgstroem, S. and Young, R., *Z. Lebensmitt. Forsch.*, **141**, 70 (1969)
30. Hornstein, I., Alford, J. A., Elliott, L. E. and Crowe, P. F., *Anal. Chem.*, **32**, 540 (1960)
31. Golovnya, R. V., Mironov, G. A. and Zhuravleva, I. L., *Zh. Anal. Khim.*, **22**, 956 (1967)
32. Gasco, L. S., Valverde, G. A. and Barrera, R. P., *Inform. Quim. Anal.*, **25**, 193, 208 (1971)
33. Keay, J. N. and Hardy, R., *J. Sci. Food Agric.*, **23**, 9 (1972)
34. Wong, N. P., Damico, J. N. and Salwin, H., *J. Ass. Off. Anal. Chem.*, **50**, 8 (1967)
35. Mendelsohn, J. M., Steinberg, M. A. and Merritt, C., *J. Food Sci.*, **31**, 389 (1966)
36. Jones, N. R., *Chemistry and Physiology of Flavors*, Avi Publishing Co. Inc., Westport, Conn., 267 (1967)
37. Hughes, R. B., *J. Sci. Food Agric.*, **15**, 290 (1964)
38. Golovnya, R. V. and Uraletz, V. P., *J. Chromatog.*, **61**, 65 (1971)
39. Golovnya, R. V. and Uraletz, V. P., *Izv. Akad. Nauk SSSR, Ser. Khim.*, **3**, 679 (1970)
40. Ronald, A. P. and Thomson, W. A, B., *J. Fish. Res. Bd, Can.*, **21**, 1481 (1964)
41. Gadbois, D. F., Mendelsohn, J. M. and Ronsivalli, L. J., *J. Food Sci.*, **32**, 511 (1967)
42. Wick, E. L., Underriner, E. and Paneras, E., *J. Food Sci.*, **32**, 365 (1967)

43. Medwadowski, B., Haley, A., van der Veen, J. and Olcott, H. S., *J. Amer. Oil Chem. Soc.*, **48**, 782 (1971)
44. Wood, G. and Hintz, L., *J. Ass. Off. Anal Chem.*, **51**, 1216 (1968)
45. Bethea, S. and Wong, N. P., *J. Ass. Off. Anal Chem.*, **51**, 1216 (1968)
46. Salwin, H. and Bond, J. F., *J. Ass. Off. Anal. Chem.*, **52**, 41 (1969)
47. Staruszkiewicz, W. F., *J. Ass. Off. Anal. Chem.*, **52**, 471 (1969)
48. Bethea, S., *J. Ass. Off. Anal. Chem.*, **53**, 468 (1970)
49. Staruszkiewicz, W. F., Bond, J. F. and Salwin, H., *J. Chromatog.*, **51**, 423 (1970)
50. Staruszkiewicz, W. F. and Starling, M. K., *J. Ass. Off. Anal. Chem.*, **54**, 773 (1971)
51. Armandola, P., *Tec. Molitoria*, **14**, 94 (1963)
52. Johansen, R. G. and Voris, S. S., *Cereal Chem.*, **48**, 576 (1971)
53. Watts, R. and Dils, R., *J. Lipid Res.*, **10**, 33 (1969)

12

Essential Oils

A. Introduction

Essential oils are volatile liquids found in various plant tissues such as fruit, leaf, flower, bark and root. They have strong odours and flavours and are used in the preparation of essences and perfumes. The introduction of flavour into food is normally achieved by addition of essences, and therefore the quality of the essential oils which produce them has a profound effect on the palatability of that food.

The quality of essential oils depends on several factors, including the part of the plant used, the plant variety and its country of origin, the method of extraction and the refining process. The judgement of this quality depended solely on organoleptic evaluation until advances in GC techniques made it possible to supplement the findings of tasting panels with analytical data.

In addition to the normal flavour compounds which are found in most foods, such as carbonyls, esters and alcohols, essential oils are characterised by their content of terpenes, sesquiterpenes and their oxygenated derivatives. Where fruit and vegetables mainly yield volatile flavour constituents which are not of the terpene family, the pertinent GC analyses are discussed in Chapter 13.

Although the terpenes and sesquiterpenes are hydrocarbons, their GC analysis has not received the same extensive attention as that of hydrocarbons in petroleum technology.

Humphrey[1] has discussed the general GC conditions for the examination of volatile oils, paying particular attention to the choice of stationary phase, column temperature, gas flow rate and column length. If a separation of all types of essential oils is to be attempted, Humphrey recommends the use of Carbowax 20M as stationary phase, since its polarity is not extreme and it will separate both polar and non-polar substances to some degree. A 15% loading is recommended and Carbowax 20M is stable up to 250°, temperature programming ideally being performed between 50 and 250°C. Apiezon L is recommended as a stationary phase for the separation of non-polar volatiles and it is also recommended that a small amount of polar phase be added to prevent tailing of alcohols if these are present.

A better approach to the analysis of essential oils is to fractionate the volatiles prior to GC and then analyse each fraction individually. Bevitt and Cheshire[2] separated alcohols by partitioning them into propylene glycol from carbon tetrachloride and separating the hydrocarbons and oxygenated compounds by silicic acid column chromatography. They used three sets of GC conditions to analyse the various fractions, and these conditions are included in *Table 12.1* together with GC details, where known, of some of the more important references in this chapter.

The essential oils have been divided into four categories: those derived from fruit, leaf or whole plant, flower and root.

B. Essential oils from fruit

1. Lemon and Meyer lemon oils

Most lemon oil is prepared by cold-pressing of the fruit in order to preserve the flavour of the essential oil. The major ingredients of lemon oil are terpenes and by far the largest terpene component is d-limonene. In 1960 Bernhard[3] examined Californian cold-pressed lemon oil for composition and identified five terpenes and 20 oxygenated compounds. In the same year Clark and Bernhard[4] concentrated their study on the hydrocarbon fractions of the same type of oil. They separated this fraction from esters, aldehydes, ketones and alcohols by distillation and found eight hydrocarbons. In an examination of the non-hydrocarbon fraction of the same oil, Clark and Bernhard[5] separated the 4% of the oxygenated compounds in that fraction from the 96% terpene hydrocarbon fraction using a modification of the deterpination method of Kirchner and Miller[6], ten oxygenated compounds being positively identified. These compounds, although minor constituents, are the most valuable from a diagnostic point of view, according to Slater[7], who examined both natural and terpeneless lemon oils. In an examination of lemon oils from California, Arizona and Texas, Ikeda *et al.*[8] showed that d-limonene was the major constituent of the seven terpenes identified. β-pinene and γ-terpinene were also found in reasonable concentration, with smaller amounts of α-pinene, sabinene, myrcene and p-cymene. Capillary column GC coupled to MS was used by MacLeod, McFadden and Buigues[9] in a comparison of the volatiles obtained from cold-pressed lemon oils from Argentina, Arizona, Australia, California, Florida, Israel and Italy. They found no significant qualitative differences and only small quantitative ones.

Stanley *et al.*[10] analysed citrus oils specifically for aldehydes by a clean-up procedure utilising the compounds formed between aldehydes and Girard T reagent. These derivatives are water-soluble and extraction with organic solvent removed non-carbonyls, the original aldehydes being regenerated by formaldehyde from the Girard T derivatives. Extraction of the resulting aldehydes with isopentane was followed by GC, which showed that citral was the major aldehyde present in lemon oil.

Glycerol extraction and column chromatography on alumina was used by Hunter and Moshonas[11] to obtain the alcohol fractions of lemon, lime, grapefruit and tangerine oils. Lemon oil analysis has also been carried out by Ikeda and Spitler[12], Guenther and Pfeiffer[13], Guenther[14] and Fischer and Still[15].

Citral, being the major aldehyde and probably the most important flavour constituent of lemon oil, exists in *cis* and *trans* forms. The ratio of geranial (*trans*-citral) to neral (*cis*-citral) of natural Argentinian lemon oil was found to lie between 3.36 and 3.50 by Montes[16]. He found that the citral extracted

Table 12.1 GC conditions used in the analysis of essential oils

Class of compound	Ref.	Column dimensions	Stationary phase	Support material	Carrier gas and conditions	Temperature or temperature programme/°C	Detector
Monoterpenes, C_1 – C_{10} aldehydes, ketones, esters, alcohols, ethers	2	2.4m x 3.6mm	20% PEG 400	60/100 Celite	75 : 25, H_2 : N_2	78	FI
Oxyterpenes, sesquiterpenes, C_8 –C_{16} aldehydes, ketones, esters, alcohols, ethers	2	1.5m x 3.6mm	20% PEG 400	60/100 Celite	75 : 25, H_2 : N_2	130	FI
Terpenes	8	10ft x 0.25in	25% DEGS	42/60 firebrick	He, 120 ml/min	75	TC
Terpenes	21	36ft x 0.5in	25% Carbowax 20M	30/60 Chromosorb P	He –	135	–
Sesquiterpenes	29	8ft x 0.25in O.D.	30% Carbowax 20M	Chromosorb P	He, 60ml/min	180	FI
Alcohols	23	20ft x 0.25in O.D.	17% Carbowax 20M	AW Chromosorb G	He, 54ml/min	150–220 at 1.1/min	–

Alcohols, esters	12	10ft x 0.25in	20% PDEAS	60/80 Chromosorb P	He, 180ml/min	136	TC
Alcohols, esters ketones	61	–	10% PEGA	60/72 AW Celite 545	Ar, 90ml/min	75	AI
Carbonyls, esters	30	20ft x 0.125in	20% Carbowax 20M	60/80 Chromosorb P	He, 60ml/min	100–225 at 1/min	TC
General	66	150ft x 0.01in I.D.	UCON 50-HB-2000	capillary	He, 20 p.s.i. inlet pressure	80 for 3 min and 80–150 at 2/min	FI
General	76	8m x 1.5mm	5% Carbowax 20M	60/80 AW Chromosorb W	N_2, 30 ml/min	150	FI
General	82	–	20% Reoplex 400	60/80 AW Chromosorb W	He, 75ml/min	170	–
General	82	–	10% XE-60	60/80 AW Chromosorb W	He, 75ml/min	150	–

from lemongrass oil had a ratio of 2.50 and since this citral can be extracted from lemongrass oil and used to fortify natural lemon oil, a study of this ratio offers a means of detecting that addition. Montes claims that a 1% addition of lemongrass citral to natural lemon oil is detectable by this means.

In order to ensure that United States Pharmacopeia (USP) cold-pressed lemon oil has been expressed by that treatment and not by any form of distillation, the sample has to conform to a UV spectrum as laid down in the USP. MacLeod, Lundin and Buigues[17] discovered an oil which failed to conform in its UV pattern, and subsequent GC analysis, supported by IR and NMR spectroscopy, showed the presence of benzyl ether. The ether appeared after elution of the sesquiterpenes on an SE-30 column and its presence indicated that the oil had either been distilled or made up artificially.

The rapidly developing lemon industry in Florida in the late 1960s caused the trade to look for other fruits similar to lemon which could be used to enlarge the oil output. The Meyer lemon is a lemon—orange hybrid and the cold-pressed oil has many of the characteristics of lemon oil. However, Meyer lemon oil has an unusual flavour; Moshonas, Shaw and Veldhuis[18] used GC to investigate its composition and found 6% thymol in addition to the 92% of limonene. It is the thymol which gives the oil a peculiar flavour, and these workers have suggested that it might be advantageous to use Meyer lemon oil for its thymol content.

2. Lime oil

The composition of lime oil has not received the analytical attention of lemon oil and the GC study has been mainly confined to the hydrocarbon and alcohol constituents. Ikeda *et al.*[8] used a column chromatographic separation process prior to GC examination of the monoterpenes in a sample of cold-pressed lime oil, the monoterpene content being almost 70% of the total volatiles, as compared with more than 80% for lemon, orange and grapefruit oils. Most of the monoterpene fraction was d-limonene, with reasonable amounts of β-pinene and γ-terpinene, and is therefore comparable to lemon oil. In an examination of nine essential oils, Jain, Varma and Bhattacharyya[19] found d-limonene and *p*-cymene in sweet lime oil. Hunter and Moshonas[11] specifically analysed lime oil for alcohols (see under Lemon Oil, p. 193)

3. Orange oil

One of the most comprehensive early GC applications to orange oil analysis was carried out by Bernhard[20], who divided the oil into terpene and terpenoid fractions. Fifteen terpenes and 37 terpenoids were tentatively identified by using two stationary phases together with IR spectroscopy, 14 of the total of 52 compounds being reported for the first time.

As with lemon oil, the largest part of orange oil consists of monoterpenes, and Ikeda *et al.*[8], in examinations of Californian and Florida oil, found that

196

almost all the monoterpene fraction consisted of d-limonene. Hunter and Brogden[21] fractionated cold-pressed orange oil in a molecular still after a two-stage centrifugation. The first distillate, comprising 94.6% of the oil, contained terpenes and low molecular weight oxygenated compounds, leaving a residue of 5.4% which contained sesquiterpenes. The oxygenated compounds were removed from the terpenes on a basic alumina column, and the terpenes and sesquiterpenes were chromatographed separately using MS to identify the components and IR spectroscopy to support that evidence. In an extension of this work Hunter and Brogden[22] found that the major constituent of Florida orange oil was the sesquiterpene valencene.

Nineteen alcohols were isolated and identified in cold-pressed Valencia orange oil by Hunter and Moshonas[23]. They partitioned the extract between carbon tetrachloride and propylene glycol and the non-alcoholic compounds were removed by the former. The propylene glycol fraction was cleaned up by alumina column chromatography prior to GC, MS and IR spectroscopy being used to identify the alcohols. *Figure 12.1* is a chromatogram of the alcohols of Valencia orange oil.

Stanley *et al.*[10] specifically looked for aldehydes in orange oil, and other workers have investigated the oil composition of orange essence[24-28].

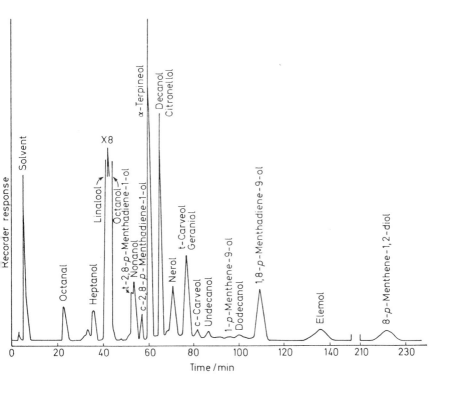

Figure 12.1 Chromatogram of 35 μl sample of alcohol fraction of Valencia orange oil (after Hunter and Moshonas[23]). For GC conditions, see Table 12.1

197

4. Grapefruit oil

Ikeda *et al.*[8] have determined the monoterpene content of a Texas grapefruit oil, and the sesquiterpene content of a cold-pressed oil has been given by Hunter and Brogden[29].

The aldehyde analysis of citrus oils carried out by Stanley *et al.*[10] included oils from white and pink grapefruit and showed $C_7 - C_{12}$ aldehydes to be present.

Twenty-two carbonyl flavour constituents of grapefruit oil have been separated and identified by Moshonas[30], who also used molecular still distillation as the separation technique. After column chromatographic clean-up, the GC examination showed 12 aldehydes, nine esters and the ketone nootkatone, which characterises grapefruit oil. Moshonas asserted that grapefruit, orange and tangerine oils have close similarities in terpene and sesquiterpene hydrocarbon content, the volatile carbonyl constituents giving the flavour differences between the three oils. The carbonyls identified in grapefruit oil all have potent and distinctive odours which probably contribute to the flavour and aroma of the oil. Nootkatone has also been found in grapefruit essence[31].

Hunter and Moshonas[11] specifically analysed grapefruit oil for alcohols (see under Lemon Oil, p. 193).

5. Tangerine and mandarin oils

The monoterpene hydrocarbons of tangerine and mandarin oils have been identified by Ashoor and Bernhard[32] using five stationary phases and IR spectroscopy. These workers were the first to report Δ^3-carene and α- and β-phellandrene in tangerine and mandarin oils. The d-limonene concentration of mandarin oil was much lower than that of tangerine oil, whereas the γ-terpinene concentration was four times greater. No Δ^3-carene was found in lemon, lime or grapefruit oils.

6. Bergamot oil

In an examination of five samples of bergamot oil by a chemical and a GC method, Mesnard and Bertucat[33] found that four contained linalool between 23 and 30% but that the fifth contained no linalool but an unidentified alcohol.

In a study of the monoterpene composition of several non-citrus essential oils, Ikeda *et al.*[34] found that bergamot oil contained 32% monoterpenes, greater than half of which was d-limonene. Thirteen per cent of γ-terpinene, 11% of β-pinene and small amounts of *p*-cymene, α-pinene, sabinene, myrcene, ocimene and α-thujene comprised the remainder of the monoterpene fraction.

7. Coriander oil and caraway oil

Coriander and caraway oils have been analysed using GC by Deryng, Strzelecka and Walewska[35] in support of their TLC findings, and they have tabulated the components.

Ikeda *et al.*[34] found that caraway oil contained 38% d-limonene and that coriander oil contained only 11% monoterpenes, the largest components being γ-terpinene, *p*-cymene, d-limonene and α-pinene.

Thirteen compounds were separated from coriander oil by Akimov and Voronin[36], and the same authors[37] used the same approach in an examination of the oxidation products of the same oil. Linalool between 12 and 89% and also citral between 6 and 84% were found in the oxidised oil.

8. Fennel and dill oils

Paukov, Rudenko and Kucherov[38] used capillary column GC to determine retention data and Kováts indices of several components of fennel oil on three stationary phases, whereas Bukala *et al.*[39] used the same technique and identified 15 components, the notable ones being anethole and fenchone.

The respective concentrations of anethole and fenchone in fennel oil, according to the ripened state of the fruit, have been determined[40] and general analysis of the oil has been carried out by Zolotovich and Hickethier[41] and by Martin and Berner[42].

In an examination of Indian dill oil, Shah, Qadry and Chauhan[43] found that the four main constituents were carvone, dihydrocarvone, limonene and dillapiole, the latter being a toxic ingredient and present at between 12 and 15%.

Betts[44] stored dill fruits in ethanol for 6 months at room temperature and determined the carvone and dillapiole contents of the extracts. The carvone content of both 'European' and 'Indian' dill was approximately the same, but whereas 'Indian' dill contained dillapiole at twice the carvone concentration, 'European' dill contained none.

9. Nutmeg oil and mace oil

Nutmeg is a very important spice and is used for flavouring many foods. The composition of nutmeg oil is consequently of equal importance, the dried seeds containing 5–15% of volatile oil and 25–40% fixed oil.

Ikeda *et al.*[34] found that two-thirds of nutmeg oil consisted of monoterpenes, the highest concentrations being those of sabinene and α- and β-pinene.

The compositions of East Indian 'Banda' oil, which is reputed to be the finest quality, and of the West Indian oil have been compared by Bejnarowicz and Kirch[45]. The two compounds safrole and eugenol were present in the East Indian but absent in the West Indian oil, and these could be important contributors to the finer flavour of East Indian oil. On the other hand, it seems more likely that an unidentified compound, which is present in the East Indian oil at between 13.9 and 25.9% and only at 1.7% in the West Indian oil, has a more profound effect on the flavour difference.

Sammy and Nawar[46] and Shulgin[47] have analysed nutmeg oil and the effects on the resulting oil on storage have been the subject of study by Sanford and Heinz[48].

Mace, being the dried aril which enwraps the nutmeg seed, yields an oil which has many similarities to oil of nutmeg. Forrest, Heacock and Forrest[49] used GC with MS, NMR and IR spectroscopy in deducing the composition of mace oil. They recorded a high level of 87.5% monoterpenes, plus 5.5% monoterpene alcohols and 6.5% aromatic ethers.

10. Cardamom oil

The chemical composition of Alleppy cardamom oil has been deduced using GC by Baruah, Bhagat and Saikia[50]. They steam-distilled cardamom seeds to give a colourless, volatile oil which was analysed on 5% Carbowax 20M and on 30% Carbowax 1000, 21 peaks being obtained. Fourteen compounds were identified by retention data on the two columns, the largest being eucalyptol and terpenyl acetate.

Ikeda *et al.*[34] found that sabinene was the largest component of the monoterpene fraction of cardamom oil, together with reasonable amounts of d-limonene, myrcene and α-pinene.

11. Black pepper oil

The monoterpene hydrocarbons in black pepper oil have been deduced by Wrolstad and Jennings[51], who isolated these compounds by vacuum distillation prior to GC on four different stationary phases. The main monoterpenes present were α-pinene, sabinene, β-pinene, limonene and Δ^3-carene. This supported the earlier findings of Ikeda *et al.*[34], who found that black pepper oil consisted of 57% of monoterpenes.

12. Apricot oil

In an attempt to discover the compounds responsible for the flavour of apricot oil, Tang and Jennings[52] analysed apricot essence after two different methods of extraction. The essence was distilled and the distillate extracted with isopentane in a continuous liquid–liquid extractor. Alternatively, the head space over the essence was passed through a charcoal column and the volatiles removed by CS_2. Terpenes and terpene alcohols were identified, but it was concluded that apricot aroma is not due to any one compound but to a mixture, in definite ratios, of a number of compounds.

13. Blackcurrant oil

Andersson and von Sydow[53] have analysed blackcurrant essential oil which was prepared by extraction of the berries with pentane, followed by steam distillation. The volatiles boiling at a temperature greater than 150°C constituted 9 p.p.m. of the fresh weight of the berries and contained many terpenes, sesquiterpenes and terpene alcohols, the major components being 25.9% of Δ^3-carene and 11.6% of caryophyllene.

14. Cranberry oil

The essential oil from American cranberries has been examined by Anjou and von Sydow[54] using GC and MS. The 89 compounds found, amounting to 83% of the oil, included 19 aliphatic alcohols, 20 aliphatic carbonyls, 19 terpene derivatives and 19 aromatic compounds. The major component was 23.7% α-terpineol, followed by 9.0% benzyl alcohol. The oil obtained from mountain cranberries was known by these workers to contain, by contrast, 40.2% benzyl alcohol and only 0.7% α-terpineol.

15. Hop oil

The analysis of hop oil is of particular importance to the brewing industry. Hops provide the bitter principle in beer, and the flavour constituents of hops are in a more concentrated form in the oil. Hop constituents in beer are discussed in Chapter 14, p. 240.

In a very early study of hop oils from 16 varieties of hop, Rigby and Bethune[55] found that the major constituents were myrcene, humulene and methyl nonyl ketone. In European oil humulene was found in higher concentration than myrcene, the reverse being the case for North American oil.

Several workers have subsequently examined the composition of hop oil[56-59].

The effect of column temperature on the separation of the four major hydrocarbons in hop oil has been studied by Roberts[60]. He experimented with a 10% Apiezon L stationary phase loading and various temperatures between 100 and 200°C, and found the best temperature was 175° for separating myrcene, farnesene, β-caryophyllene and humulene. Although Roberts[61] found that these were the major constituents of hop oil, he realised that the main flavouring compounds lay in the oxygenated fraction. *Figure 12.2* is a chromatogram of part of this oxygenated fraction and shows the major components to be 2-methylbutylisobutyrate (peak 3), methyl geranate (peak 12) and an unknown (peak 14). Other components tentatively identified were methyl pelargonate (peak 8), linalool (peak 9) and the combined methyl nonyl ketone and methyl caprate (peak 10).

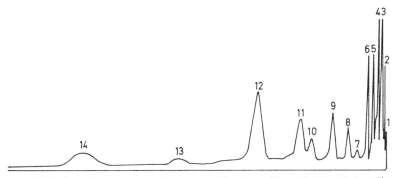

Figure 12.2 Gas chromatogram of oxygenated fraction of hop oil (after Roberts[61]). For GC conditions, see Table 12.1

In an examination of the unsaponified constituents of the oxygenated fraction, Roberts[62] identified, in addition to those already mentioned, 2-methylbutanol, methyl undecyl ketone, geraniol, β-caryophyllene epoxide and humulene epoxide.

Likens and Nickerson[63] analysed oils from several hop varieties and obtained sufficient differences between chromatograms to enable differentiation of the varieties. Buttery and Ling[64] continued this work and evolved a scheme for differentiation of American hops through the GC analysis of the oils. Chromatograms produced by different hop oils have also been compared by Silbereisen and Krueger[65] using capillary column GC.

C. Essential oils from leaf or whole plant

1. Peppermint, rue, pennyroyal, mint and spearmint oils

Most study of the oils of the mint family has concerned peppermint oil. Its persistent, distinctive flavour has been used to advantage in the sugar confectionery industry, and the strength of its flavour in masking unpleasant ones has been exploited by the pharmacist.

Peppermint oil is made up of a few major constituents and numerous minor ones, and for this reason Cieplinski and Averill[66] used capillary column GC with temperature programming for its analysis. *Figure 12.3* is a chromatogram of

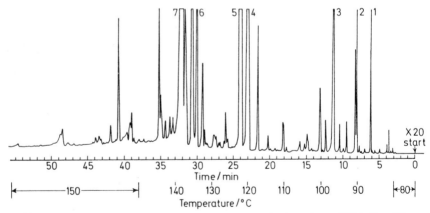

Figure 12.3 Peppermint oil, Yakima Valley (after Cieplinski and Averill[66]). For GC conditions, see Table 12.1

Yakima Valley peppermint oil, giving the programming detail and showing the predominance of menthol (peak 7), menthone (peak 4), menthofuran (peak 5) and eucalyptol (peak 3).

In experiments using boric acid as part of the stationary phase to remove primary and secondary alcohols, Ikeda, Simmons and Grossman[67] isolated menthol from peppermint oil. They used an orthodox column of 20% Carbowax 20M backed up by a column of the same material plus 3% boric acid. With menthol removed in this way, the analysis of the remaining constituents was facilitated.

In the Report of the Food Standards Committee of 1965 on Flavouring Agents[68] it was considered that rue oil should be one of 16 substances prohibited for use in foods. Rue oil is an abortifacient. Jansen and van der Kolk[69] examined preparations on sale in the Netherlands which possibly contained rue oil admixed with peppermint oil. Their method could be adapted for the detection of rue oil in other oils extracted from food, and simply consists of the addition of rue oil to the sample and seeing if any existing peaks are enhanced. Two of the six peaks were enhanced in the analysis of one of the preparations, where added peppermint oil had no effect on these but enhanced two other peaks. By measuring peak areas of sample and also the sample with standard additions of the two oils, it was possible to estimate the concentration of rue oil in peppermint oil.

In the Report already referred to[68] pennyroyal oil is recommended to be prohibited for use in foods, but little GC study has been made of it. Manfredini and Montes[70] examined pennyroyal oil and also the oils of peppermint, spearmint and Japanese mint, and have listed the various components.

Preparative GC, separating mint oil into five fractions, was performed by Fedeli and Pedrinella[71] prior to analytical GC of each fraction. Components identified were α-pinene, β-pinene, sabinene, limonene, camphene, eucalyptol, γ-terpinene and menthol.

It has always been considered that peppermint oil may be distinguished from other mint oils by its unique menthofuran content. Nigam and Levi[72], however, have found small amounts of menthofuran in mint oil samples from Argentina, Brazil, Formosa, India, Japan and South Africa. They also found menthofuran in some samples of spearmint, pennyroyal and other mint oils.

In an examination of spearmint oil, Burks and Gjerstad[73] found 19 GC peaks in the chromatogram, carvone being by far the major one. Also identified among the more substantial peaks were eucalyptol, α-pinene and linalool.

2. Tansy oil

The Food Standards Committee Report already referred to[68] recommended the prohibition of tansy oil as an ingredient of food. The oil is produced from the whole plant and is characterised by being made up largely of isothujone. A study was made of a commercial oil and a local tansy oil by von Rudloff[74] in order to assess the terpene contents. He used four stationary phases to confirm the identities of isothujone, thujone, camphor, sabinene, eucalyptol, carvone, α-pinene and camphene. The commercial oil contained 13.9% of camphor but only traces were found in the oil made from local plants.

3. Sage and thyme oils

Brieskorn and Wenger[75] separated the terpenes of sage oil by fractional distillation and identified several of these by GC. Thujone accounted for half of the terpenes, and eucalyptol, camphor, borneol, α-pinene, β-pinene and bornyl acetate together accounted for two-fifths. More recently, Rasmussen, Rasmussen and Svendsen[76] identified thujone, camphor, β-terpineol and borneol by volatilising the essential oil straight into the GC injection block. Five to ten mg of sample was placed in a cage which was pushed into the hot zone of the injection block. At a temperature of 200°C the sage oil was swept by nitrogen on to the column and the monoterpene hydrocarbons were separated.

The monoterpene hydrocarbon composition of thyme oil has been deduced by Ikeda *et al.*[34] Almost a third of the oil consists of monoterpenes, *p*-cymene accounting for more than half of the fraction. Nineteen per cent γ-terpinene was also found, together with small amounts of myrcene, α-terpinene, α-pinene, α-thujene, camphene, d-limonene, β-pinene, α-phellandrene and Δ³-carene.

4. Rosemary oil

Ikeda *et al.*[34] found that 61% of rosemary oil consisted of monoterpenes, α-pinene being the major constituent. Other terpenes identified were camphene, β-pinene, d-limonene, myrcene, γ-terpinene, Δ³-carene, β-phellandrene, α-phellandrene and α-terpinene.

5. Cassia oil

The qualitative composition of cassia oil has been deduced by ter Heide[77]. He used a chemical fractionation of the oil prior to GC and used MS, NMR spectroscopy, IR spectroscopy and TLC to aid identification of the 35 components.

6. Tea oil

Fresh tea leaves can be extracted with diethyl ether to yield an essential oil. This oil was analysed by Yamanishi *et al.*[78], who isolated 27 compounds, including isopentyl acetate, salicylaldehyde, benzyl acetate, indole and skatole, all hitherto unreported in tea leaf oil. The neutral fraction of the oil was analysed by Yamanishi *et al.*[79] using GC, IR spectroscopy and also odour assessment. Forty-two compounds were isolated, including linalool oxide *cis*-jasmone and hex-3-en-1-yl acetate.

In an examination of black tea essential oil, Saijo[80] found that the first compounds eluting from the GC column were the most important in characterising the oil. Ethyl acetate, ethanol and myrcene increased in concentration as fermentation of green tea occurred, and β-ocimene was only found in the completely manufactured black tea oil.

D. Essential oils from the flowers of lavender and rose

In addition to the GC analysis of fennel oil, Zolotovich and Hickethier[41] separated terpene hydrocarbons and other constituents from lavender and rose oils. After trying several stationary phases, these workers concluded that the most efficient was hèxa-*O*-(2-cyanoethyl)mannitol. Linalool, linalyl acetate, α-pinene, limonene, eucalyptol, borneol, pentanol and camphor were identified in lavender oil, and the main constituents of rose oil were found to be geraniol, rhodinol, nerol, nonanal and 2-phenylethanol. Kolsek and Maticic[81] also made a study of the different types of stationary phases for separating linalool, linalyl acetate, camphor and other constituents of lavender oil, and concluded that the best was 15% Castorwax.

E. Essential oils from the roots of ginger, wild ginger and sassafras

GC has been used to investigate the composition of ginger and wild ginger oils, particularly the minor constituents. Nigam *et al.*[82] used IR spectroscopy to aid the identification of nine constituents hitherto unreported in ginger oil, and Nigam and Levi[83] specifically looked for sesquiterpenes in that oil after experimenting with standard sesquiterpene mixtures in order to obtain the best GC conditions.

The composition of wild ginger oil has been studied by Nigam and Levi[84], who fractionated the oil by alumina column chromatography after freezing out the zerumbone. The compositions of wild ginger and ginger oils are given in *Table 12.2*.

Chinese sassafras oil has been analysed by Moryashchev and Voronin[85], who found safrole, eugenol, furfuraldehyde, camphor and other members of the terpene family.

Table 12.2 The % w/w compositions of ginger and wild ginger oils (after Nigam and co-workers[82, 84])

	Ginger oil	Wild ginger oil
α-and β-Zingiberene	35.6	−
α-Humulene	−	25.2
Camphene	1.1	21.9
Zerumbone	−	21.2
ar-Curcumene	17.7	0.8
Sesquiterpene alcohols	16.7	−
Unidentified	5.6	9.9
Farnesene	9.8	−
Humulene epoxide	−	4.6
Camphor	−	4.2
α-Pinene	0.4	3.4

(continued overleaf)

Table 12.2 (*continued*)

	Ginger oil	Wild ginger oil
Borneol	2.2	–
Borneol plus α-terpineol	–	0.6
Eucalyptol	1.3	1.8
β-Caryophyllene	–	1.6
Limonene	1.2	1.5
Citral a	1.4	–
γ-Selinene	1.4	–
Linalool	1.3	0.9
β-Phellandrene	1.3	–
Δ³-Carene	–	1.3
β-Elemene	1.0	–
Citral b	0.8	–
β-Pinene	0.2	0.6
Humulene dioxide	–	0.3
Alcohol (unidentified)	0.2	–
β-Bisabolene	0.2	–
Decyl aldehyde	0.2	–
2-Nonanol	0.2	–
Alcohol (unidentified)	0.1	–
Bornyl acetate	0.1	–
p-Cymene	0.1	–
Geraniol	0.1	–
Methyl heptenone	0.1	–
Myrcene	0.1	–
Nonyl aldehyde	0.1	–
Cumene	less than 0.1	–
2-Heptanol	less than 0.1	–

References

1. Humphrey, A. M., *Flavour Ind.*, **1**, 163 (1970)
2. Bevitt, R. N. and Cheshire, J. D., *J. Soc. Cosmet. Chem.*, **14**, 173 (1963)
3. Bernhard, R. A., *J. Chromatog.*, **3**, 471 (1960)
4. Clark, J. R. and Bernhard, R. A., *Food Res.*, **25**, 389 (1960)
5. Clark, J. R. and Bernhard, R. A., *Food Res.*, **25**, 731 (1960)
6. Kirchner, J. G. and Miller, J. M., *Ind. Eng. Chem.*, **44**, 318 (1952)
7. Slater, C. A., *J. Sci. Food Agric.*, **12**, 257 (1961)
8. Ikeda, R. M., Stanley, W. L., Rolle, L. A. and Vannier, S. H., *J. Food Sci.*, **27**, 593 (1962)
9. MacLeod, W. D., McFadden, W. H. and Buigues, N. M., *J. Food Sci.*, **31**, 591 (1966)
10. Stanley, W. L., Ikeda, R. M., Vannier, S. H. and Rolle, L. A., *J. Food Sci.*, **26**, 43 (1961)
11. Hunter, G. L. K. and Moshonas, M. G., *J. Food Sci.*, **31**, 167 (1966)
12 Ikeda, R. M. and Spitler, E. M., *J. Agric. Food Chem.*, **12**, 114 (1964)
13. Guenther, H. and Pfeiffer, W., *Reichstoffe Arom. Koerperpflegemitt.*, **18**, 52 (1968)
14. Guenther, H., *Dt. Lebensmitt. Rdsch.*, **64**, 104 (1968)
15. Fischer, R. and Still, F., *Pharm. Zentralhalle Dtl.*, **108**, 97 (1969)
16 Montes, A. L., *An. Asoc. Quim. Arg.*, **50**, 111 (1962)
17. MacLeod, W. D., Lundin, R. E. and Buigues, N. M., *J. Food Sci.*, **29**, 802 (1964)
18. Moshonas, M. G., Shaw, P. E. and Veldhuis, M. K., *J. Agric. Food Chem.*, **20**, 751 (1972)
19. Jain, T. C., Varma, K. R. and Bhattacharyya, S. C., *Perfum. Essent. Oil Rec.*, **53**, 678 (1962)
20. Bernhard, R. A., *J. Food Sci.*, **26**, 401 (1961)
21. Hunter, G. L. K. and Brogden, W. B., *J. Food Sci.*, **30**, 1 (1965)
22. Hunter, G. L. K. and Brogden, W. B., *J. Food Sci.*, **30**, 383 (1965)
23. Hunter, G. L. K. and Moshonas, M. G., *Anal. Chem.*, **37**, 378 (1965)
24. Wolford, R. W., Alberding, G. E. and Attaway, J. A., *J. Agric. Food Chem.*, **10**, 297 (1962)
25. Wolford, R. W. and Attaway, J. A., *J. Agric. Food Chem.*, **15**, 369 (1967)
26. Attaway, J. A., Wolford, R. W. and Edwards, G. J., *J. Agric. Food Chem.*, **10**, 102 (1962)
27. Attaway, J. A., Wolford, R. W., Alberding, G. E. and Edwards, G. J., *J. Agric. Food Chem.*, **12**, 118 (1964)
28. Coleman, R. L. and Shaw, P. E., *J. Agric. Food Chem.*, **19**, 520 (1971)
29. Hunter, G. L. K. and Brogden, W. B., *Anal. Chem.*, **36**, 1122 (1964)
30. Moshonas, M. G., *J. Agric. Food Chem.*, **19**, 769 (1971)
31. Moshonas, M. G. and Shaw, P. E., *J. Agric. Food Chem.*, **19**, 119 (1971)
32. Ashoor, S. H. M. and Bernhard, R. A., *J. Agric. Food Chem.*, **15**, 1044 (1967)
33. Mesnard, P. and Bertucat, M., *Ann. Falsif. Exp. Chim.*, **54**, 389 (1961)
34. Ikeda, R. M., Stanley, W. L., Vannier, S. H. and Spitler, E. M., *J. Food Sci.*, **27**, 455 (1962)
35. Deryng, J., Strzelecka, H. and Walewska, E., *Farmacja Pol.*, **22**, 7 (1966)
36. Akimov, Y. A. and Voronin, V. G., *Zh. Prikl. Khim., Leningr.*, **41**, 2561 (1968)
37. Akimov, Y. A. and Voronin, V. G., *Zh. Prikl. Khim., Leningr.*, **41**, 2344 (1968)
38. Paukov, V. N., Rudenko, B. A. and Kucherov, V. F., *Izv. Akad. Nauk SSSR, Ser. Khim.*, **1**, 15 (1968)
39. Bukala, M., Arct, J., Dul, M. and Pasternak, A., *Farmacja Pol.*, **24**, 689 (1968)
40. Betts, T. J. *J. Pharm. Pharmacol.*, **20**, 469 (1968)
41. Zolotovich, G. and Hickethier, R., *Compt. Rend. Acad. Bulg. Sci.*, **16**, 661 (1963)
42. Martin, E. and Berner, C., *Mitt. Geb. Lebensmitt. Hyg.*, **63**, 127 (1972)
43. Shah, C. S., Qadry, J. S. and Chauhan, M. G., *J. Pharm. Pharmacol.*, **23**, 448 (1971)
44. Betts, T. J., *J. Pharm. Pharmacol.*, **21**, 259 (1969)
45. Bejnarowicz, E. A. and Kirch, E. R., *J. Pharm. Sci.*, **52**, 988 (1963)
46. Sammy, G. M. and Nawar, W. W., *Chem & Ind.*, 1279 (1968)
47. Shulgin, A. T., *Nature*, **197**, 379 (1963)
48. Sanford, K. J. and Heinz, D. E., *Phytochemistry.*, **10**, 1245 (1971)
49. Forrest, J. E., Heacock, R. A. and Forrest, T. P., *J. Chromatog.*, **69**, 115 (1972)

50. Baruah, A. K. S., Bhagat, S. D. and Saikia, B. K., *Analyst,* **98,** 168 (1973)
51. Wrolstad, R. E. and Jennings, W. G., *J. Food Sci.,* **30,** 274 (1965)
52. Tang, C. S. and Jennings, W. G., *J. Agric. Food Chem.,* **15,** 24 (1967)
53. Andersson, J. and von Sydow, E., *Acta Chem. Scand.,* **18,** 1105 (1964)
54. Anjou, K. and von Sydow, E., *Acta Chem. Scand.,* **21,** 2076 (1967)
55. Rigby, F. L. and Bethune, J. L., *J. Inst. Brew.,* **63,** 154 (1957)
56. Cassuto, Mme., *Brasserie,* **15,** 40, 64 (1960)
57. Jahnsen, V. J., *Nature,* **196,** 474 (1962)
58. Buttery, R. G., McFadden, W. H., Teranishi, R., Kealy, M. P. and Mon, T. R., *Nature,* **200,** 435 (1963)
59. Buttery, R. G., McFadden, W. H., Black, D. R. and Kealy, M. P., *Proc. Amer. Soc. Brew. Chem.,* 137 (1964)
60. Roberts, J. B., *Nature,* **193,** 1071 (1962)
61. Roberts, J. B., *J. Inst. Brew.,* **68,** 197 (1962)
62. Roberts, J. B., *J. Inst. Brew.,* **69,** 343 (1963)
63. Likens, S. T. and Nickerson, G. B., *J. Agric. Food Chem.,* **15,** 525 (1967)
64. Buttery, R. G., and Ling, L. C., *J. Agric. Food Chem.,* **15,** 531 (1967)
65. Silbereisen, K. and Krueger, E., *Mschr. Brau.,* **20,** 184 (1967)
66. Cieplinski, E. W. and Averill, W., *Gas Chromatography Applications,* No. GC-AP-002, Perkin-Elmer Corporation (1962)
67. Ikeda, R. M., Simmons, D. E. and Grossman, J. D., *Anal. Chem.,* **36,** 2188 (1964)
68. *Food Standards Committee Report on Flavouring Agents,* Ministry of Agriculture, Fisheries and Food (1965)
69. Jansen, H. and van der Kolk, H., *Med. Sci. Law,* **3,** 77 (1963)
70. Manfredini, T. A. A. and Montes, A. L., *An. Asoc. Quim. Arg.,* **50,** 198 (1962)
71. Fedeli, E. and Pedrinella, L., *Riv. Ital. Sostanze Grasse,* **47,** 872 (1970)
72. Nigam, I. C. and Levi, L., *J. Pharm. Sci.,* **53,** 1008 (1964)
73. Burks, T. F. and Gjerstad, G., *J. Pharm. Sci.,* **53,** 964 (1964)
74. von Rudloff, E., *Can. J. Chem.,* **41,** 1737 (1963)
75. Brieskorn, C. H., and Wenger, E., *Arch. Pharm., Berlin,* **293,** 21 (1960)
76. Rasmussen, K. E., Rasmussen, S. and Svendsen, A. B., *Sci. Pharm.,* **39,** 159 (1971)
77. ter Heide, R., *J. Agric. Food Chem.,* **20,** 747 (1972)
78. Yamanishi, T., Kiribuchi, T., Sakai, M., Fujita, N., Ikeda, Y. and Sasa, K., *Agric. Biol. Chem. Japan,* **27,** 193 (1963)
79. Yamanishi, T., Kiribuchi, T., Mikumo, Y., Sato, H., Ohmura, A., Mine, A. and Kurata, T., *Agric. Biol. Chem. Japan,* **29,** 300 (1965)
80. Saijo, R., *Agric. Biol. Chem. Japan,* **31,** 1265 (1967)
81. Kolsek, J. and Maticic, M., *J. Chromatog.,* **14,** 331 (1964)
82. Nigam, M. C., Nigam, I. C., Levi, L. and Handa, K. L., *Can. J. Chem.,* **42,** 2610 (1964)
83. Nigam, I. C. and Levi. L., *J. Chromatog.,* **23,** 217 (1966)
84. Nigam, I. C. and Levi, L., *Can. J. Chem.,* **41,** 1726 (1963)
85. Moryashchev, A. K. and Voronin, V. G., *Zh. Anal. Khim.,* **18,** 401 (1963)

13

Fruits, Vegetables and Non-alcoholic Beverages

A. Introduction

Fruits and vegetables, being natural foods, are made up of some complex and of some simple organic constituents. When the simple constituents are in reasonable concentration, analysis has been achieved by classical macro-organic techniques. However, analytical problems arise when constituents occur in small concentrations and also when there is a desire to separate and analyse complicated mixtures of either simple or complex compounds at any level of concentration. The organic chemist has, therefore, made great use of GC in the analysis of such classes as fixed acids, amino acids, sugars, pectins, and aroma and flavour constituents of fruits and vegetables.

The compounds associated with the flavour differences between fruits and between vegetables are frequently present in only small concentrations. Having established the identities of these compounds in a particular fruit or vegetable, their ratios or trends in concentration can be used to assess quality and also the degree of maturity, fermentation or any other process designed to produce the typical flavour.

The terpenes and their related compounds are the important flavour constituents of essential oils, and since many of these oils form an integral part of fruit and vegetable flavours, it follows that many terpenes are found in these foods. Chapter 12 deals exclusively with essential oil analysis and, therefore, indirectly with the terpene composition of certain fruits and vegetables.

All other aspects pertaining to the GC of fruits, vegetables and beverages made from cocoa, coffee, tea and vanilla are found in the three sections of this chapter.

B. Fruits

In an academic study to find the best GC conditions for the analysis of the fruit acids, Hautala[1] confined his work to the separation of citric, malic and tartaric acids. He found that NPGS was the best stationary phase to separate methyl citrate and methyl malate but that tartrate separation was not satisfactory. Hautala[2] continued this academic study with a separation of the methyl esters of citric, malic, tartaric, oxalic, fumaric, benzoic and succinic acids, and con-

Table 13.1 GC conditions used in the analysis of fruits, vegetables and non-alcoholic beverages

Class of compound	Ref.	Column dimensions	Stationary phase	Support material	Carrier gas and conditions	Temperature or temperature programme/°C	Detector
Methyl esters of organic acids	2	2ft x 5mm I.D.	20% DEGS	100/110 Anakrom ABS	N₂, 10 p.s.i. inlet pressure	100 or 120 or 140	FI
TMS derivatives of acids and sugars	4,6	6ft x 0.25 in O.D.	3.8% SE-30	60/80 silanised Diatoport S	He, 50ml/min	90–240 at 6/min (acids) 160–170 at 4/min (sugars)	TC
TMS derivatives of sugars	50	6ft x 0.25in	5% NPGS	60/80 Chromosorb W	N₂, 30ml/min	110–220 at 5/min	FI
TMS derivatives of sugars	102	6ft x 0.125in	5% SE-30	80/100 Hi-perform-ance AW DMCS Chromosorb G	N₂, 30 ml/min	140–260 at 2/min	FI
TMS derivatives of flavanols	115	8ft x 4mm	3% OV-1	60/80 Gas Chrom Q	He, 200ml/min He, 187ml/min	250 (catechin, epichatecin, epigallocatechin) 300 (epichatechin gallate, epigallocatechin gallate)	FI

Compound	No.	Column	Stationary phase	Support	Carrier gas	Temperature	Detector
TFA derivatives of butyl esters of amino acids	5	6ft x 0.25in O.D.	0.65% EGA	60/80 AW Chromosorb W	He, 45ml/min	80–205 at 4/min	FI
TFA derivative of tyramine	45	4ft x 0.125in	6% QF-1 plus 4% SE-30	60/80 AW Chromosorb W	—	100–200 at 6/min	FI
Pyrazines	101	8ft x 0.06in	10% DEGA including 2% H_3PO_4	80/100 Gas Chrom A	N_2, 30ml/min	60–190 at 2/min	FI
Caffeine	118	6ft x 4mm I.D.	10% DC200 oil	80/100 Gas Chrom Q	N_2, —	190	AFI
Capsaicin	66	1.5m x 2mm	3% SE-52	silanised Gas Chrom Z	N_2, 40ml/min	150–270 at 3.5/min	FI
TMS derivative of capsaicin	67	2m x 3mm	1% or 3% JXR	100/120 Gas Chrom Q	N_2, 35ml/min	150–230 at 6/min	FI
Vanillin, ethyl vanillin	119	6ft x 0.25in O.D.	20% Carbowax 20M	60/80 Gas Chrom P	He, 90ml/min	200	FI
Sulphur-containing compounds	78	50ft x 0.02in I.D.	Carbowax 20M	capillary	He, 6ml/min	50–175 at 2/min	FI
General	33	– x 0.125in	5% Carbowax 20M	60/80 Gas Pack-F	—	50–230	FI
General	95	1.2m x 4mm	10% di-n-decyl phthalate	firebrick	Ar, 95ml/min	50	AI

cluded that this separation on a 20% DEGS stationary phase could be applied to fruits. *Table 13.1* gives the full GC conditions and also includes the conditions of some of the more important references in this chapter.

Harvey, Hale and Ikeda[3] analysed tobacco, fruit, fruit products and vegetables for malic, citric and oxalic acids by conversion to their methyl esters and extraction of these into chloroform for GC examination. Citric acid was found at more than 5% in samples of lemon and lime juice and also in fresh tomatoes. A dried apple sample contained 1.5% malic acid and a sample of frozen spinach contained 5.5% oxalic acid.

Fernandez-Flores, Kline and Johnson[4] made ethanolic extracts of fruits, precipitated the acids as their lead salts and converted these into their TMS derivatives. Results for glycolic, succinic, fumaric, malic, tartaric, citric, syringic and quinic acid concentrations of 26 types and varieties of fruits have been tabulated. An examination of the chromatographic profile of the TMS derivatives of the various acids can aid the identification of a fruit or its variety.

The amino acid content of fruits can also be used as an index for determining their authenticity. Fernandez-Flores *et al.*[5] determined the concentrations of 15 amino acids in 22 varieties of fruits and have tabulated the results. The fruit juice or fruit extracted in water was treated with alcohol and the amino acids were cleaned up on an ion exchange resin before conversion to their *N*-TFA-n-butyl derivatives for GC. As little as 0.5 mg amino acid per 100 g fruit can be detected and the method offers another means of fruit identity. Apple, blackberry and cranberry contain only minimal amounts of amino acids, whereas fresh fig, prune and cantaloupe have relatively large amounts of amino acids. Very few fruits examined contained hydroxyproline but all contained aspartic acid.

Kline, Fernandez-Flores and Johnson[6] have also analysed large numbers of fruit samples for sugars. Interfering acids were precipitated as their lead salts and the sugars were converted into their TMS derivatives for GC. These workers tabulated results for the determination of fructose, glucose, sucrose and sorbitol in 28 samples, including 20 different fruits. Fruits normally contained equal concentrations of glucose and fructose, except apple and pear, where fructose was dominant, and cranberry, which contained more glucose than fructose. There was a relatively high concentration of sucrose in banana, cantaloupe, nectarine, orange, peach and pineapple, whereas most berry fruits contained none. Prunes, because of their partial desiccation, contained relatively high concentrations of fructose and sucrose, and the sorbitol content, ranging between 9 and 16%, was by far the highest for any of the fruits examined.

The examination of a fruit or a fruit product for fixed acids, amino acids and sugars by GC, as outlined by these American Food and Drug Authority personnel, gives the analyst three sets of data which could prove invaluable where there is a suspicion that adulteration by another fruit or any other substance has occurred. Analytical chemistry applied to fruits and vegetables has become toothless since the advent of the unscrupulous additions of substances which have served as analytical parameters, such as potash, phosphate and certain amino acids, and the aforementioned composite GC application offers a different approach to the problem.

Fitelson[7] made a qualitative study of the carbohydrates in fruit juices and used the information to detect adulteration. Acids were removed as lead salts, and after evaporation of the solution, the carbohydrate residue was dissolved in hot pyridine and the TMS derivatives were made for GC. All fruit juices showed a major fructose peak followed by two smaller glucose peaks. Cherry, prune, apple and pear juices also contained sorbitol, in contrast to the small berry fruit juices – strawberry, blackberry and raspberry. The characteristic profiles of several pure fruit juices have been established and several commercial fruit concentrates that were judged to be adulterated on the basis of abnormal sugar patterns had this confirmed by other parameters, such as ash, potash and phosphate. The adulteration of a fruit juice, both with another juice and by addition of sugar syrup, can be detected by these means.

Zamorani, Roda and Lanzarini[8] pyrolysed pectins by depositing them from solution on to a platinum spiral and drying until approximately 1 mg of dry substance was accumulated. Following a 6 s pyrolysis the products were analysed by GC, each pectin producing a characteristic pyrogram. The conditions of drying and subsequent pyrolysis are critical; otherwise this might offer another useful GC method for the confirmation of the identity of fruits, although there remains the problem of removing the pectins from the samples.

1. Apple

Malic acid has been shown to be the dominant acid of apple[4] and the total amino acid content of 3.3 mg per 100 g fruit is low in comparison with other fruits[5]. Winesap and Golden Delicious varieties contained twice as much fructose as glucose[6].

In the carbohydrate analysis of apple juice, Fitelson[7] found sorbitol and sucrose as well as glucose and fructose.

In an early application of GC in fruit analysis, Meigh[9] studied the evolution of ethylene from stored apples. Ethylene can be used to stimulate the ripening of fruits and is naturally emitted by apples. Meigh studied the concentration of ethylene in different concentrations of atmospheric carbon dioxide, and found that high levels of carbon dioxide retarded ethylene emission.

Apple flavanol glycosides and aglycones have been studied[10], as also have the phenolic amines[11].

The aroma and flavour constituents of both the apple[12-18] and its juice[19-24] have received considerable attention. Many esters, alcohols and aldehydes are considered to be important flavour and aroma constituents, and the isolation of these volatiles prior to GC analysis has involved head space, distillation and solvent extraction techniques, either singly or in combination.

2. Pear

The fruit acid concentration for a Bartlett variety pear showed malic and citric acids to be dominant[4], and at 23.3 mg aspartic acid per 100 g fruit this amino acid accounted for most of the total amino acid content[5].

The sugar profile of pear juice is similar to that of apple. Fitelson[7] found that pear juice contained some sorbitol and sucrose as well as glucose and fructose. Kline et al.[6] analysed Bosc and Bartlett varieties of pear for sugars and found that the glucose content was only a third or less of that of the fructose content. The sorbitol content of 2.6% for both pear varieties is significantly higher than that of the two apple samples similarly examined.

213

In the flavonoid analysis described by Duggan[10] where the TMS derivatives of the flavanol aglycones were chromatographed, pears contained isorhamnetin together with a small amount of quercetin.

The study of pear volatiles has been made chiefly on the Bartlett variety. Jennings and Sevenants[25] examined this variety by chromatographing the essence, and found that hexyl acetate and methyl *trans*-2:*cis*-4 decadienoate were flavour contributors. The latter ester was used by Heinz, Creveling and Jennings[26] as an indicator of Bartlett pear maturity. The juice from the same variety of pear has also been analysed[27,28].

Romani and Ku[29] placed Bartlett pears in gallon jars at 20°C and passed water-saturated air continually over the fruit. The jars were sealed daily for 0.5 h and head space samples were taken for GC analysis. Ethylene, acetaldehyde and simple alkyl acetates were tentatively identified and the concentrations of these used to assess the degree of maturity of the fruit. This method has the advantage that it is non-destructive and can be used as a monitoring process until the correct moment for marketing or processing is reached. Ethyl acetate concentration continued to increase with ripening, whereas ethylene and carbon dioxide reached a maximum before ripening. Irradiation of the fruit suppressed the total volatile content.

Pribela[23] has described methods for the isolation of free fatty acids, carbonyls and alcohols from pear concentrate, and Gasco *et al.*[24] have listed the volatiles found in pear juice on three different stationary phases.

3. Grape

Fernandez-Flores *et al.*[4] found that malic and tartaric acids were the dominant ones in grape, and it is significant that of all the fruits analysed by these workers only grape contains tartaric acid. Harvey *et al.*[3] found that grape juice contained 0.29% of malic acid, but did not determine tartaric acid.

Fernandez-Flores *et al.*[5] analysed black, red and white grape samples for total amino acids, and the sample of black grapes was noteworthy in that it contained 27 mg threonine per 100 g fruit, whereas the other samples did not contain any. Very few fruits contain hydroxyproline, but the black, red and white grapes contained 1.8, 1.5 and 1.6 mg per 100 g fruit, respectively.

The same three types of grape were analysed for sugar content by the method described previously by Kline *et al.*[6] All samples contained approximately equivalent concentrations of glucose and fructose, and small concentrations of sucrose and sorbitol were also present.

Far more attention has been paid to the use of GC in the analysis of wine volatiles than to its use in the analysis of the grape itself (see also Chapter 14 under Wine). Pribela[23] isolated free fatty acids, carbonyls and alcohols from grape concentrate, and Gasco *et al.*[24] have listed the volatiles of grape juice as found by different extraction methods using three different stationary phases.

Mattick *et al.*[30] determined methyl anthranilate in Concord grape juice by a GC method which was found to be more sensitive than the existing chemical method, which involved a diazotisation and coupling reaction. Methyl anthranilate is considered to be a major flavour constituent of Concord grape juice, and the presence and concentration of it is an important factor to vintners in an assessment of grape quality. Forty-seven samples of juice gave a range of nil to 6.8 p.p.m. methyl anthranilate.

4. Peach and nectarine

Malic and citric acids were found to be the dominant acids of peach[31], and quinic acid appears to be present in significant concentration[4].

Most of the total amino acid content of a peach sample analysed by Fernandez-Flores et al.[5] was aspartic acid and no lysine was found.

A peach sample analysed for sugars by Kline et al.[6] showed a large sucrose content relative to the other sugars.

The volatile components important to peach flavour have been reported by Jennings and Sevenants[32]. They prepared an essence from Red Globe variety peaches by vacuum distillation and extracted the essence with diethyl ether for GC. IR spectroscopy was used to support identification of benzaldehyde, benzyl

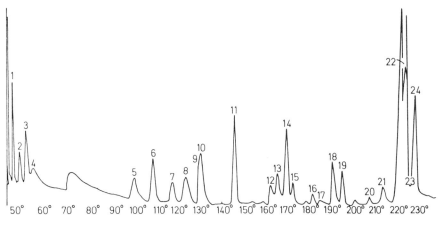

Figure 13.1 Chromatogram of Red Globe peach essence (after Sevenants and Jennings[33]). For GC details, see Table 13.1. 1, Acetaldehyde; 2, methyl acetate; 3, ethyl acetate; 4, ethyl alcohol; 5, hexyl formate; 6, hexyl acetate; 7, trans-2-hexenyl acetate; 8, hexyl alcohol; 9, acetic acid; 10, trans-2-hexene-1-ol; 11, benzaldehyde; 12, isovaleric acid; 13, ethyl benzoate; 14, γ-caprolactone; 15, benzyl acetate; 16, γ-heptalactone; 17, caproic acid; 18, benzyl alcohol; 19, γ-octalactone; 20, γ-nonalactone; 21, hexyl benzoate; 22, γ-deca-lactone; 23, α-pyrone; 24, δ-decalactone

alcohol, γ-caprolactone, γ-octalactone, γ-decalactone and δ-decalactone. In a second investigation Sevenants and Jennings[35] found a total of 24 volatiles in a similar essence and concluded that no one compound was responsible for the odour and flavour of peach but that typical peach flavour was formed by a mixture of several volatiles. *Figure 13.1* is a chromatogram showing these 24 volatiles.

The analysis of nectarine for fruit acids[4], amino acids[5] and sugars[6] gave similar results to those of peach.

5. Apricot

Molina, Soler and Cambronero[34] studied the aroma components of apricot by capillary column GC and determined the octa-, nona-, deca- and dodecalactone contents.

6. Prune

Fernandez-Flores *et al.*[4] showed that prunes contained a high malic acid content of almost 2% and a lack of citric acid. Harvey *et al.*[3], in their analysis of acids in fruit juices, found that prune juice contained 0.5% malic acid and no citric acid.

The comparatively large aspartic acid composition of prunes is reflected in their high total amino acid content[5].

As might be expected of a dried fruit, the total sugar content is much higher than most other fruits listed. The glucose content is higher than that of fructose but the sucrose content is either nil or very small. The sorbitol content, ranging from 9 to 16%, is by far the highest concentration in all the fruits examined. It follows that a significant addition of sucrose to prune juice would be detected by this type of GC analysis, and Flynn and Wendt[35] removed organic acids from prune juice by precipitation of their lead salts and converted the sugars to their TMS derivatives in a search for such an addition. The code of Federal Regulations, USA, requires that prune juice must contain not less than 18.5% soluble solids, and it is possible that sucrose could be added to supplement that figure.

7. Plum

Fernandez-Flores *et al.*[4] found that nearly all the acid contribution in plums was from the 1.5% malic acid. The same authors[5], in their amino acid analysis of fruits, found that of two samples of purple plums and one of yellow the latter gave a much higher total amino acid content.

8. Cherry

Fernandez-Flores *et al.*[4] included cherry in their analysis of fruit acids and showed that the major acid is syringic, followed by malic and citric acids. In his examination of fruit juices for carbohydrates, Fitelson[7] found that cherry juice contained sorbitol as well as the more common sugars.

9. Pineapple

Pineapple was one of the few fruits found to contain hydroxyproline on analysing for amino acids[5]. Sugar analysis showed sucrose at 6.6% to be the major contributor[6].

The flavour and aroma characteristics of pineapple have received the attention of several workers[36-41]. Among important contributors to pineapple flavour are allyl hexanoate, ethyl acetate, the methyl and ethyl esters of β-methyl-thiopropionic acid and 2,5-dimethyl-4-hydroxy-3($2H$)-furanone.

10. Strawberry

Most of the fruit acid of strawberry is citric, comprising 940 mg per 100 g fruit[4]. Aspartic acid is the major amino acid component[5].

In his examination of fruit juices for possible adulteration with sugars, Fitelson[7] found that pure strawberry juice did not contain sorbitol; Kline *et al.*[6] found only 0.2% sucrose in the fruit, out of a total of 6.8%; and the bulk of the carbohydrate content of strawberry is evenly divided between glucose and fructose.

In the examination of certain fruits for flavonoids by Duggan[10], already described in this chapter, it was found that strawberry contained kaempferol and quercetin.

Apart from the usual esters, aldehydes and ketones found as fruit volatiles, acetals have been found in strawberries[42-44].

11. Raspberry

Fitelson[7] found that raspberry juice, like strawberry juice, contained no sorbitol and little or no sucrose. Therefore the presence of these sugars in raspberry juice would indicate adulteration.

The tyramine content of raspberry and other fruits was determined by Coffin[45] using GC of the TFA derivative after clean-up of the fruit extract on a cation exchange resin. Fifteen samples of fresh raspberries gave tyramine in the range 12.8–92.5 μg per g fruit, and raspberry jam samples contained this amine up to 38.4 μg per g jam. Relative to other fruits, such as orange, lemon, banana, plum, apricot and avocado, raspberry contains a high concentration of tyramine but the content is too variable for use as a quantitative measure of raspberry content in mixed fruit preparations.

12. Blackberry

Fernandez-Flores *et al.*[5] found that alanine and aspartic acid were the main amino acids in a blackberry sample. Kline *et al.*[6], in their carbohydrate analysis of fruits, found approximately equivalent amounts of glucose and fructose, with no sucrose or sorbitol, in blackberries.

Scanlan, Bills and Libbey[46] analysed commercial blackberry essence for flavour components. The concentrated juice was saturated with salt and extracted with ethyl chloride and GC was carried out on a BDS capillary column. Sixteen volatiles were identified with the aid of MS and included acetals, esters, alcohols, ketones, terpenes and 3,4-dimethoxyallylbenzene, which has a musty odour and is thought to be an important contributor to blackberry flavour.

13. Blueberry

The dominant fruit acid of blueberry is citric[4], and its total amino acid content, at 1.5 mg per 100 g fruit, is remarkably low, aspartic acid being absent[5].

The sugar content of blueberry is similar to that of blackberry, showing that blueberry contains approximately equivalent amounts of glucose and fructose and no sorbitol or sucrose. The presence of these last-named sugars in blueberry juice would therefore indicate adulteration.

14. Cranberry

Cranberry has a relatively high acidity, mainly due to the 1.5% of quinic acid, although malic and citric acids are both present at over 1%. Since this quinic acid content is considerably higher than that in other fruits analysed, this might be used to give an assessment of cranberry in a mixed fruit product. Harvey *et al.*[3] in the determination of some acids in fruit juices did not look for quinic acid but found 0.88% citric acid and 0.83% malic acid in a sample of cranberry juice.

Kline *et al.*[6] have shown that there is a significant excess of glucose over fructose in cranberry, sucrose and sorbitol being absent.

15. Blackcurrant

Blackcurrant concentrate was analysed for volatile constituents by Gasco *et al.*[24], who used a circulation-type evaporator to concentrate blackcurrant juice. A portion of the concentrate was caused to react with 2,4-dinitrophenylhydrazine and the phenylhydrazones were extracted into chloroform for GC analysis. Another portion of the concentrate was saturated with salt and extracted with either *o*-dichlorobenzene, 2-chloroethyl ether or diethylphthalate for analysis. Head space analyses were also performed and the compiled information is given for volatile composition of blackcurrant concentrate using three different stationary phases.

16. Orange

Fernandez-Flores *et al.*[4] and Primo, Sanchez and Alberola[47] have determined the fruit acids in oranges, using GC methods.

Gee[48] examined dried orange powder for amino acids by GC after preparing the trifluoroacetates of the methyl esters. Ten milligrams proline per g powder was the major amino acid found, with aminobutyric acid the second major one at half that concentration. Using the same amino acid derivatives for GC, Alberola and Primo[49] analysed orange juices from a number of fruit varieties and found the major components to be proline, hydroxyproline and aspartic acid. This is in contrast to the findings of Fernandez-Flores *et al.*[5], who did not find hydroxyproline in oranges. Alberola and Primo found no valine, leucine or threonine after examination of a variety of Spanish and American juices and only small amounts of phenylalanine, tyrosine and serine. Citric acid and sucrose, which have both been used to adulterate orange juice, gave small peaks corresponding to some of the amino acids present in the juice. Three unidentified peaks were also found in the sucrose, and if these unidentified substances were always present in sucrose, the authors suggest that this analysis offers a means of detecting adulteration of orange juice by sugar addition.

Orange juice was examined for the presence of phenolic amines by Coffin[11]. The juice was cleaned up on an ion exchange resin and the purified phenolic amines converted to their TFA derivatives for GC. All ten samples of orange juice contained synephrine in the range 12.5−23.4 mg/l and another unidentified phenolic amine at a similar concentration.

When orange juice is prepared, the pasteurisation under inherently acid conditions produces more glucose and fructose by hydrolysis of sucrose. Stepak and Lifshitz[50] analysed Israel citrus juices for glucose, fructose, galactose and sucrose, and pointed out that if an unpasteurised juice contained glucose and fructose in significantly unequal proportion, the hydrolysis of sucrose under the pasteurisation conditions would give a more equal concentration of the two monosaccharides in the fruit product. This could be utilised when adulteration of a fruit juice by sucrose was suspected. A typical chromatogram of the TMS derivatives of the sugars present in Shamouti orange juice is shown in *Figure 13.2.*

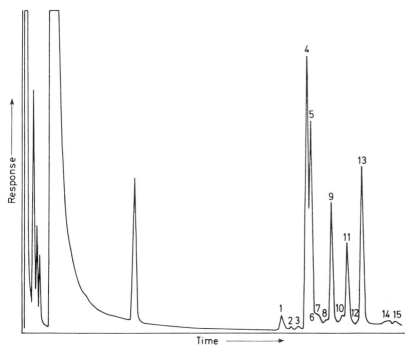

Figure 13.2 Gas chromatogram of silylated sugars in Shamouti orange juice as separated on 5% neopentyl glycol adipate on 60–80 mesh Chromosorb W: 2, rhamnose; 3, xylose; 4 and 5, fructose; 8, galactose; 9, α-glucose; 11, β-glucose; 13, sucrose; 15, trehalose (after Stepak and Liftshitz[50]). For full GC details, see Table 13.1

17. Lemon and lime

Among the fruit acids in lemon and lime there is a predominance of citric acid [3,4]. Fernandez-Flores *et al.*[5] have determined the amino acid content, and Kline *et al.*[6] the carbohydrate content of lemon and lime.

18. Grapefruit

Grapefruit has been analysed for fruit acids[4] and amino acids[5], the dominant acids being citric and aspartic, respectively.

19. Melon and cantaloupe

The analyses of fruit acids[4], amino acids[5] and sugars[6] of both melon and cantaloupe have been recorded by GC methods.

Melon is an important fresh market food in the USA, but little attention has been paid to its aroma constituents. Kemp, Stoltz and Knavel[51] used vacuum distillation followed by GC of the distillate and supported by MS in order to assess the volatile components of musk melon, and found that this fruit is characterised by a series of alcohols and aldehydes, each comprising nine carbon atoms. *cis*-6-nonen-1-ol was found and was previously unreported as a naturally occurring compound.

20. Banana

The analyses of fruit acids[4], amino acids[5] and sugars[6] of banana have been recorded by GC methods.

Banana aroma constituent analysis by GC has received some attention, and McCarthy *et al.*[52] have attempted to correlate volatile constituents with banana flavour. A head space method was used and the volatiles found were placed in one of three categories according to odour: banana-like; fruity; green, woody or musty. The banana-like odiferous volatiles were found to be amyl acetate, propionate and butyrate and also isoamyl acetate.

Issenberg and Wick[53], McCarthy, Wyman and Palmer[54] and Tressl *et al.*[55] are among workers who have studied the composition of banana volatiles.

21. Avocado

The small amount of investigation of avocado composition using GC methods has included amino acid analysis[5] and the estimation of C_{17} oxygenated aliphatic compounds, which contribute to the unpleasant flavour of the unripe fruit[56].

22. Fig

Fig has been the subject of analysis for amino acids[5] and sugars[6] using GC methods.

23. Tomato

Malic, citric and oxalic acids were determined in several fruit juices and vegetables by Harvey *et al.*[3] and included a sample of fresh tomatoes: 5.6% citric acid was present and also 0.5% of both malic and oxalic acids.

Gee[48] analysed a tomato powder for amino acid content via GC of the methyl esters of the trifluoroacetates and found that the powder contained 27% glutamic acid, 13% γ-aminobutyric acid and 13 other amino acids at 6% or less.

The volatile constituents of tomato have been analysed by several workers[57-61].

24. Cucumber

The flavour of cucumbers has been investigated using GC by Forss *et al.*[62] The cucumbers were diced, macerated and vacuum distilled and the volatiles extracted into petroleum ether prior to chromatography with IR spectroscopic examination of fractions to confirm identities. The important flavour constituent was found to be nona-2,6-dienal, which has the typical cucumber flavour. n-Hexanal, hex-2-enal and non-2-enal were carbonyls also found, identities being further confirmed by column chromatographic separation of the 2,4-dinitrophenylhydrazones.

25. Hop

The major GC contribution in the elucidation of the aroma and flavour of hops is by the analysis of the oil (Chapter 12, p. 20).

Hops and hop extracts were analysed for volatile sulphur compounds by Baerwald and Miglio[63]. The sample was held under nitrogen in a sealed ampoule for 30 min at 50°C and the head space analysed by GC using FPD. Ethanethiol, dimethyl sulphide, dimethyl disulphide and diethyl disulphide were detected.

26. Capsicum

The main interest in capsicum, from the GC analytical point of view, is in capsaicin (the vanillyl amide of isodecylenic acid), which is the pungent principle. Capsicum oil was examined for capsaicin by Hollo, Kurncz and Bodor[64], who tested various GLC stationary phases and recommended a low loading of a mixture of NPGS and DEGS on silanised glass powder. The capsaicin content of three oils was in the range 1.3–1.9%.

The GC estimation of capsaicin was performed by Todd and Perun[65] in order that successive batches of paprika would, after blending, contain the same capsaicin concentration and therefore the same pungency. The capsaicin was hydrolysed and the methyl ester of vanillic acid was chromatographed.

Quaglio, Romagnolo and Cavicchi[66] determined capsaicin by prolonged extraction of the dried ripe fruit with dichloromethane. The capsaicin was determined directly by GC with confirmation of results being afforded by silica gel TLC.

Mueller-Stock, Joshi and Buechi[67] made a study of capsaicin and the related compounds dihydrocapsaicin, nordihydrocapsaicin and *N*-(4-hydroxy-3-methoxybenzyl)nonanamide, all found as constituents of capsicum. The powdered fruit was extracted with methanol and the extract evaporated to give an oil which was dissolved in 70% acetic acid. The acid solution was eventually extracted into dichloromethane, cleaned up on activated carbon and then trimethysilylated for GC analysis. The relative proportions of the four compounds in capsicum were 46–77% capsaicin, 20–40% dihydrocapsaicin, 2.4–12% nordihydrocapsaicin and 1.5–4.6% *N*-(4-hydroxy-3-methoxybenzyl)nonanamide. This last-named compound has been added to powdered capsicum as an adulterant and the GC conditions were modified when only this constituent was determined.

C. Vegetables

Less GC study has been made of vegetables than of fruits, and the main interest has concerned the analysis of aroma and flavour constituents, particularly to discover whether or not stored vegetables maintain their original flavour characteristics.

1. Potato, carrot and swede

One of the first applications of a head space sampling method for GC with a dual FI system was made by Buttery and Teranishi[68] in the examination of dehydrated potato, carrot and pear for their aroma constituents. Dehydrated potato was mixed with water, the mixture was heated and the head space was analysed, giving a chromatographic profile of the particular sample. These workers showed that potato or carrot granules stored at 75°F for 1 year in an oxygen atmosphere produced more peaks than a sample stored at −30°F in a nitrogen atmosphere for a similar length of time. This was attributed to the production of off-flavour compounds. The same workers[69] developed their style of examining foods for aroma constituents by head space methods, and considered that the production of a chromatogram of such constituents, which they termed 'aromagram', was a more consistent and less time-consuming method than organoleptic evaluation. Reconstituted dehydrated potato was examined by GC before and after storage, and the hexanal produced by oxidation of linoleate was measured. 2-Methyl propanal plus 2- and 3-methylbutanal were found in potato which had become oxidised and had 'browned', and these aldehydes were thought to be associated with that state.

Using a similar approach, Self[70] analysed the volatiles of potato and swede. Experiments have been conducted to find the changes in potato volatiles on storage by Self, Rolley and Joyce[71], and Gumbmann and Burr[72] specifically analysed potatoes for sulphur-containing compounds.

Although many flavourings can be used in potato crisps, it is important that the combined potato and oil flavour does not deteriorate on storage. Darkening and oil decomposition have been observed in potato crisps which affect flavour, and Dornseifer and Powers[73] have investigated the production of carbonyls in stored crisps. Steam distillation was followed by formation of the carbonyl 2,4-dinitrophenylhydrazones, regeneration of the carbonyls, GC analysis and IR spectroscopic analysis to support the GC retention time identification data. Ethanal, propenal, 2-propanone, n-butanal, 2-pentenal, 2,3-butanedione, 2-hexenal, n-heptenal and 2-heptenal were tentatively identified. It was also found that ethanal and 2,3-butanedione decreased with storage of the potato crisps, whereas 2-propenal increased. Such changes of carbonyls produced by autoxidation of the 35—50% oil in potato crisps were thought to have an important effect on the change in their flavour.

2,5-Dimethylpyrazine has been identified by GC in potato chips as an important volatile constituent by Deck and Chang[74]. The chips were fried and a water slurry was made of them which was vacuum distilled and the distillate was extracted into diethyl ether for GC. IR spectroscopy and MS were used to confirm the identity of the pyrazine, which has a potato chip-like flavour with a threshold of 2 p.p.m. in oil and 1 p.p.m. in water.

When carrots are stored in the presence of oxygen, the β-carotene becomes oxidised, resulting in loss of carrot colour, and the vegetable develops an off-odour described as 'violets'. Ayers *et al.*[75] investigated the problem with GC and identified β-ionone 5,6-epoxide as the major constituent of the off-odour, with smaller amounts of α- and β-ionone.

2. Onion, garlic and chive

The main interest in the application of GC to the analysis of members of the onion family has centred on the aroma composition and on the analysis of sulphur-containing compounds in particular.

Carson and Wong[76] vacuum distilled an aqueous onion slurry and extracted it with isopentane. GC analysis was supported by IR spectroscopy and also by chemical derivatives to classify the volatile components.

The aroma of freeze-dried onion was studied in detail by Behun and Hrivnak[77] using a head space sampling procedure after making an aqueous extract of the onion. The GLC analysis was performed using a stationary phase of 10% bis-(2-ethylhexyl) sebacate.

A comprehensive study of the flavour components of onion oil has been carried out by Brodnitz, Pollock and Vallon[78]. They used GC coupled to MS for the analysis with IR and NMR spectroscopy to support their findings. In an extension of this study, Brodnitz and Pollock[79] identified nine important sulphur-containing compounds, the major one being di-1-propyldisulphide, which was found in the range 30—42% in the 17 oils analysed. The ratio of this compound to the combined di-1-propyldisulphide and 1-propyl-propenyl disulphide has been used to determine whether any other oil similar to onion oil, e.g. garlic oil, had been added or whether there had been addition of any of the various sulphides. Garlic oil was found to be rich in allyl disulphide, and chive oil to have a relatively high concentration of dimethyl disulphide.

Three asymmetric disulphides were found in garlic by Jacobsen *et al.*[80] using the head space of the chopped tissue for GC with IR spectroscopy to confirm identities. Head space sampling and vacuum distillation were used to isolate the volatiles of garlic oil by Bernhard *et al.*[81], who used GC analysis to show that garlic oil contained allyl monosulphide and allyl alcohol.

Because the EC detector is relatively sensitive compared with the FI detector in its response to di- and trisulphides, a dual-channel system, splitting the effluent equally to both detectors, was used in the determination of sulphur-containing components of garlic by Oaks, Hartmann and Dimick[82]. Hexane extracts and also head space samples of garlic were analysed, and the EC to FI response ratios of the di- and trisulphides were significantly greater than unity.

The volatiles from chives have been determined by Wahlroos and Virtanen[83] using GC with MS detection.

3. Cabbage, broccoli and spinach

A survey of the volatiles which boil at less than 120°C in various green and root vegetables was conducted by Self, Casey and Swain[84]. Capillary column GC was used in the survey, and H_2S, acetaldehyde, dimethyl sulphide, methanol and methanethiol were commonly found.

The importance of sulphur-containing compounds as flavour and aroma constituents of cabbage has been confirmed by GC analysis[85,86].

In their determination of organic acids in fruits and vegetables, Harvey *et al.*[3] found that frozen broccoli contained 0.5% malic acid and frozen spinach contained 5.5% oxalic acid.

4. Pea and bean

The volatile components present in the steam above peas in a blancher were analysed by Ralls *et al.*[87] After adsorption on to charcoal, the volatiles were solvent extracted, concentrated by distillation and fractionated by GC prior to analysis of each fraction, using capillary column GC with MS to aid identification. Eleven alcohols, ten acetals, six aldehydes, eight esters and five sulphur-containing compounds were tabulated.

Bengtsson and Bosund[88] studied the formation of volatiles in peas stored at various temperatures, and Murray *et al.*[89] have examined unblanched frozen green peas for alcohols, which are important flavour contributors. Pyrazines, also important flavour constituents, were analysed in these vegetables[90].

The inositol content of Californian small white and pinto beans was determined by Seifert[91], using the hexa-acetate derivative for GC after reaction with acetic anhydride and toluene-4-sulphonic acid. Average amounts of 71 and 215 p.p.m. inositol were found in the small white and pinto beans, respectively.

5. Mushroom

Holtz[92] analysed mushrooms for sugar content, using GC of the TMS derivatives. Mushrooms were homogenised in 20% aqueous methanol, filtered and freeze-dried and the derivatives made. The major component was mannitol at a concentration of 11.5 mg per g mushroom with 2.2 mg per g mushroom of total glucose. Fructose and sucrose were present at approximately 0.5 mg per g mushroom.

Some important flavour constituents have been determined in dried mushroom by Thomas[93], and those volatiles which develop on heating have been investigated by Picardi and Issenberg[94].

D. Non-alcoholic beverages

The non-alcoholic beverages have been included in this chapter because they are essentially of vegetable origin and much of the application of GC to their analysis arises from the flavour and aroma composition of the beans or leaves.

There is great commercial interest, particularly in coffee and tea, and many attempts have been made to discover the best way to put quality control on a more scientific basis than organoleptic testing. The combinations of GC with either MS or IR spectroscopy have been the most common techniques applied in the analysis of the volatile constituents of non-alcoholic beverages, and where the problems and data accruing become too complex, e.g. total coffee volatiles as a means of defining geographical origin, the attachment of a computer becomes a necessity.

1. Cocoa

Several workers have studied the flavour and aroma composition of roasted and unroasted cocoa beans, not only because of their use as the basis for a beverage but also because the nib portion of the bean is an essential ingredient of chocolate couverture.

Bailey *et al.*[95] examined the cocoa varieties Bahia, Sanchez, Accra, Arriba and Trinidad, and found that they contained the same volatiles but in different ratios. Roasted beans yielded more of the same volatiles than did unroasted ones, and the major component was isovaleraldehyde, which was found at 42 mol % in roasted Bahia beans. Isobutyraldehyde and propionaldehyde were major components at concentrations of 15.4 and 13.0 mol %, respectively, and *Figure 13.3* shows the increase in the concentration of these aldehydes with increase in roasting temperature. Head space samples were analysed by GC with MS to aid identification.

Head space samples were also taken by van Praag, Stein and Tibbetts[96] in an examination of roasted cocoa nibs for aroma components. Steam distillation was another method of isolation, and the distillate was divided into basic, acidic and neutral fractions. Using these two approaches, a total of 56 compounds were identified by GC using MS to aid identification, and the most important aroma compounds were acetaldehyde, isobutyraldehyde, isovaleraldehyde, benzaldehyde, phenylacetaldehyde, 5-methyl-2-phenyl-2-hexenal, 2-furaldehyde, methyl disulphide, acetic acid, isopentyl acetate and 11 alkyl-substituted pyrazines. The cocoa nib is the part of the bean which is used to give the liquor for chocolate and therefore these volatiles are important to the flavour of that confectionery. The precursors of chocolate aroma have been investigated by Rohan[97] using GC. He found that aroma compounds increased with time of fermentation of cocoa beans and reached a maximum concentration after 4—5 days.

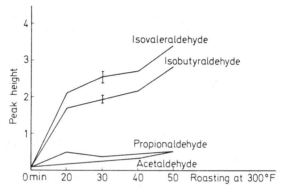

Figure 13.3 Change of gas chromatogram peak heights of four volatile components of cocoa beans with roasting period (after Bailey et al.[95])

Among many references to the application of GC to aroma constituent analysis of cocoa, the work of Marion *et al.*[98], van der Wal *et al.*[99] and Flament, Willhalm and Stoll[100] is noteworthy.

In the last few years the importance of pyrazines in heat-treated foods has been recognised for their contribution to flavour and aroma. Therefore roasted cocoa beans were studied for the presence of these aroma constituents by Reineccius, Keeney and Weissberger[101]. Ground roasted beans were mixed with Celite in a column and extracted into diethyl ether and this extract was shaken with decinormal acid solution to remove the pyrazines. These were made insoluble by making the solution slightly alkaline and the pyrazines were dissolved in dichloromethane for GC analysis. Two stationary phases plus MS identification were used, and the two largest components were found to be tetramethyl and trimethyl pyrazines. *Figure 13.4* shows a chromatogram of pyrazines found in roasted Ghanaian cocoa beans. Reineccius *et al.* examined several varieties of roasted beans and found that the total pyrazine concentration lay in the range 142—698 mg per 100 g bean. In fermented beans pyrazine concentration increased rapidly during roasting to reach a maximum which did not change with extended roasting. As pyrazines were formed, amino acid and sugar concentrations decreased, although some amino acids were present at the end of roasting, thus indicating that it was the sugar concentration which was the decisive factor in pyrazine production. Tetramethyl pyrazine was the only pyrazine found in unroasted beans. Coupled with this investigation of pyrazine concentration, Reineccius *et al.*[102] determined the sugars in cocoa beans. Twenty per cent aqueous methanol was used to extract the sugars and lead acetate was added to precipitate polyhydric phenols. The TMS derivatives were made for GC analysis, and fructose, glucose, sorbose, sucrose, inositol, mannitol and two unidentified sugars were found. Relative concentrations varied with beans of different geographical origin. Fermented varieties of cocoa bean contained 2–16 times more fructose than glucose, the preferential consumption of the latter during fermentation having an important effect on the development of chocolate flavour. Unfermented Sanchez beans contained 1% sucrose, whereas fermented Bahia and Ghana types contained 0.05–0.12%. This work is a good example of the use of MS in classifying compounds followed by GC to identify, by retention times, individuals within each class.

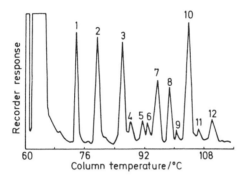

Figure 13.4 Gas chromatographic separation (adipate column) of alkylpyrazines in the basic fraction of Ghana cocoa beans (roasted at 150°C for 30 min). 1: Pyrazine (internal standard); 2: acetoin and methylpyrazine; 3: mixture of 2,5-dimethyl, 2,6-dimethyl and ethyl pyrazines; 4: 2,3-dimethylpyrazine; 5: 2-ethyl-6-methylpyrazine; 6: 2-ethyl-5-methylpyrazine; 7: trimethylpyrazine; 8: 2,5-dimethyl-3-ethylpyrazine; 9: 2,3-dimethyl-6-ethylpyrazine; 10: tetramethylpyrazine; 11 and 12: contamination. (after Reineccius et al.[101]). For GC conditions see Table 13.1

2. Coffee

Since the production of coffee is of great commercial importance to countries where it can be satisfactorily cultivated, it is not surprising that a great deal of research has been conducted into the flavour and aroma of different varieties of roasted and unroasted coffees. With the advent of GC the analysis of these flavour constituents became easier, and when coupled to MS and computers to sort out and evaluate the complex nature of the chromatograms, most of the compounds responsible for the unique aroma of coffee were deduced.

As early as 1960, Rhoades[103] extracted seven trade varieties of coffee and found 19 volatiles. He conducted experiments using different temperatures of roasting and found that the ratio of biacetyl to acetyl propionyl increased almost linearly with increase of roasting temperature; this is shown in *Figure 13.5*. Rhoades suggested that it might be possible to use this ratio to determine the degree of roast and that a ratio of less than 1:1 was typical of a light roast,

227

one of between 1:1 and 1:3 was typical of a medium roast, and a ratio greater than 1:3 was typical of a dark roast. H_2S increased in concentration as the roasting temperature was raised to 350°F, but decreased thereafter.

Green bean volatiles were also determined and were found to be the same as those of roasted beans but in lesser concentration. Data were also given for the volatiles found in fresh coffee, 6 day old coffee and instant coffee showing diminished concentrations with age or processing.

Several other workers have analysed coffee for its aroma constituents by GC methods[104-108].

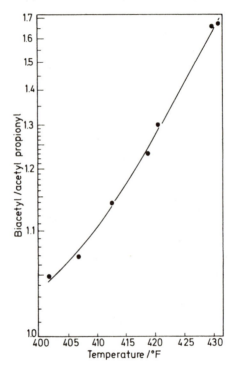

Figure 13.5 Relationship between degree of roast and the ratio of biacetyl to acetyl propionyl (after Rhoades[103])

Powers and Keith[109] prepared steam distillates of four coffees, each with different flavour characteristics, and analysed them by GC. The calculation of ratios of peak heights was achieved by computer programming using 80 variables. This discriminant analysis procedure was established to calculate important ratios in the first instance and subsequently to calculate those ratios which would yield less fruitful information. The complex chromatograms obtained in aroma research are best analysed by this approach, particularly if a magnetic tape feed is incorporated between the GC and computer. Biggers, Hilton and Gianturco[110] used a similar method when differentiating between the two varieties of coffee *Coffea Arabica* and *Coffea robusta*. The finest coffee is reputed to be *Coffea Arabica*, which is in short supply relative to *Coffea robusta*, and therefore the two varieties are blended. Although tasting is the method of evaluating the blends, these workers consider that the GC profiles, albeit complex, can be used in conjunction with organoleptic tests, provided that the GC information can be analysed by computer. A chromatogram is shown with 404 peaks or shoulders, indicating the complexity of coffee aroma, but 32 were chosen as the most significant, and with computer discriminant analysis of peak ratios, it was possible to deduce the geographical origins of coffee, which can be subsequently used in blending.

Although caffeine is normally determined using classical chemical methods, its presence in coffee can be rapidly estimated by the method of Vitzthum, Barthels and Kwasny[111], who utilised a nitrogen-sensitive detector.

The isostatic permeation of volatile aroma constituents of food through packaging films was studied by Weintjes, Maarse and van Straten[112]. They applied their study to the diffusion of the aroma constituents of roasted coffee through polyethylene film, by concentrating and condensing the permeating gases and then heating the condensate to produce volatiles for GC analysis.

3. Tea

As with cocoa and coffee, the aroma and flavour constituents of tea have received attention of the gas chromatographer[106,107,113,114].

Green tea is of considerable commercial importance in Japan and China, and its flavour character is determined to some extent by the presence of flavanols. Pierce *et al.*[115] extracted iced powdered tea with pyridine and formed the TMS derivatives of catechin, epicatechin, epigallocatechin, epicatechin gallate and epigallocatechin gallate, which were the flavanols present in fresh green leaf and in Japanese green tea. The chromatograms were determined isothermally at two different temperatures and showed that fresh green leaf contained a total of 19% flavanols, with epigallocatechin gallate the major contributor at 10%. The Japanese tea sample contained 14% total flavanols, of which 7.3% came from epigallocatechin gallate.

Flavanols have also been determined in green and black tea by Collier and Mallows[116,117].

Caffeine was quantitatively determined in instant tea by Newton[118]. The presence of numerous free amino acids resulting from protein breakdown makes it necessary to use basic and acidic column clean-up procedures. The caffeine is finally extracted into chloroform and GC carried out using a potassium chloride tipped AFID which, in being capable of detecting 1 ng caffeine, is 100 times as sensitive as the FID. The method agreed well with the standard UV spectroscopic method and, in a collaborative study involving seven laboratories, three brands of instant tea gave average results of 3.05, 4.40 and 5.94% caffeine.

4. Vanilla

Vanilla is a substance obtained from the pods of the Mexican orchid *Vanilla planifolia,* and alcoholic extracts are made from it, producing vanilla essence, which is a widely used flavouring. The quality of the essence depends to a large extent on its vanillin content, which has been determined by Martin, Feeny and Scaringelli[119]. They separated vanillin and also ethyl vanillin by GC and found that the vanillin content of nine samples was in the range 0.11 — 0.44 g per 100 ml vanilla essence. The presence of ethyl vanillin would indicate adulteration, but none was found in the nine samples. More recently Schlack and Dicecco[120] have produced a method for the determination of vanillin in beverages and other foods.

There is a possibility that adulteration of vanilla essence could occur by the addition of combinations of the following compounds: *p*-cresol, 4-phenylbut-3-en-2-one, γ-octanolactone, dihydrocoumarin, vanitrope and ethyl anisate. These volatiles can enhance the flavour of vanilla, and Potter[121] used a GC method to detect their presence in vanilla essence. The main analytical problems were the large concentrations of ethanol and vanillin. Isopentane was used as extractant and GC was carried out on a non-polar phase to minimise the interference of these two major volatiles.

As early as 1959, Burchfield and Paul[122] examined vanilla extracts for non-carbonyl volatiles. The vanilla extracts were treated with 2,4-dinitrophenyl-hydrazine to precipitate the carbonyls, including the major component, vanillin. Isopentane was used to extract the remaining volatiles, and GC fractions were compared in a qualitative study of different varieties of vanilla bean.

References

1. Hautala, E., *J. Ass. Off. Anal. Chem.,* **49,** 619 (1966)
2. Hautala, E., *J. Ass. Off. Anal. Chem.,* **50,** 287 (1967)
3. Harvey, W. R., Hale, R. W. and Ikeda, R. M., *Tobacco Sci.,* **14,** 141 (1970)
4. Fernandez-Flores, E., Kline, D. A. and Johnson, A. R., *J. Ass. Off. Anal. Chem.,* **53,** 17 (1970)
5. Fernandez-Flores, E., Kline, D. A., Johnson, A. R. and Leber, B. L., *J. Ass. Off. Anal. Chem.,* **53,** 1203 (1970)
6. Kline, D. A., Fernandez-Flores, E. and Johnson, A. R., *J. Ass. Off. Anal. Chem.,* **53,** 1198 (1970)
7. Fitelson, J., *J. Ass. Off. Anal. Chem.,* **53,** 1193 (1970)
8. Zamorani, A., Roda, G. and Lanzarini, G., *Ind. Agrar.,* **9,** 35 (1971)
9. Meigh, D. F., *J. Sci. Food Agric.,* **11,** 381 (1960)
10. Duggan, M. B., *J. Ass. Off. Anal. Chem.,* **52,** 1038 (1969)
11. Coffin, D. E., *J. Ass. Off. Anal. Chem.,* **52,** 1044 (1969)
12. Grevers, G. and Doesburg, J. J., *J. Food Sci.,* **30,** 412 (1965)
13. Jakob, M. A., Hippler, R. and Luethi, H. R., *Mitt. Geb. Lebensmitt. Hyg.,* **60,** 223 (1969)
14. Nishimura, K. and Hirose, Y., *Agric. Biol. Chem. Japan,* **28,** 1 (1964)
15. Paillard, N., *Fruits,* **20,** 189 (1965)
16. Paillard, N., *Fruits,* **23,** 385 (1968)
17. Drawert, F., Heimann, W., Emberger, R. and Tressl, R., *Chromatographia,* **2,** 57 (1969)
18. Drawert, F., Heimann, W., Emberger, R. and Tressl, R., *Chromatographia,* **2,** 77 (1969)
19. Mueller, W., *Wiss. Z. Tech. Univ., Dresden,* **12,** 589 (1963)
20. MacGregor, D. R., Sugisawa, H. and Matthews, J. S., *J. Food Sci.,* **29,** 448 (1964)
21. Koch, J. and Schiller, H., *Z. Lebensmitt. Forsch.,* **125,** 364 (1964)
22. Barrera, R. P., Gasco, L. S. and de la Cruz, F. C., *An. R. Soc. Esp. Fis. Quim.,* **64,** 517 (1968)

23. Pribela, A., *Prumysl Potravin*, **18**, 51 (1967)
24. Gasco, L. S., Barrera, R. P. and de la Cruz, F. C., *J. Chromat. Sci.*, **7**, 228 (1969)
25. Jennings, W. G. and Sevenants, M. R., *J. Food Sci.*, **29**, 158 (1964)
26. Heinz, D. E., Creveling, R. K. and Jennings, W. G., *J. Food Sci.*, **30**, 641 (1965)
27. Heinz, D. E. and Jennings, W. G., *J. Food Sci.*, **31**, 69 (1966)
28. Heinz, D. E., Sevenants, M. R. and Jennings, W. G., *J. Food Sci.*, **31**, 63 (1966)
29. Romani, R. J. and Ku, L., *J. Food Sci.*, **31**, 558 (1966)
30. Mattick, L. R., Robinson, W. B., Weirs, L. D. and Barry, D. L., *J. Agric. Food Chem.*, **11**, 334 (1963)
31. Li, K. C. and Woodroof, J. G., *J. Agric. Food Chem.*, **16**, 534 (1968)
32. Jennings, W. G. and Sevenants, M. R., *J. Food Sci.*, **29**, 796 (1964)
33. Sevenants, M. R. and Jennings, W. G., *J. Food Sci.*, **31**, 81 (1966)
34. Molina, .P., Soler, A. and Cambronero, J., *An. Bromatol.*, **26**, 51 (1974)
35. Flynn, C. and Wendt, A. S., *J. Ass. Off. Anal. Chem.*, **53**, 1067 (1970)
36. Connell, D. W., *Aust. J. Chem.*, **17**, 130 (1964)
37. Rodin, J. O., Himel, C. M., Silverstein, R. M., Leeper, R. W. and Gortner, W. A., *J. Food Sci.*, **30**, 280 (1965)
38. Silverstein, R. M., Rodin, J. O., Himel, C. M. and Leeper, R. W., *J. Food Sci.*, **30**, 668 (1965)
39. Rodin, J. O., Coulson, D. M., Silverstein, R. M. and Leeper, R. W., *J. Food Sci.*, **31**, 721 (1966)
40. van den Dool, H., Hansen, A. and van der Puijl, I., *Z. Lebensmitt. Forsch.*, **138**, 272 (1968)
41. Working Party No. IV of the Commission of Experts of the Association Internationale des Fabricants de Confiserie, *Choc. Confis. Fr.*, **280**, 7 (1972)
42. McFadden, W. H. and Teranishi, R., *Nature*, **200**, 329 (1963)
43. McFadden, W. H., Teranishi, R., Corse, J., Black, D. R. and Mon, T. R., *J. Chromatog.*, **18**, 10 (1965)
44. Teranishi, R., Corse, J. W., McFadden, W. H., Black, D. R. and Morgan, A. I., *J. Food Sci.*, **28**, 478 (1963)
45. Coffin, D. E., *J. Ass. Off. Anal. Chem.*, **53**, 1071 (1970)
46. Scanlan, R. A., Bills, D. D. and Libbey, L. M., *J. Agric. Food. Chem.*, **18**, 744 (1970)
47. Primo, E. Y., Sanchez, J. P. and Alberola, J., *Rev. Agroquim. Technol. Alimentos*, **3**, 349 (1963)
48. Gee, M., *Anal. Chem.*, **39**, 1677 (1967)
49. Alberola, J. and Primo, E., *J. Chromat. Sci.*, **7**, 56 (1969)
50. Stepak, Y. and Lifshitz, A., *J. Ass. Off. Anal. Chem.*, **54**, 1215 (1971)
51. Kemp, T. R., Stoltz, L. P. and Knavel, D. E., *J. Agric. Food Chem.*, **20**, 196 (1972)
52. McCarthy, A. I., Palmer, J. K., Shaw, C. P. and Anderson, E. E., *J. Food Sci.*, **28**, 379 (1963)
53. Issenberg, P. and Wick, E. L., *J. Agric. Food Chem.*, **11**, 2 (1963)
54. McCarthy, A. I., Wyman, H. and Palmer, J. K., *J. Gas Chromat.*, **2**, 121 (1964)
55. Tressl, R., Drawert, F., Heimann, W. and Emberger, R., *Z. Naturf.*, **24b**, 781 (1969)
56. Brown, B. I., *J. Chromatog.*, **86**, 239 (1973)
57. Schormueller, J. and Grosch, W., *Z. Lebensmitt. Forsch*, **118**, 385 (1962)
58. Schormueller, J. and Grosch, W., *Z. Lebensmitt. Forsch*, **126**, 188 (1965)
59. Pyne, A. W. and Wick, E. L., *J. Food Sci.*, **30**, 192 (1965)
60. Buttery, R. G. and Seifert, R. M., *J. Agric. Food Chem.*, **16**, 1053 (1968)
61. Schormueller, J. and Kochmann, H-J., *Z. Lebensmitt. Forsch.*, **141**, 1 (1969)
62. Forss, D. A., Dunstone, E. A., Ramshaw, E. H. and Stark, W., *J. Food Sci.*, **27**, 90 (1962)
63. Baerwald, G. and Miglio, G., *Mschr. Brau.*, **23**, 288 (1970)
64. Hollo, J., Kuruez, E. and Bodor, J., *Lebensmitt.-Wiss. Technol.*, **2**, 19 (1969)
65. Todd, P. H. and Perun, C., *Food Technol.*, **15**, 270 (1961)
66. Quaglio, M. P., Romagnolo, M. and Cavicchi, G. S., *Farmaco, Ed. Prat.*, **26**, 349 (1971)
67. Mueller-Stock, A., Joshi, R. K. and Buechi, J., *J. Chromatog.*, **63**, 281 (1971)
68. Buttery, R. G. and Teranishi, R., *Anal. Chem.*, **33**, 1439 (1961)
69. Buttery, R. G. and Teranishi, R., *J. Agric. Food Chem.*, **11**, 504 (1963)

Fruits, Vegetables and Non-alcoholic Beverages

70. Self. R., in *Recent Advances in Food Science* (J. M. Leitch and D. N. Rhodes, Eds.), Butterworths, London, 170 (1963)
71. Self, R., Rolley, H. L. J. and Joyce, A. E., *J. Sci. Food Agric.*, **14**, 8 (1963)
72. Gumbmann, M. R. and Burr, H. K., *J. Agric. Food Chem.*, **12**, 404 (1964)
73. Dornseifer, T. P. and Powers, J. J., *Food Technol.*, **17**, 1330 (1963)
74. Deck, R. E. and Chang, S. S., *Chem. & Ind.*, 1343 (1965)
75. Ayers, J. E., Fishwick, M. J., Land, D. G. and Swain, T., *Nature*, **203**, 81 (1964)
76. Carson, J. F. and Wong, F. F., *J. Agric. Food Chem.*, **9**, 140 (1961)
77. Behun, M. and Hrivnak, J., *Prumysl Potravin*, **16**, 658 (1965)
78. Brodnitz, M. H., Pollock, C. L. and Vallon, P. P., *J. Agric. Food Chem.*, **17**, 760 (1969)
79. Brodnitz, M. H. and Pollock, C. L., *Food Technol.*, **24**, 78 (1970)
80. Jacobsen, J. V., Bernhard, R. A., Mann, L. K. and Saghir, A. R., *Arch. Biochem. Biophys.*, **104**, 473 (1964)
81. Bernhard, R. A., Saghir, A. R., Jacobsen, J. V. and Mann, L. K., *Arch. Biochem. Biophys.*, **107**, 137 (1964)
82. Oaks, D. M., Hartmann, H. and Dimick, K. P., *Anal. Chem.*, **36**, 1560 (1964)
83. Wahlroos, O. and Virtanen, A. L., *Acta Chem. Scand.*, **19**, 1327 (1965)
84. Self, R., Casey, J. C. and Swain, T., *Chem. & Ind.*, 863 (1963)
85. Bailey, S. D., Bazinet, M. L., Driscoll, J. L. and McCarthy, A. I., *J. Food Sci.*, **26**, 163 (1961)
86. MacLeod, A. J. and MacLeod, G., *J. Sci. Food Agric.*, **19**, 273 (1968)
87. Ralls, J. W., McFadden, W. H., Seifert, R. M., Black, D. R. and Kilpatrick, P. W., *J. Food Sci.*, **30**, 228 (1965)
88. Bengtsson, B. and Bosund, I., *Food Technol.*, **18**, 773 (1964)
89. Murray, K. E., Shipton, J., Whitfield, F. B., Kennett, B. H. and Stanley, G., *J. Food Sci.*, **33**, 290 (1968)
90. Murray, K. E., Shipton, J. and Whitfield, F. B., *Chem. & Ind.*, 897 (1970)
91. Seifert, R. M., *J. Ass. Off. Anal. Chem.*, **55**, 1194 (1972)
92. Holtz, R. B., *J. Agric. Food Chem.*, **19**, 1272 (1971)
93. Thomas, A. F., *J. Agric. Food Chem.*, **21**, 955 (1973)
94. Picardi, S. M., and Issenberg, P., *J. Agric. Food Chem.*, **21**, 959 (1973)
95. Bailey, S. D., Mitchell, D. G., Bazinet, M. L. and Weurman, C., *J. Food Sci.*, **27**, 165 (1962)
96. van Praag, M., Stein, H. S. and Tibbetts, M. S., *J. Agric. Food Chem.*, **16**, 1005 (1968)
97. Rohan, T. A., *J. Food Sci.*, **32**, 402 (1967)
98. Marion, J. P., Mueggler-Chavan, F., Viani, R., Bricout, J., Reymond, D. and Egli, R. H., *Helv. Chim. Acta*, **50**, 1509 (1967)
99. van der Wal, B., Sipma, G., Kettenes, D. K. and Semper, A. T. J., *Rec. Trav. Chim. Pays-Bas Belg.*, **87**, 238 (1968)
100. Flament, I., Willhalm, B. and Stoll, M., *Helv. Chim. Acta*, **50**, 2233 (1967)
101. Reineccius, G. A., Keeney, P. G. and Weissberger, W., *J. Agric. Food Chem.*, **20**, 202 (1972)
102. Reineccius, G. A., Andersen, D. A., Kavanagh, T. E. and Keeney, P. G., *J. Agric. Food Chem.*, **20**, 199 (1972)
103. Rhoades, J. W., *J. Agric. Food Chem.*, **8**, 136 (1960)
104. Viani, R., Mueggler-Chavan, F., Reymond, D. and Egli, R. H., *Helv. Chim. Acta*, **48**, 1809 (1965)
105. Weeren, R. D., Werner, H. and Wurziger, J., *Dt. Lebensmitt.-Rdsch.*, **62**, 13 (1966)
106. Heins, J. T., Maarse, H., ten Noever de Brauw, M. C. and Weurman, C., *J. Gas Chromat.*, **4**, 395 (1966)
107. Reymond, D., Mueggler-Chavan, F., Viani, R., Vuataz, L. and Egli, R. H., *J. Gas Chromat.*, **4**, 28 (1966)
108. Stoffelsma, J. and Pypker, J., *Rec. Trav. Chim. Pays-Bas Belg.*, **87**, 241 (1968)
109. Powers, J. J. and Keith, E. S., *J. Food Sci.*, **33**, 207 (1968)
110. Biggers, R. E., Hilton J. J. and Gianturco, M. A., *J. Chromat. Sci.*, **7**, 453 (1969)
111. Vitzthum, O. G., Barthels, M. and Kwasny, H., *Z. Lebensmitt. Forsch.*, **154**, 135 (1974)
112. Wientjes, A. G., Maarse, H. and van Straten, S., *Mitt. Geb. Lebensmitt. Hyg.*, **58**, 61 (1967)

113. Brandenberger, H. and Mueller, S., *J. Chromatog.,* **7,** 137 (1962)
114. Wickremasinghe, R. L. and Swain, T., *J. Sci. Food Agric.,* **16,** 57 (1965)
115. Pierce, A. R., Graham, H. N., Glassner, S., Madlin, H. and Gonzalez, J. G., *Anal. Chem.,* **41,** 298 (1969)
116. Collier, P. D. and Mallows, R., *J. Chromatog.,* **57,** 29 (1971)
117. Collier, P. D. and Mallows, R., *J. Chromatog.,* **57,** 19 (1971)
118. Newton, J. M., *J. Ass. Off. Anal. Chem.,* **52,** 653 (1969)
119. Martin, G. E., Feeny, F. J. and Scaringelli, F. P., *J. Ass. Off. Agric. Chem.,* **47,** 561 (1964)
120. Schlack, J. E. and Dicecco, J. J., *J. Ass. Off. Anal. Chem.,* **57,** 329 (1974)
121. Potter, R. H., *J. Ass. Off. Anal. Chem.,* **54,** 39 (1971)
122. Burchfield, H. P. and Paul, E. A., *Contributions from Boyce Thompson Institute,* **20,** 217 (1959)

14

Alcoholic Beverages and Vinegar

A. Introduction

Ever since fermented liquor was produced, the prime interest has been its intoxicating effect. This has generated a desire to know what compound or compounds were responsible for this effect, and so the determination of ethanol in alcoholic beverages has become the most important single analytical feature.

In general, the measurement of ethanol in an alcoholic beverage has depended either on the difference in specific gravity of ethanol and the beverage which contains it or on the oxidation of the ethanol to acetaldehyde or acetic acid. More recently ethanol has been separated and determined by GC, and the specificity of the technique has probably produced a more accurate assessment of ethanol content than the older methods.

Apart from the differences in ethanol content between the various types of beverage, there are obvious differences in flavour and aroma. Prior to GC, the analysis of flavour and aroma volatiles was an extremely difficult and tedious operation and the analyst had to be content with the total concentration of a particular organic class, e.g. total fusel oil or total esters. The separation and determination of the various alcohols which make up fusel oil has been achieved by GC and this has been of particular assistance to the distiller. The isolation of the whole range of volatiles present in beer and wort, and especially the sulphur-containing compounds and hop oil constituents, has helped the brewery chemist to understand more fully the changes that take place during the fermentation process. The vintner, too, has used GC in numerous ways, including the study of the flavour changes which occur during maturation of wine.

It must be pointed out that GC has not replaced organoleptic testing of alcoholic beverages, but GC evidence is used alongside the findings of the taste panel.

Sugars and amino acids are precursors of fusel oil and other flavour components of alcoholic beverages, and therefore, in this context, their determination in beer wort or in wine must is often desired. The GC analysis of sugars via their TMS derivatives and of amino acids by their N-TFA methyl esters are most satisfactory ways of achieving this.

234

Oxidation, alkaline oxidation, iodine and ester values provide some of the data in the assessment of vinegar types and their quality. Some of the substances which contribute to these values have been isolated and determined by GC methods, e.g. ethyl acetate and biacetyl.

Some of the more important applications of GC to the analysis of alcoholic beverages and vinegars are given in *Table 14.1*.

B. Beer and wort

Beer is the fermented liquor prepared from malted barley and often flavoured with hops. Wort is the name given to the liquor while it is undergoing fermentation. Through the ages the prime analytical interest in beer has been its ethanol content, the determination of which has been made, in recent times, by distillation and specific gravity measurement methods. Having established the conditions, the GC determination of ethanol in beer or wort is a relatively simple matter, since no preparation of the sample is necessary. Silbereisen *et al.*[1] determined ethanol and also carbon dioxide in beer using direct injection on to a Porapak Q column. This column material was also used by Trachman[2], who compared the GSC method with that of the established distillation and gravity method and obtained good agreement. Butanol was used as internal standard and, by using temperature programming, the last peak was eluted after 12.5 min, thus providing a quick method. Lie, Haukeli and Gether[3] also used Porapak Q as stationary phase-cum-support when determining ethanol in beer. Equal volumes of degassed beer and 5% aqueous acetone were mixed and injected on to the chromatograph, which was operated isothermally at 135°C.

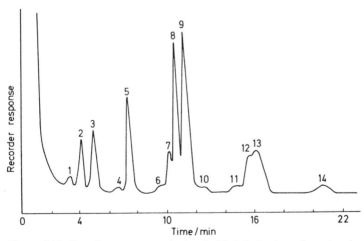

Figure 14.1 Gas chromatogram of trimethylsilyl derivatives of wort carbohydrates and added phenyl-β-D-glucopyranoside: 1, α-fructose; 2, α-glucose; 3, β-glucose; 4, phenyl-α-D-glucopyranoside (impurity); 5, phenyl-β-D-glucopyranoside; 6, unidentified; 7, sucrose; 8, α-maltose; 9, β-maltose; 10, isomaltose; 11, unidentified; 12, α-maltotriose; 13, β-maltotriose; 14, maltotetraose (after Clapperton and Holiday[5]). For GC conditions, see Table 14.1

Table 14.1 GC conditions used in the analysis of alcoholic beverages and vinegar

Class of compound	Ref.	Column dimensions	Stationary phase	Support material	Carrier gas and conditions	Temperature or temperature programme/°C	Detector
Methanol	89	1.5m x 4mm I.D.	100/120 Porapak	100/120 Porapak	N_2, 15ml/min	between 70 and 125	FI
Ethanol	2	6ft x 0.19in	50/80 Porapak Q	50/80 Porapak Q	He, 60ml/min	100–120 at 10/min	FI
Fusel oil, ethyl acetate	99	40ft x 0.125in O.D.	3% Carbowax 400	100/120 Chromosorb W	He, 34ml/min	82	FI
2-Phenylethanol	63	1.22m x 4.5mm I.D.	0.2% Apiezon L	60 mesh glass microbeads	N_2/H_2, 1/1, 15ml/min	80	FI
Glycerol	9	6ft x 0.25in O.D.	80/120 Par 1	80/120 Par 1	He, 60ml/min	195	FI
Sorbitol acetate	133	6ft x 4mm I.D.	10% DC-200	100/120 Gas Chrom Q	N_2, 120ml/min	200	FI
TMS derivatives of sugars	5	3ft x 0.125in O.D.	1.5% silicone gum rubber	80/100 AW, DMCS Chromosorb W	N_2, 25ml/min	100–350 at 8/min	FI
TMS derivatives of sugars	8	1m x 0.125 in	3% SE-52	60/80 silanised Chromosorb W	N_2, 30ml/min	150–350 at 4/min	FI
N-TFA methyl esters of amino acids	12	10ft x 0.25in	0.75% Carbowax 1540 plus 0.25% dimethyl-propane-1,3-diol succinate	Chromosorb W	N_2, 40ml/min	80–200 at 4/min	FI

Aldehydes	11	6ft x 0.25in	20% Carbowax 4000	60/80 Chromosorb W	He, 75ml/min	75	FI
Vicinal diketones	37	3m x 0.125 in	10% Carbowax 20M	60/80 AW, DMCS Chromosorb W	N_2, 20ml/min	70	EC
TMS derivatives of acids	70	6ft x 0.25in O.D.	3.8% SE-30	Diatoport S	He, 50ml/min	130	FI
TMS derivatives of phenolic acids	17	5ft x 0.125in	3% OV-1	60/80 Chromosorb G-HP	N_2, 65ml/min	70–110 at 4/min 110–280 at 6/min	FI
Esters	79	50ft x 0.02in I.D.	UCON LB550X	capillary	He, 7ml/min	32 for 3 min 32–150 at 2/min	FI
Ethyl esters of fatty acids	107	6ft x 0.125in O.D.	10% FFAP	60/80 AW, DMCS Chromosorb G	He, 60ml/min	100–225 at 7.5/min	FI
General	117	10ft x 0.125in O.D. 10ft x 0.125in O.D.	10% Tergitol NPX 10% Surfonic N-300	100/120 AW Chromosorb W 100/120 AW Chromosorb W	He, 38ml/min	60–135 at 3/min	FI
Sulphur-containing compounds	56	20ft x 0.25in O.D.	10% Triton X-305	80/100 AW DMCS Chromosorb G	N_2, 60ml/min	40–150 at 6/min	FP
Carbon dioxide	86	9ft x 0.125in O.D.	60/80 charcoal	60/80 charcoal	He, 50ml/min	40	TC

237

Although the presence of methanol in beer is not the problem it can be in distilled alcoholic beverages, a collaborative study of a GC method based on that applied by the AOAC for fusel oil determination was organised by Dyer[4]. Samples of beer containing between 0.01 and 0.35% methanol were sent to the collaborators and good agreement was obtained.

Before the advent of GC, it was impossible to quantitatively analyse all the sugars present in beer and wort. Since sugars are one of the basic ingredients of alcoholic beverages, any technique which can differentiate and determine the individual members in the starting material, the fermenting liquor and the final product is of importance to the brewery chemist. A sugar analysis of worts and beers was made by Clapperton and Holliday[5] using TMS derivatives. *Figure 14.1* shows a chromatogram of the TMS derivatives of wort sugars with phenyl-β-D-glucopyranoside as internal standard. Fructose, glucose, sucrose, maltose and maltotriose figures for five worts showed that the dominant sugar was maltose, which lay in the range 2.68—4.80 g per 100 ml wort. The importance of the distribution of the sugars in wort on beer quality was emphasised by Tuning[6], who determined the proportions of fructose, glucose, maltose, maltotriose, maltotetraose, maltopentaose and sucrose in wort samples as their TMS derivatives. The same derivatives were used by Marinelli and Whitney[7] in the examination of wort, beer and brewing syrup for carbohydrates. Wort contained maltose and maltotriose as major sugar constituents; the maltose content decreased as fermentation proceeded; and the major sugar constituents of the final beer were maltotriose and maltotetraose. Similar results were obtained by Otter and Taylor[8], who also examined the sweeter types of beer in which the added sucrose becomes inverted to give a significant glucose plus fructose content. These workers listed the sugar compositions of various ales, stouts and lagers.

Feil and Marinelli[9] determined the glycerol content of lager by direct injection of the sample using Par 1 as column material. Par 1 is a macroreticular resin. It is a polymer of styrene with divinyl benzene cross-linking and is particularly suitable for glycerol analysis, which is difficult to achieve on conventional column materials. Eleven lagers were analysed and averaged 1.58 mg glycerol per litre of lager with a range of 1.40—1.83 mg per litre. The range of glycerol found in Canadian beer by Parker and Richardson[10] was 1.14—2.50 mg per litre. They used the TMS derivative of glycerol, and included maltose, isomaltose, maltotriose, maltotetraose and dextrin in the analysis. They concluded that although the normal glycerol content was below the taste threshold, the presence of glycerol was necessary to give a proper balance to the flavour constituents.

As fermentation proceeds, the analysis of amino acids in wort can yield valuable information regarding its proteolytic changes. Olsen[11] estimated the valine, alanine, leucine and isoleucine content of wort because amino acid concentration has an effect not only on fusel oil production but also on the formation of diketones. For example, a shortage of valine in the wort may cause a high level of biacetyl in the final product, which, if above the flavour threshold, would be undesirable. The wort was saturated with salt, norleucine was added as internal standard and ninhydrin was added to convert the amino acids to their respective aldehydes at 70°C. Head space samples were taken for GC analysis after 30 min, and because the reaction was not completed, critical parameters which had to be standardised were the amount of ninhydrin used, the reaction temperature and the time of reaction. Under standard conditions the method was simple and rapid for the analysis of valine as methyl propanal.

Kurosky and Bars[12] determined the amino acid content of beer and wort using the GC of the *N*-TFA methyl esters. The amino acids were separated by adsorption on Dowex 50W-X8 resin and eluted with dilute ammonia solution, and the derivatives were formed and dissolved in dichloromethane for injection. Of the 13 amino acids detected, proline was the major constituent in all the samples and was completely dominant in the British lagers. The Canadian beers had significant contributions from alanine, phenyl alanine, valine and leucine. *Figure 14.2* is a chromatogram of the *N*-TFA methyl esters of the amino acids in a sample of ale.

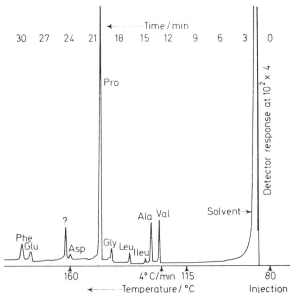

Figure 14.2 Chromatogram of an ale sample: Val, valine; Ala, alanine; Ileu, isoleucine; Leu, leucine; Gly, glycine; Pro, proline; Asp, aspartic acid plus asparagine; Glu, glutamic acid plus glutamine; Phe, phenylalanine (after Kurosky and Bars[12]). For GC conditions see Table 14.1

A special nitrogen-selective thermionic detector was used by Baerwald and Prucha[13] to determine the amino acids of wort and beer by the GC of their *N*-TFA butyl esters. The detector can be used in the presence of large concentrations of carbohydrates.

Arkima[14, 15] has examined beer for up to 16 volatile acids, and Harrison and Collins[16] have analysed beer for α-keto acids, notably pyruvic acid. The non-volatile phenolic acids present in beer have also been determined[17,18].

The volatile components in beer and wort, particularly the fusel oils, esters, diketones and sulphur-containing compounds, have been the subject of much investigation since the advent of GC. Although ethanol content is still an important factor from a merchandising aspect, the understanding of flavour composition of beer largely brought about by the use of GC has meant that brewers have found it easier to standardise their products.

The fusel oils and 2-phenylethanol are important contributors to the aroma and flavour of beer. As early as 1958, 2-methyl-1-propanol and 3-methyl-1-butanol were detected in beer distillates by Fleischmann and Daghetta[19], and these alcohols were also detected by Bitter[20], who concluded that they gave beer a harsh flavour. Sihto and Arkima[21] examined Finnish beers for fusel oil by extracting with 2:1 diethyl ether:pentane. The solvent was evaporated to low bulk and chromatographed, and results showed the major component was generally 3-methyl-1-butanol with some 2-methyl-1-butanol plus 2-phenylethanol. Some beers contained an abnormally high level of 3-methyl-1-butyl acetate, which had a deleterious effect on the palatability of the beer.

Fusel oil concentration has been determined in Finnish and other lagers[22], and *Table 14.2* shows the fusel oil, 2-phenylethanol, ethyl acetate and isoamyl acetate contents of stout, pale ale, brown ale and lager[23].

Table 14.2 The composition in beer of fusel oil and some other volatiles/p.p.m. (after Morgan[23])

	Stout (10)*	Pale ale (4)	Brown ale (4)	Lager (4)
Propanol	13–60	31–48	17–29	5–10
2-Methyl-1-propanol	11–98	18–33	11–33	6–11
2-Methyl-1-butanol	9–41	14–19	8–22	8–16
3-Methyl-1-butanol	33–169	47–61	28–77	32–57
2-Phenylethanol	20–55	36–53	19–44	25–37
Ethyl acetate	11–69	14–23	0–18	8–14
3-Methyl-1-butyl acetate	0–4.9	0–4.2	0.4–2.6	0–2.0

* Number of samples examined is given in parentheses.

Engan[24] considered 2-phenylethanol to be the most important aromatic alcohol found in beer and that the fermentation temperature had a greater influence on its formation than on that of the aliphatic alcohols. He analysed seven Norwegian pilsner beers, and 2-phenylethanol was found to lie within the range 5.6–27 p.p.m. 2-Phenylethanol has also been determined in American beer[25], and with tyrosol and tryptophol in beer and wine[26].

The influence of hop oil constituents on beer or wort volatiles has been studied by GC methods[27-30].

There has been an interest in the concentration of diketones in beer[31-37], and it is considered that biacetyl and pentan-2,3-dione are important flavour constituents. The concentrations of acetaldehyde and other aldehydes have been determined by Ronkainen, Arkima and Suomalainen[38].

General analyses of volatile components of beer or wort have been performed by several workers[39-53].

Hashimoto and Kuroiwa[54] used the technique of flash exchange GC to determine the sulphur volatiles of wort and beer. Volatile sulphur compounds were swept from the sample by nitrogen into mercuric chloride solution and the resulting complexes were heated with toluene-3,4-dithiol in a sample vaporiser and the gases led directly on to the GC column. Thioformaldehyde, dithioform-aldehyde and thioacetone were reported in beer and wort, and no mercaptans were found. It was suggested that mercaptans, if found in beer, were probably artefacts.

As the FPD has become commercially available there has been an increasing interest in the sulphur-containing constituents of food. Since most of these constituents are sufficiently volatile to appear in head space vapour, this method of sampling has been combined with FPD by the majority of chromatographers who have latterly studied the sulphur-containing volatiles of beer.

Harrison and Coyne[55] used this combination after incubation of beer at 35°C for 1 h, and found that the only significant compound was dimethyl sulphide, which was within the range 2–6 μg/l. 50°C for 1 h were in the incubation conditions used by Richardson and Mocek[56] in examining Canadian ales and lagers by the same method. They found that lager contained more sulphur-containing compounds than ale and agreed that dimethyl sulphide was the major component in each. The ales contained between 9 and 31 μg/l and the lagers between 18 and 6.5 μg/l. In the lager samples trace amounts of methyl n-butyl sulphide and dimethyl disulphide were also found.

Several other workers have examined beer for sulphur-containing compounds[57–61].

The gases oxygen, nitrogen and carbon dioxide in beer have been sampled by the head space method. Drawert, Postel and Kurer[62] used an empty column separating two columns which contained silica gel and molecular sieve 13X.

C. Cider and perry

Cider is the fermented juice of apples which are specially selected for their high tannin content. Perry is made in similar fashion from a particular type of pear. Cider is not the universal drink that beer is, and very little investigation of its flavour constituents has been carried out by GC.

Kieser et al.[63] determined the 2-phenylethanol content of yeasted and un-yeasted ciders using a distillation and extraction method to isolate the alcohol: 50 ml cider was distilled to give 25 ml distillate and 25 ml water was added to the distillate and the mixture was redistilled to give a second 25 ml distillate. This procedure was repeated and the final distillate was extracted with ethylene chloride and this solution used for GC. Many previously reported methods for the GC determination of 2-phenylethanol produced a retention time of greater than an hour, but in this method the retention time is 8 min. The 3,5-dinitro-benzoate was made to verify the presence of the alcohol. Results showed that ciders produced by addition of commercial yeast for fermentation gave between 7 and 15 p.p.m. 2-phenylethanol, whereas cider produced from the natural yeast flora contained approximately 100 p.p.m.

The fusel oil composition of cider and perry was investigated by Pollard et al.[64]. The alcohols were extracted either by vapour cycling through cold traps or by distillation. The extracts were placed in containers such that there was a fivefold excess of head space which was sampled after incubation of the samples at 65°C for 15 min. Concentrations of propanol, 2-methyl-1-propanol, butanol and 2- and 3-methyl-1-butanol were tabulated for six ciders extracted by vapour cycling, seven ciders extracted by distillation and three perries also extracted by distillation. There was little to choose between the two methods of extraction, and the major cider components were 2- and 3-methyl-1-butanol within the range 74–142 p.p.m. and 2-methyl-1-propanol within the range 14–82 p.p.m. The comparative figures for the perry samples were 86–104 p.p.m. 2- and 3-methyl-1-butanol and 36–43 p.p.m. 2-methyl-1-propanol.

Reinhard[65] saturated cider with salt and extracted the mixture with 2:1 diethyl ether:pentane in order to isolate the volatiles. He found that cider made from apple syrup yielded lower concentrations of butanol and hexanol than that made from fresh fruit.

Biacetyl has been determined in cider by Koch, Hess and Gruss[66]. They used a head space sampling method after incubation of the sample at 30°C for 15 min and introduction of nitrogen at 0.3 atm.

D. Wine and must

Wine is the product of the fermentation of grape juice, and the yeast used for the fermentation has a higher ethanol tolerance than has the beer yeast. Therefore wine has a higher ethanol content than beer, and some wines, such as sherry and port, being fortified with grape spirit, have even higher ethanol contents. Must is the name given to the liquor before the wine has fully fermented.

Ethanol in wine can be determined using any of the methods used for beer (see p.235) by simply diluting the wine to give a suitable ethanol content for the analysis. When a large number of samples are required to be analysed on a routine basis, an automated system is an advantage, and Stockwell and Sawyer[67] have described diagramatically an automated system for analysing wines, essences and tinctures. Mixed 2-propanol-1-propanol internal standard was blended automatically with the sample and the resulting solution automatically injected and chromatographed.

Sugars in wine are normally determined by methods based on the reduction of cupric to cuprous ions before and after inversion of the sample. These methods become difficult to operate quantitatively in mixtures of three or more sugars. Capella and Losi[68] determined glucose, fructose and sucrose in wine by the GC of the TMS derivatives. Wine was decolorised with activated carbon, cleaned up with a mixed ion exchange resin and filtered. The solution was evaporated to dryness under vacuum at less than 60°C after addition of ethanol and the TMS derivatives prepared.

Sake, a fermented beverage made from rice, was analysed for α-ethyl glucoside and sugar alcohols by Imanari and Tamura[69]. After extraction and clean-up by ion exchange and gel chromatography, analysis was carried out by GC supported by MS. Sake was found to contain approximately 2 mg α-ethyl glucoside per ml and smaller amounts of erythritol, arabinol and mannitol.

Brunelle, Schoeneman and Martin[70] determined the fixed acids in wines by GC of their TMS derivatives after precipitation of the lead salts to remove sugars, which do not form insoluble lead salts. Succinic, malic, tartaric and citric acids were determined in this way, and grape wines and other fruit wines contained malic acid, whereas only grape wines contained tartaric acid.

Formic, acetic, propionic, 2-methyl-1-propionic, butyric, valeric, hexanoic, heptanoic and octanoic acids were the volatile ones found in apple wine by Sugisawa, Matthews and MacGregor[71] either by GC of the free acids on four

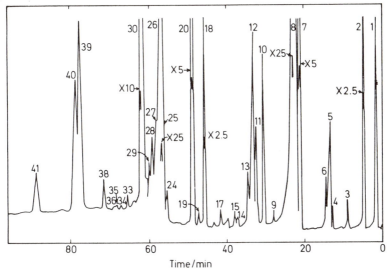

Figure 14.3 Chromatogram of extract of Riesling wine volatiles. Column: Perkin-Elmer support-coated open tubular (50 ft x 0.02 in I.D.) coated with Ucon Oil LB 550X stationary phase. Helium flow rate, 7 ml/min; temp. 32°C for 3 min, then programmed at 2°C/min to 150°C. For full GC conditions see Table 14.1. (After Hardy and Ramshaw[79].)

Peak No.	Identity	Peak No.	Identity
1	Solvent	20	3-Methylbutyl hexanoate
2	Ethyl acetate	21	Di-ethyl succinate
3	Unknown, apparent mol.	22	Propyl octanoate
	wt. 88, base peak 87	23	Unknown
4	Isobutyl acetate	24	Isobutyl octanoate
5	Isobutyl alcohol	25	2-Phenylethyl acetate
6	Ethyl butyrate	26	2-Phenylethyl alcohol
7	3-Methylbutyl acetate	27	Hexyl hexanoate
8	3-Methylbutyl alcohol	28	Ethyl decanoate
9	3-Methylbutyl pro-	29	Ethyl decenoate[a]
	pionate	30	3-Methylbutyl octanoate
10	Ethyl hexanoate	32	Heptyl hexanoate
11	Hexyl acetate	33	3-Methylbutyl ethyl
12	Hexyl alcohol		succinate
13	3-Methylbutyl butyrate	34	2-Phenylethyl butyrate
14	Valerate ester	35	Isobutyl decanoate
15	Butyl valerate	36	Isobutyl decenoate[a]
16	Propyl hexanoate	37	2-Phenylethyl ester
17	Isobutyl hexanoate,	38	Hexyl octanoate
	3-methylbutyl valerate	39	3-Methylbutyl decanoate
18	Ethyl octanoate	40	3-Methylbutyl decenoate[a]
19	Unknown, apparent mol.	41	2-Phenylethyl hexanoate
	wt. 138, base peak 82		

[a] *From MS data only.*

different stationary phases or by utilising the methyl esters and two other sets of GC conditions.

Prior to GC it was a difficult analytical task to assess the levels of the individual components of fusel oil, although it was known that these were very important flavour and aroma constituents. Fleischmann and Daghetta[19] determined 2-methyl-1-propanol and 3-methyl-1-butanol in wine after distillation, and Webb and Kepner[72] similarly isolated propanol, 2-methyl-1-propanol, 2-methyl-1-butanol and 3-methyl-1-butanol. The latter workers investigated variations in column length, amount and nature of the stationary phase, the support and the temperature with respect to the efficiency of separation. Burgundy, Montrachet, Jerez, Muscat Raisin and Zinfandel wines were analysed for fusel oil composition, and in all cases the major component was 3-methyl-1-butanol.

Direct injection of wine was used by Prillinger and Horwatitsch[73] and the peaks due to ethyl acetate, methanol, propanol and butanol were identified. The butanol concentration of wines made from other fruits was higher than that from grapes.

Nykanen *et al.*[26] determined the aromatic alcohols tyrosol and tryptophol in French red and white wines, Algerian red wine, Yugoslavian white wine and Finnish red and white berry wines. Tyrosol was found in the range 8–43 mg per litre of wine and typtophol in the range 0.2–3.1 mg per litre of wine.

Several workers have detected or determined volatile components of table wine[74-82] and sherry[83,84].

The favoured extractants for volatiles are diethyl ether, n-pentane and iso-pentane, used either singly or in combination. Hardy and Ramshaw[79] experimented with the use of trichlorofluoromethane as extractant since it does not extract water or ethanol but does extract higher alcohols, carbonyls and esters. The trichlorofluoromethane solution was extracted with propylene glycol, which removed the higher alcohols, leaving behind the carbonyls and esters. Coupled GC and MS confirmed the identities of 41 components which were mainly esters, ethyl 3-methylbutyl succinate being reported for the first time. *Figure 14.3* shows a chromatogram of Riesling wine volatiles with a legend for the identified peaks.

Benzaldehyde was determined in port wine by simple GC and TLC methods by Ramos and Gomes[85].

Still wine has been defined in the United States as a wine containing less than 0.256 g carbon dioxide per 100 ml. Since the tax on carbonated wine was 15–20 times greater than that on still wine, it was important to accurately determine the carbon dioxide content of wines. Ashmead, Martin and Schmit[86] have described a GSC method for this determination using a column of 60/80 mesh charcoal. Although results were higher than those obtained manometrically, it was not a statistically significant difference, and the fact that sulphur dioxide did not interfere in the analysis made the method a viable one.

E. Spirits

Spirits are the distilled and refined products of fermented beverages. The type of spirit obtained depends on the ingredients used to prepare the fermented liquor and, in the UK, it is an offence to sell whisky, brandy, rum and gin at less than 65° proof spirit. The determination of ethanol is therefore an important legal consideration, and established methods based on corrected direct specific gravity, specific gravity measurement after distillation and oxidation with dichromate have been used effectively. The GC methods described for beer (see p. 235) can be adapted for the analysis of ethanol in spirits by merely diluting the sample. Most GC methods use an internal standard, but Pietsch, Oehler and Kasprick[87] used direct injection of spirit samples without addition of internal standard and calibrated each new column. More recently an automated method has been devised[88].

The analysis of methanol in spirits is important because small amounts can be distilled from a fermented beverage, particularly if the wrong distillation conditions are used. The UK Research Committee on the analysis of potable spirits has reported on the estimation of methanol in spirits[89]. After a collaborative exercise, the Committee cited three GC methods suitable for the analysis of methanol and these methods gave results which were shown to be in close agreement with the AOAC colorimetric method based on the oxidation of methanol to formaldehyde and its colour reaction with chromotropic acid. A similar comparison was made by Dyer[90].

1. Brandy and wine distillate

Brandy is made by distillation of wine and therefore contains, apart from a higher ethanol content, the most volatile of the flavour constituents of wine, the predominant substances being fusel oil and the esters of the fusel oil alcohols.

The methanol content of brandy was shown by Dyer[90] to be in the range 0.018–0.067%. Brandies made from other fruit wines were also analysed for methanol, and pear, apricot, plum and cherry brandies contained 0.35, 0.32, 0.37 and between 0.17 and 0.31%, respectively.

Much interest has been shown in the fusel oil composition of spirits and brandy in particular. Several workers have determined the fusel oil concentration of brandy[73, 91-95].

The sum of the concentrations of the higher alcohols from C_3 to C_5 can be taken as the fusel oil content of brandy, according to Singer and Stiles[96]. GC analysis showed that the alcohols greater than C_5 were present in insignificant amount. Total higher alcohol concentrations of 15 brandies were determined and found to be in the range 230–510 mg/l. Singer[97] continued this work, comparing brandy with various whiskies, and concluded that spirituous beverages were distinguished from each other by the presence of secondary constituents which generally arise during fermentation and which change in character and proportion in subsequent distillation and maturation. Samples were adjusted to 40% v/v ethanol, pentanol was added as internal standard and direct injection was used for chromatography. 3-Methyl-1-butanol, 2-methyl-1-propanol, propanol and 2-butanol were determined; cheaper brandies contained more than 5 mg of the latter per 100 ml, whereas more expensive brandies invariably contained less. The proportions of some of these alcohols were also used as a guide to brandy quality. For example, good-quality brandy has a propanol to 2-methyl-1-propanol ratio of between 0.37 and 0.47, whereas poor-quality brandy has a ratio of between 0.52 and 0.66. In a further phase of this study, Singer[98] evaluated stationary phases suitable for the separation of 2-methyl- and 3-methyl-1-butanol, and found that 10% diethyl tartrate and 20% PEG200 were the most suitable. Analysis of 20 good-quality brandies gave a ratio of 2-methyl- to 3-methyl-1-butanol of between 0.19 and 0.24, whereas the same ratio for five inferior brandies was 0.24–0.26.

Direct injection of brandy with butanol as internal standard was used by Martin and Caress[99] in the analysis for fusel oil and ethyl acetate. Three per cent Carbowax 400 was the stationary phase chosen, particularly since it gave reasonable separation of 2-methyl- and 3-methyl-1-butanols.

The important flavour and aroma compound 2-phenylethanol, and higher esters, were determined in brandy and wine distillate by Reinhard[100]. Ethyl octanoate concentration increased during storage of wine distillate, and the ratios of various esters were used to characterise different-quality brandies.

Kahn and Conner[101] determined 2-phenyl ethanol in a number of brandies, whiskies, rums and tequilas, using direct injection on to a FFAP column. The three brandies contained between 8 and 10 p.p.m. 2-phenyl ethanol.

In a study of the fusel oil composition of distilled muscat raisin wine, Kepner and Webb[102] separated the volatiles into 'low' boilers, viz. 3-methyl-1-butanol, 2-methyl-1-butanol, 2-methyl-1-propanol and propanol, and 'high' boilers, viz. ethyl decanoate and other esters. The 'low' boilers, of which 3-methyl-1-butanol was the major constituent, comprised 85% of the total volatiles.

Esters are also important flavour constituents of brandy and several workers have detected or determined them[78,103–106].

Ethyl octanoate, decanoate and dodecanoate were the most important of the esters found in brandies and wine distillates by Guymon and Crowell[107]. Samples were diluted with water to an ethanol concentration of 20% v/v and extracted four times with methylene chloride. The combined extracts were dried and concentrated for GC. Wine distillates, still containing suspended yeast cells, yielded 2–3 mg ethyl octanoate, 5–10 mg ethyl decanoate and 5–8 mg dodecanoate per 100 ml distillate. Distillates containing no yeast cells contained lesser concentrations of these esters. Hexanoate, with its fragrant and intense odour and suggestive of banana oil, octanoate, with its pungent, fusel oil odour, decanoate, with its milder tallowy odour, and dodecanoate, with its waxy, candle-like odour, are all considered to be very important contributors to the bouquet of matured spirits.

2. Whisky and bourbon

There are essentially two types of whisky: malt, produced by the distillation of fermented malted barley; and grain, produced by the distillation of fermented unmalted maize. The bulk of marketed whisky in the UK consists of blends of the two types. Bourbon is a whisky made from rye or maize as a basis and has a large market in the US and Canada. (For convenience, the variant spelling 'whiskey' will not be used in this section.)

Ethanol has been measured in whisky by specific gravity methods, either after distillation or directly and applying a correction for the small amounts of soluble solids present. It can be determined after suitable dilution, by direct injection GC methods, using an internal standard such as 1-propanol or 2-propanol. In an industry where technology is still developing, GC is used to monitor ethanol in stillage. Simpson[108] reviewed quality control in the distilling industry and highlighted the increasing use of GC in measuring ethanol, methanol, higher alcohols and flavour constituents as a supplement to the established organoleptic testing.

Methanol has been determined in whisky by the colour reaction between formaldehyde and chromotropic acid after oxidation of the methanol. Dyer[90] used a GC method and found that whisky contained between 0.005 and 0.007% methanol.

Since the fusel oil contents of spirits are generally derived from the sugars and nitrogen-containing compounds, the qualitative and quantitative GC approach is the same for whisky as it is for brandy. Chemically, the fusel oil constituents can be estimated by colour reaction with either *p*-dimethylamino-benzaldehyde or 4-hydroxybenzaldehyde-3-sulphonic acid. These methods give the total higher alcohol content, but if individual alcohols are to be determined, GC would be the chosen technique.

Martin *et al.*[92] gave the 3-methyl-1-butanol, 2-methyl-1-propanol and propanol concentration ranges for whisky as 31–223, 6–45 and 3–40 g/100 l, respectively, whereas Bober and Haddaway[93] showed that Scotch whisky with between 620 and 800 p.p.m. 2-methyl-1-propanol was significantly higher in that alcohol than Canadian whisky, which contained between 50 and 100 p.p.m. Bourbon was distinguished from other whiskies by having a higher level of 3-methyl-1-butanol and of 2-methyl-1-butanol at 270 and 80 p.p.m., respectively.

Singer and Stiles[96] found that, for most practical purposes, the sum of the C_3–C_5 alcohols gave the total higher alcohol content. Three samples of whisky were analysed and the total higher alcohol content was in the range 200–250 mg per 100 ml. In a similar study, Singer[97] examined many samples of whisky for fusel oil content in an endeavour to find differences between the various types of product. Between 162 and 265 mg 3-methyl-1-butanol per 100 ml malt whisky was found, whereas only between 0.8 and 19 mg per 100 ml grain whisky was measured. Eleven samples of blended whisky contained between 75 and 94 mg 3-methyl-1-butanol per 100 ml and indicated that such blends were not far from being 1:1 mixtures. US bourbon whisky gave high 3-methyl-1-butanol values, and so did a sample of Irish whisky.

The ratio of 2-methyl-1-butanol to 3-methyl-1-butanol in whiskies of various origin has been given by Singer[98]. Whereas the ratio is useful in differentiating brandies, there is too big an overlap of values for it to be useful in differentiating whiskies. However, malt whisky has higher concentrations of both alcohols than blended whisky, and US bourbon whisky has even higher concentrations. Dutch and Canadian whiskies are low in both alcohols. It should therefore be possible to distinguish the following by GC: malt from blended whisky; US bourbon whisky from any other whisky; and Scotch whisky from either Dutch or Canadian whisky.

Several other workers have detected or determined fusel oil components[109-112], 2-phenylethanol[101], esters[113,114] and carbonyls[115] in whisky and related spirits.

Duncan and Philp[116] used three different sets of GC conditions in the analysis of Scotch whisky volatiles, each set to perform a particular analysis. Direct GC of whisky on one column gave most of the alcohols, phenol and the higher esters, whereas a second column gave ethyl acetate, diethyl acetal, 3-methyl-1-butyl acetate and butanol. A third column was used specifically for the analysis of carbonyls after formation of the bisulphite compounds and regeneration of the carbonyls with sodium carbonate. The volatiles of grain whisky were compared with those of malt whisky using the three columns, and the qualitative and quantitative differences between the two types of whisky are shown in *Table 14.3*.

Table 14.3 The volatile composition of unmatured malt and grain whisky/p.p.m. (after Duncan and Philp[116])

Volatile component	Malt whisky	Unmatured grain whisky
Acetaldehyde plus acetal	37–39	5 7
Diethyl acetal	14–17	6–7
Acetone	less than 0.2	nil
Isobutyraldehyde	14–15	0.4–0.5
Isovaleraldehyde	20–22	nil
Propanol	270–320	360–440
2-Methyl-1-propanol	490–530	560–650
Butanol	5–7	nil
3-Methyl-1-butanol	1 560–1 680	nil
Ethyl acetate	420–450	nil
3-Methyl-1-butyl acetate	28–32	nil
Ethyl lactate	9.7–11	nil
Ethyl octanoate	18–19	nil
Ethyl decanoate	54–58	nil
Ethyl dodecanoate	43–50	nil
Ethyl tetradecanoate	3.4–3.9	nil
Ethyl hexadecenoate	25–30	nil
Ethyl hexadecanoate	34–40	nil
Furfural	12–13	nil
2-Phenylethyl acetate	18–21	nil
2-Phenyl ethanol	27–32	nil

This approach highlights the difficulties encountered when attempting to include in one chromatogram all the volatiles of a food irrespective of their different polarities. Kahn, La Roe and Conner[117] devised a system of two GC columns in series, one polar and one non-polar, in order to detect compounds with different polarities in distilled alcoholic beverages. Direct injection was used when compounds were present in sufficient quantity, and diethyl ether: pentane extraction was used if sensitivity was lacking. MS confirmed identities of 35 volatiles which included 1,1,3-triethoxy propane, a compound formed between ethanol and acrolein which is synthesised by certain bacteria in corn, rye and barley malt grains.

Eighty-seven volatile components of bourbon have been listed by Kahn *et al.*[118], who used a diethyl ether:pentane extraction method. Five stationary phases were utilised to separate the volatiles, some of these compounds being identified for the first time.

Malt whisky is characterised by its 'smoky' flavour, which is thought to be derived from the drying of malt in peat smoke. Macfarlane[119] conducted an interesting study on peat smoke condensate using a GC technique. Malt dried over a peat fire contained a proportion of phenols, and the smoke condensate contained phenol, furfuraldehyde and traces of cresol and guiaicol.

3. Rum

Rum is the spirit distilled from sugar cane juice or molasses, and most rum is processed in the Caribbean countries. Whereas it is higher alcohols which are the dominant flavour volatiles of brandy and whisky, it is esters which largely contribute to rum flavour.

Fusel oil and higher alcohol analyses have been performed on rum, usually as a comparative exercise with other spirits. Martin *et al.*[92] found that the major fusel oil component of rum was 3-methyl-1-butanol, which was found up to 144 g/100 l. Propanol and 2-methyl-1-propanol were found in the ranges 28–100 and 4–36 g/100 l, respectively.

In an analysis of whisky and rum for ethyl acetate and fusel oil, Martin and Caress[99] found 13.8 g ethyl acetate per 100 l rum and 12.4 g 3-methyl-1-butanol per 100 l rum. Smaller amounts of propanol and 2-methyl-1-propanol were also present.

Three Jamaican rums analysed by Bober and Haddaway[93] contained more than 2000 p.p.m. 1-propanol. These workers found that Virgin Islands rum could be distinguished from other rums by its high concentration of 2-propanol at approximately 50 p.p.m.

Deluzarche *et al.*[120] detected fusel oil constituents in a variety of liqueurs, rum and geneva. The significance in their analysis of rum was the absence of methanol, butanol, 2-butanol and 2-pentanol. In their examination of spirits for total C_3 –C_5 alcohols, Singer and Stiles[96] found that two samples of rum contained 5 and 20 mg per 100 ml.

Kahn and Conner[101] examined spirits for 2-phenylethanol and did not detect any of this alcohol in four rums. This analysis could be utilised if there was a suspicion that rum had been sophisticated with wine or other spirits.

Maurel and Lafarge[121], Maurel[122] and Maurel, Sansoulet and Giffard[123] have made complete studies of the volatiles of rum. In contrast to Deluzarche *et al.*[120], they found traces of methanol, 1-butanol and 2-butanol, and average concentrations of propanol, 2-methyl-1-propanol and total C_5 alcohols were 300, 570 and 1300 p.p.m., respectively. A fine grade of rum contained tertiary butanol and isovaleric acid, which are not normally found in ordinary-grade rums.

Head space sampling of rum vapour was used for the quantitative determination of volatiles by Kepner, Maarse and Strating[44]. Thirteen simple volatiles were found and the effects on the volatiles entering the head space after dilution of the rum were discussed.

The quality of rum is largely dependent on the amounts of the ethyl esters of the fatty acids of even carbon number between 8 and 16. This was the conclusion of Stevens and Martin[124], who examined rum by GC after chloroform extraction of the diluted sample and vacuum evaporation of the extract.

4. Gin and geneva

Because gin is a rectified spirit made from barley or rye, it is almost devoid of flavour constituents other than those present by virtue of the addition of juniper, angelica, coriander, etc. It is therefore not surprising that very little study has been made of gin volatiles by GC.

Seventeen samples of gin were examined by the UK Government Chemist[125] and found to have different chromatographic profiles from those of authentic samples. The 17 gins had two very small peaks, whereas the authentic samples showed three larger peaks in the same region. It was considered that the 17 samples were not genuine gins.

In the study of Deluzarche *et al.*[120] of liqueurs and spirits for higher alcohols, geneva, which is a gin principally manufactured in Europe, was found to contain 3-methyl-1-butanol, 2-methyl-1-propanol and propanol. No methanol was found.

5. Tequila and mescal

Tequila and mescal are spirits made in Mexico. Kahn and Conner[101] included tequila in the spirits in which they determined 2-phenylethanol concentration. The four samples examined contained this alcohol in the range 14—37 p.p.m.

The volatile components of tequila and mescal were determined by Moreno and Llama[126] using columns of Carbowax 600 and Carbowax 1540 in series.

F. Confectionery containing ethanol

Liqueur chocolates are chocolate capsules containing alcoholic beverages — either spirits, liqueurs or fortified wines. Normally the ethanol content lies between 0.4 and 1.3 proof gallons per 100 lb chocolate.

This type of confectionery item has been analysed for ethanol content by Chaveron[127]. He made a 10% solution or filtered dispersion of the confectionery item in 0.5% aqueous propanol and injected the mixture, calculating the ethanol content with the aid of propanol as internal standard. Chocolate, sugar and oleoresin of capsicum did not interfere with the analysis, although sugar contaminated the injector. A removable stainless steel grid was placed in the injector so that non-volatile substances were deposited on this grid, which could be readily exchanged.

G. Vinegar

Vinegar is the product of the alcoholic and acetous fermentation of a sugar-containing solution, and may be flavoured with spices. Spirit vinegar is produced by fermentation and distillation, and will have a significant ethanol content. Morgantini[128] used a GC method to determine the ethanol content of such vinegar by direct injection on to a column of 20% PEGS, and compared peak heights of ethanol in standards and sample.

In the chemical tests carried out to distinguish brewed and non-brewed vinegars, biacetyl in the brewed product makes a significant contribution to the alkaline oxidation value and also the ester value. Cesari, Cusmano and Boniforti[129] determined biacetyl and also 3-hydroxy-butan-2-one in vinegar by GC. They prepared facsimile vinegars containing these volatiles to test out their method, and found it to be accurate and sensitive.

Alcohols, esters, acids and also 3-hydroxy-butan-2-one in spirit and cider vinegars were determined by Kahn, Nickol and Conner[130]. They produced chromatograms which showed the complex pattern of the cider vinegar constituents as compared with those of commercial distilled vinegars and those obtained by the acetous fermentation of denatured ethanol. Some of the cider vinegar volatiles were shown to arise from the parent beverage.

Twenty-five volatiles were found by Aurand *et al.*[131] in cider and wine vinegars, including ethanol, ethyl acetate, acetaldehyde and acetone. The vinegars produced from natural sources showed a greater content of higher alcohols and esters than distilled ones, and such differences in the chromatographic profiles of these two types of vinegar could be used to differentiate them.

The aroma constituents of spirit vinegar were determined by Suomalainen and Kangasperko[132]. They extracted these volatiles with 2:1 diethyl ether: pentane, and found that ethyl acetate was the only volatile present in any significant concentration.

Wine and cider vinegars which are made with grapes and cider apples, respectively, as the basis of the fermentations, were analysed by Hundley[133] for sorbitol. The neat vinegar was acetylated and extracted with chloroform and the evaporated extract injected on to a 10% DC-200 column. No sorbitol was found in authentic wine vinegars but the 20 cider vinegars analysed had sorbitol contents ranging from 112 to 638 mg per 100 ml. Adulteration of wine vinegar by as little as 10% of cider vinegar could therefore be detected by this sorbitol analysis.

References

1. Silbereisen, K., Krueger, E., Schubert, B. and Anthon, F., *Mschr. Brau.*, **20,** 121 (1967)
2. Trachman, H., *Communs. Wallerstein Labs.*, **32,** 111 (1969)
3. Lie, S., Haukeli, A. D. and Gether, J. J., *Brygmesteren*, **27,** 281 (1970)
4. Dyer, R. H., *J. Ass. Off. Anal. Chem.*, **55,** 564 (1972)
5. Clapperton, J. F. and Holliday, A. G., *J. Inst. Brew.*, **74,** 164 (1968)
6. Tuning, B., *Proc. Eur. Brew. Conv.*, 191 (1971)
7. Marinelli, L. and Whitney, D., *J. Inst. Brew.*, **72,** 252 (1966)
8. Otter, G. E. and Taylor, L., *J. Inst. Brew.*, **73,** 570 (1967)
9. Feil, M. F. and Marinelli, L., *Proc. Amer. Soc. Brew. Chem.*, **29 (1969)**
10. Parker, W. E. and Richardson, P. J., *J. Inst. Brew.*, **76,** 191 (1970)
11. Olsen, A., *Communs. Wallerstein Labs.*, **33,** 19 (1970)
12. Kurosky, A. and Bars, A., *J. Inst. Brew.*, **74,** 200 (1968)
13. Baerwald, G. and Prucha, J., *Brauwissenschaft*, **24,** 397 (1971)
14. Arkima, V., *Mschr. Brau.*, **18,** 121 (1965)
15. Arkima, V., *Mschr. Brau.*, **21,** 247 (1968)
16. Harrison, G. A. F. and Collins, E., *Proc. Amer. Soc. Brew. Chem.*, 101 (1968)
17. Dallos, F. C. and Koeppl, K. G., *J. Chromat. Sci.*, **7,** 565 (1969)
18. Dallos, F. C., Lautenbach, A. F. and West, D. B., *Proc. Amer. Soc. Brew. Chem.*, 103 (1967)
19. Fleischmann, L. and Daghetta, A., *La Ric. Sci.*, **28,** 2286 (1958)
20. Bitter, H., *Brauwissenschaft*, **19,** 278 (1966)
21. Sihto, E. and Arkima, V., *J. Inst. Brew.*, **69,** 20 (1963)
22. Arkima, V., *Mschr. Brau.*, **21,** 25 (1968)
23. Morgan, K., *J. Inst. Brew.*, **71,** 166 (1965)
24. Engan, S., *Brygmesteren*, **26,** 23 (1969)
25. van der Kloot, A. P. and Wilcox, F. A., *Proc. Amer. Soc. Brew. Chem.*, 93 (1963)
26. Nykanen, L., Puputti, E. and Suomalainen, H., *J. Inst. Brew.*, **72,** 24 (1966)
27. Likens, S. T. and Nickerson, G. B., *Proc. Amer. Soc. Brew. Chem.*, 5 (1964)
28. Nickerson, G. B. and Likens, S. T., *J. Chromatog.*, **21,** 1 (1966)
29. Buttery, R. G., Black, D. R., Lewis, M. J. and Ling, L., *J. Food Sci.*, **32,** 414 (1967)
30. Hartley, R. D., *J. Inst. Brew.*, **74,** 550 (1968)
31. Harrison, G. A. F., Byrne, W. J. and Collins, E., *J. Inst. Brew.*, **71,** 336 (1965)
32. Harrison, G. A. F., Byrne, W. J. and Collins, E., *Proc. Eur. Brew. Conv.*, 352 (1965)
33. Drews, B., Baerwald, G. and Niefind, H. J., *Mschr. Brau.*, **21,** 96 (1968)
34. Steffen, P., *Nahrung*, **13,** 697 (1969)
35. Baerwald, G., *Mschr. Brau.*, **23,** 351 (1970)
36. Spaeth, G., Niefind, H. J. and Martina, M., *Schweizer BrauRdsch.*, **82,** 121 (1971)
37. Haukeli, A. D. and Lie, S., *J. Inst. Brew.*, **77,** 538 (1971)
38. Ronkainen, P., Arkima, V. and Suomalainen, H., *J. Inst. Brew.*, **73,** 567 (1967)
39. Harrison, G. A. F., *Proc. Eur. Brew. Conv.*, 247 (1963)
40. Maule, D. R., *J. Inst. Brew.*, **73,** 351 (1967)
41. Drews, B., Baerwald, G. and Niefind, H. J., *Mschr. Brau.*, **21,** 131 (1968)
42. van der Kloot, A. P. and Wilcox, F. A., *Proc. Amer. Soc. Brew. Chem.*, 113 (1960)
43. Krueger, E. and Neumann, L., *Mschr. Brau.*, **23,** 269 (1970)
44. Kepner, R. E., Maarse, H. and Strating, J., *Anal. Chem.*, **36,** 77 (1964)
45. Dalgliesh, C. E., *Brewers J. (London)*, 31 (1967)
46. Maendl, B., Wullinger, F., Binder, W. and Piendl, A., *Brauwissenschaft*, **22,** 477 (1969)
47. Jahnsen, V. J. and Horn, G. B., *Proc. Amer. Soc. Brew. Chem.*, 194 (1965)
48. Powell, A. D. G. and Brown, I. H., *Proc. 9th Conv. Inst. Brewing, Australian Section*, 257 (1966)
49. Trachman, H. and Saletan, L. T., *Proc. Amer. Soc. Brew. Chem.*, 19 (1969)
50. Engan, S., *Brygmesteren*, **28,** 191 (1971)
51. van Gheluwe, J. E. A., Belleau, G., Jamieson, A. and Buday, A., *Proc. Amer. Soc. Brew. Chem.*, 49 (1966)
52. von Szilvinyi, A. and Puspok, J., *Brauwissenschaft*, **16,** 204 (1963)
53. Kepner, R. E., Strating, J. and Weurman, C., *J. Inst. Brew.*, **69,** 399 (1963)

54. Hashimoto, H. and Kuroiwa, Y., *Proc. Amer. Soc. Brew. Chem.*, 121 (1966)
55. Harrison, G. A. F. and Coyne, C. M., *J. Chromatog.*, 41, 453 (1969)
56. Richardson, P. J. and Mocek, M. *Proc. Amer. Soc. Brew. Chem.*, 128 (1971)
57. Jansen, H. E., Strating, J. and Westra, W. M., *J. Inst. Brew.*, 77, 154 (1971)
58. Drews, B., Baerwald, G. and Niefind, H. J., *Mschr. Brau.*, 22, 140 (1969)
59. McCowen, N. M., Palamand, S. R. and Hardwick, W. A., *Proc. Amer. Soc. Brew. Chem.*, 136 (1971)
60. Sinclair, A., Hall, R. D. and Burns, D. T., *Proc. Eur. Brew. Conv.*, 427 (1969)
61. Sinclair, A., Hall, R. D., Burns, D. T. and Hayes, W. P., *J. Sci. Food Agric.*, 21, 468 (1970)
62. Drawert, F., Postel, W. and Kurer, C., *Brauwissenschaft*, 23, 217 (1970)
63. Kieser, M. E., Pollard, A., Stevens, P. M. and Tucknott, O. G., *Nature*, 204, 887 (1964)
64. Pollard, A., Kieser, M. E., Stevens, P. M. and Tucknott, O. G., *J. Sci. Food Agric.*, 16, 384 (1965)
65. Reinhard, C., *Dt. LebensmittRdsch.*, 64, 251 (1968)
66. Koch, J., Hess, D. and Gruss, R., *Z. Lebensmitt. Forsch.*, 146, 143 (1971)
67. Stockwell, P. B. and Sawyer, R., *Anal. Chem.*, 42, 1136 (1970)
68. Capella, P. and Losi, G., *Ind. Agrar.*, 6, 7 (1968)
69. Imanari, T. and Tamura, Z., *Agric. Biol. Chem., Japan*, 35, 321 (1971)
70. Brunelle, R. L., Schoeneman, R. L. and Martin, G. E., *J. Ass. Off. Anal. Chem.*, 50, 329 (1967)
71. Sugisawa, H., Matthews, J. S. and MacGregor, D. R., *J. Food Sci.*, 27, 435 (1962)
72. Webb, A. D. and Kepner, R. E., *Amer. J. Enol. Viticult.*, 12, 51 (1961)
73. Prillinger, F. and Horwatitsch, H., *Mitt. Rebe Wein, Obstb. Fruchteverwert.*, 15, 72 (1965)
74. Webb, A. D., Riberau-Gayon, P. and Boidron, J-N., *Bull. Soc. Chim. France*, 1415 (1964)
75. Lemperle, E. and Mecke, R., *Z. Anal. Chem.*, 212, 18 (1965)
76. Drawert, F. and Rapp, A., *Chromatographia*, 1, 446 (1968)
77. Bayer, E., *J. Gas Chromat.*, 4, 67 (1966)
78. Koch, J., Hess, D. and Gruss, R., *Z. Lebensmitt. Forsch.*, 147, 207 (1971)
79. Hardy, P. J. and Ramshaw, E. H., *J. Sci. Food Agric.*, 21, 39 (1970)
80. Bertrand, A., *Chim. Anal.*, 53, 577 (1971)
81. Schreier, P. and Drawert, F., *Z. Lebensmitt. Forsch.*, 154, 273 (1974)
82. Schreier, P., Drawert, F. and Junker, A., *Z. Lebensmitt. Forsch.*, 154, 179 (1974)
83. Webb, A. D. and Kepner, R. E., *Amer. J. Enol. Viticult.*, 13, 1 (1962)
84. Suomalainen, H. and Nykanen, L., *Acta Chem. Fenn.*, B39, 252 (1966)
85. Ramos, M. da C. and Gomes, L. G., *Anais Inst. Vinho Porto*, 23, 11 (1969)
86. Ashmead, H. L., Martin G. E. and Schmit, J. A., *J. Ass. Off. Agric. Chem.*, 47, 730 (1964)
87. Pietsch, H. P., Oehler, R. and Kasprick, D., *Nahrung*, 12, 885 (1968)
88. Venturella, V. S., Graves, D. and Lang, R. E., *J. Ass. Off. Anal. Chem.*, 57, 118 (1974)
89. Research Committee on the Analysis of Potable Spirits, *J. Ass. Publ. Anal.*, 10, 49 (1972)
90. Dyer, R. H., *J. Ass. Off. Anal. Chem.*, 54, 785 (1971)
91. Bouthilet, R. J. and Lowrey, W., *J. Ass. Off. Agric. Chem.*, 42, 634 (1959)
92. Martin, G. E., Caggiano, G. and Schlesinger, H., *J. Ass. Off. Agric. Chem.*, 46, 294 (1963)
93. Bober, A. and Haddaway, L. W., *J. Gas Chromat.*, 1, 8 (1963)
94. Egorov, I. A., Rodopulo, A. K. and Pisarnitskii, A. F., *Dokl. Akad. Nauk SSSR*, 151, 729 (1963)
95. Drawert, F. and Rapp, A., *Z. Lebensmitt. Forsch.*, 126, 406 (1965)
96. Singer, D. D. and Stiles, J. W., *Analyst*, 90, 290 (1965)
97. Singer, D. D., *Analyst*, 91, 127 (1966)
98. Singer, D. D., *Analyst*, 91, 790 (1966)
99. Martin, G. E. and Caress, E. A., *J. Sci. Food Agric.*, 22, 587 (1971)
100. Reinhard, C., *Dt. Lebensmitt.-Rdsch.*, 67, 349 (1971)
101. Kahn, J. H. and Conner, H. A., *J. Ass. Off. Anal. Chem.*, 55, 1155 (1972)

102. Kepner, R. E. and Webb, A. D., *Amer. J. Enol. Viticult.*, **12,** 159 (1961)
103. de Vries, M. J., *S. Afr. J. Agric. Sci.*, **5,** 395 (1962)
104. Mecke, R. and de Vries, M. J., *Z. Anal. Chem.*, **170,** 326 (1959)
105. Prillinger, F. and Horwatitsch, H., *Mitt. Rebe Wein, Obstb. Fruchteverwert.*, **16,** 115 (1966)
106. Reinhard, C., *Dt. Lebensmitt.-Rdsch.*, **65,** 223 (1969)
107. Guymon, J. F. and Crowell, E. A., *Amer. J. Enol. Viticult.*, **20,** 76 (1969)
108. Simpson, A. C., *Process Biochem.*, 18 (February) (1973)
109. Kahn, J. H., Trent, F. M., Shipley, P. A. and Vordenberg, R. A., *J. Ass. Off. Anal. Chem.*, **51,** 1330 (1968)
110. Research Committee on the Analysis of Potable Spirits, *J. Ass. Publ. Anal.*, **8,** 81 (1970)
111. Jones, K. and Wills, R., *J. Inst. Brew.*, **72,** 196 (1966)
112. Sihto, E., Nykanen, L. and Suomalainen, H., *Tekn. Kem. Aikl.*, **19,** 753 (1962)
113. Martin, G. E., Schoeneman, R. L. and Schlesinger, H. L., *J. Ass. Off. Agric. Chem.*, **47,** 712 (1964)
114. Brunelle, R. L., *J. Ass. Off. Anal. Chem.*, **51,** 915 (1968)
115. Ronkainen, P., Brummer, S. and Suomalainen, H., *J. Chromatog.*, **28,** 270 (1967)
116. Duncan, R. E. B. and Philp, J. M., *J. Sci. Food Agric.*, **17,** 208 (1966)
117. Kahn, J. H., La Roe, E. G. and Conner, H. A., *J. Food Sci.*, **33,** 395 (1968)
118. Kahn, J. H., Shipley, P. A., La Roe, E. G. and Conner, H. A., *J. Food Sci.*, **34,** 587 (1969)
119. Macfarlane, C., *J. Inst. Brew.*, **74,** 272 (1968)
120. Deluzarche, A., Maillard, A., Maire, J-C., Sommer, J-M. and Wagner, M., *Ann. Falsif. Exp. Chim.*, **60,** 173 (1967)
121. Maurel, A. J. and Lafarge, J-P., *Compt. Rend. Acad. Agric. France*, **49,** 332 (1963)
122. Maurel, A. J., *Compt. Rend. Acad. Agric. France*, **50,** 52 (1964)
123. Maurel, A. J., Sansoulet, O. and Giffard, Y., *Ann. Falsif. Exp. Chim.*, **58,** 291 (1965)
124. Stevens, R. K. and Martin, G. E., *J. Ass. Off. Agric. Chem.*, **48,** 802 (1965)
125. Report of the Government Chemist, HMSO, London, 19 (1965)
126. Moreno, A. M. and Llama, M., *Rev. Soc. Quim. Mex.*, **13,** 1A (1969)
127. Chaveron, II., *Choc. Confis. Fr.*, 23 (1968)
128. Morgantini, M., *Boll. Lab. Chim. Prov.*, **13,** 117 (1962)
129. Cesari, A., Cusmano, A. M. and Boniforti, L., *Boll. Lab. Chim. Prov.*, **15,** 444 (1964)
130. Kahn, J. H., Nickol, G. B. and Conner, H. A., *J. Agric. Food Chem.*, **14,** 460 (1966)
131. Aurand, L. W., Singleton, J. A., Bell, T. A. and Etchells, J. L., *J. Food Sci.*, **31,** 172 (1966)
132. Suomalainen, H. and Kangasperko, J., *Z. Lebensmitt. Forsch.*, **120,** 353 (1963)
133. Hundley, H. K., *J. Ass. Off. Anal. Chem.*, **51,** 1272 (1968)

15
Sugar Products and Syrups

A. Introduction

Many sections of the food industry require sugar or sugar syrup as a raw material, and the control of the composition of these sugar ingredients is vital. The analysis of single sugars is relatively simple, being carried out by established techniques, such as polarimetry and titrimetry. Mixtures of sugars can be a complex analytical problem using these techniques, and GC has been developed for use in the separation and determination of the sugar compositions of syrups, juices and molasses.

Before it is possible to apply a GC method to a product containing several sugars, the conditions of separation, possible derivative formation and analysis have to be evaluated for the components. In 1961 Greenwood, Knox and Milne[1] investigated the possibility of eliminating derivative formation in the analysis of sugars and analysed the products of their thermal decomposition. The pyrolysis products of sucrose, maltose, cellulose, starch, jute hemicellulose and alginic acid were shown in chromatograms, 10 of the 11 ensuing volatiles being identified. This approach does not appear to have been followed by other workers, possibly because it has been more convenient to produce volatile sugar derivatives for chromatography.

Sugars are relatively non-volatile and it is essential that volatile derivatives are made for GC analysis. The first derivatives to be chromatographed were, naturally, those already used for chemical analysis of sugars. Kircher[2] methylated the α- and β-pyranosides of arabinose, xylose, glucose, mannose and galactose, and resolved them on a column of methylated hydroxyethylcellulose. He applied this separation to the analysis of starch, cellulose, guar gum and dextran. Kircher[3] later made the methyl glycosides of glucose, mannose and xylose, and then formed their acetyl derivatives using acetic anhydride in pyridine. Because glycodisation usually gives anomers of sugars, Kircher manipulated the GC conditions of the analysis of the acetylated glycosides such that the anomers were chromatographed together, thus simplifying the resulting chromatogram.

Hause, Hubicki and Hazen[4] formed the hexa-acetate of sorbitol by refluxing this sugar alcohol with acetic anhydride. The acetate was chromatographed on a fluoroalkyl silicone stationary phase using mannitol, similarly acetylated, as internal standard. Sorbitol was found in a sample of shredded coconut at a concentration of 0.5% using this method.

255

Sugar Products and Syrups

The sensitivity obtained with EC detection of halogenated compounds was exploited by Tamura and Imanari[5], who made the TFA derivatives of glucose, galactose and mannose using trifluoroacetic anhydride in tetrahydrofuran and heating at 40–50°C for 10 min. They used several different stationary phases and showed that these derivatives were most suitable for sugar analysis by GC.

Luke[6] also used the TFA derivatives of sucrose and lactose since these were better separated than the corresponding TMS derivatives and the analysis of these sugars in milk chocolate was facilitated. Luke found commercial samples of milk chocolate to contain between 41 and 44% sucrose and between 6.3 and 6.9% lactose. Chocolate syrups, semisweet chocolate and baking chocolate were also analysed for these two sugars.

By far the most popular of the volatile compounds of sugars used for GC purposes are the TMS derivatives. Sweeley *et al.*[7] were the first to produce a viable method based on the formation of these derivatives for carbohydrates, and many sugar chemists have used that method, often with some modification to suit the particular analysis. These derivatives were utilised to determine the sugar and sugar alcohol contents of a variety of foods by Mueller and Goeke[8].

In the analysis of carbohydrates present in substances of animal origin, such as glycoproteins, mucopolysaccharides and glycolipids, it is desirable to determine the small concentrations of *N*-acetylated amino sugars. The most common of these compounds are glucosamine (2-amino-2-deoxy-D-glucose) and galactosamine (2-amino-2-deoxy-D-galactose), and their determination is made difficult by the presence of much larger concentrations of the normal simple hexoses. Karkkainen and Vihko[9] decided to make the TMS derivatives but found that the sole use of TMCS did not silylate the amino group. However, incorporation of BSA effected complete silylation of all the functional groups in the molecule, although this method was found less suitable in quantitative analysis of amino sugars owing to the lability of the Si–N bond and also the overlapping of glucosamine and galactosamine peaks on many stationary phases. These workers used combined GC and MS to classify and differentiate these isomers.

The analysis of the carbohydrate content of milk, dairy products, fruits and vegetables and their products, cereals and flour confectionery products is discussed in the relevant chapters of this book. Chapter 14 gives details of sugar determinations on alcoholic beverages and on the pre-fermented liquors, which has particular significance since sugars are precursors of fusel oils found in distilled beverages.

Some of the more important GC conditions of analysis are given in *Table 15.1*.

B. Cane sugar products

Sugar cane, and sugar beet, are the major sources of sucrose. The first sugar solution in the manufacture of sucrose or syrup from sugar cane is the raw juice and this has been analysed for mono-and di-saccharides by Vidaurreta, Fournier and Burks[10]. The juice was dried under reduced pressure and the residue was trimethylsilylated using myo-inositol as internal standard. An analysis of Louisiana cane juice gave 16% sucrose, 0.74% fructose and 0.40% glucose.

Table 15.1 GC conditions used in the analysis of sugar products and syrups

Class of compound	Ref.	Column dimensions	Stationary phase	Support material	Carrier gas and conditions	Temperature or temperature programme/°C	Detector
TFA derivatives of sugars	6	6ft x 4mm I.D.	20% SE-30	80/100 AW Chromosorb W	N$_2$, 20 ml/min	200	FI
TMS derivatives of sugars	7	6ft x 0.25in O.D.	3% SE-52	80/100 AW silanised Chromosorb W	–,15–20 p.s.i. inlet pressure	210	FI
TMS derivatives of sugars, sugar alcohols	8	3m x 0.125in	3% OV-1	100/120 Gas Chrom Q	N$_2$, 30ml/min	180 for 5 min, 180–300 at 4/min	FI
N-TFA derivatives of butyl esters of amino acids	13	6ft x 0.25in O.D.	0.65% EGA	60/80 Chromosorb W	He, 45ml/min	80 for 5 min, 80–205 at 4/min	FI
TMS derivative of aconitic acid	15	6ft x 0.125in O.D.	10% SE-30	80/100 S810	He, 50ml/min	140 for 16 min, 140–240	FI
Acetaldehyde	16	3ft x 4mm I.D.	15% Carbowax 20M	100/120 HMDS Chromosorb W	N$_2$, 40ml/min	75	FI
TMS derivatives of sugars	21	2ft x 0.25in O.D.	0.25% SE-52	60/80 glass beads (Glassport M)	He, 80ml/min	75–245 at 6.4/min	FI
General	31	4ft x 0.25in O.D.	20% Carbowax 20M	60/80 AW Chromosorb W	He, 50ml/min	50 for 4 min, 50–240 at 3.5/min	TC

257

After removal of some of the water, cane juice becomes molasses or syrup, which is a viscous liquid containing approximately 30% uncrystallisable sucrose and 20% total reducing sugars. Walker[11], using the TMS derivatives for GC, found large concentrations of sucrose and smaller concentrations of glucose and fructose in cane molasses which he examined.

Mahoney and Lucas[12] asserted that although polarimetry was widely adopted for many analytical purposes throughout the sugar industry, it was not an accurate technique when measuring sugars in mixtures of unknown composition. Although attempts have been made to apply corrections to the formulae used to calculate sugar contents by polarimetry, little or no allowance has been made for substances present in sugar syrups which are not removed at the clarification stage. These authors found that the GC analysis using TMS derivatives gave a more accurate analysis of sucrose in the liquors from massecuites. Trehalose was added as internal standard to the liquor which was dissolved in 7:1 pyridine: dimethyl sulphoxide. The mixture was dehydrated with calcium hydride, trimethylsilylated and the derivatives separated by GC. Samples of liquor contained sucrose in the range 58–66%.

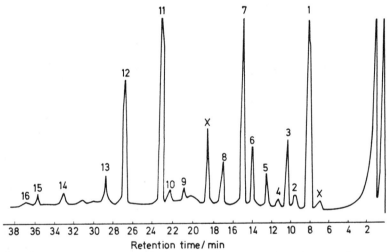

Figure 15.1 Gas chromatographic pattern of amino acids in Iowa sorghum cane syrup: 1, alanine; 2, valine; 3, glycine; 4, isoleucine; 5, leucine; 6, threonine; 7, proline and serine; 8, aminobutyric acid; 9, hydroxyproline; 10, phenylalanine; 11, aspartic acid; 12, glutamic acid; 13, tyrosine; 14, lysine; 15, tryptophan; 16, arginine; X, not identified (after Johnson et al.[13]). For GC details see Table 15.1

Some interest has been shown in the nitrogenous matter found in molasses and syrups, and Johnson, Corliss and Fernandez-Flores[13] have separated and determined the amino acids of cane syrups. The amino acids were concentrated from the syrups by an ion exchange method and GC carried out on the N-TFA derivatives of the butyl esters. Sixteen amino acids were found; *Figure 15.1* is a chromatogram of these amino acids in a sample of Iowa sorghum cane syrup showing the major contributor to be aspartic acid. The sorghum cane is closely allied to the sugar cane and yields a similar syrup.

Aconitic acid has been separated from other acids in table syrups by Johnson and Fernandez-Flores[14]. This acid is the predominant organic acid in sugar cane and sorghum cane syrups, and its concentration is important since high levels can cause problems with clarification and liming operations in syrup production. The analytical method depends on precipitation of acids as their lead salts and trimethylsilylation of the lead salts. Besides aconitic acid, cane syrup contained smaller concentrations of succinic, malic, citric and syringic acids, whereas, in addition to aconitic acid, sorghum syrup contained malic and citric acids with very small concentrations of succinic and fumaric acids.

This method has been extended by Mehltretter and Otten[15], who completely recovered aconitic acid from aqueous solution of the syrup by precipitation as lead aconitate. The lead salt was dried and silylated for GC using tartaric acid as internal standard. These workers found 0.54, 1.93 and 0.99% aconitic acid in Egyptian final molasses, Louisiana A molasses and sorghum syrup, respectively.

C. Beet sugar products

Sugar beet is an important source of sucrose, and it is therefore desirable to know the composition of the extracted juice and of the molasses. Walker[11] used TMS derivatives of carbohydrates in his analysis of beet molasses and found that the presence of the large concentration of sucrose tended to render difficult the quantitative examination of the other sugars. He detected glucose, fructose, inositol, galactinol, raffinose and ketoses in the molasses. The presence of raffinose in a syrup or molasses sample would indicate that sugar beet was its origin and not sugar cane.

Interest has been shown in the lactic acid and other organic acid contents of beet juice and molasses. The level of these acids in the beet products is significant in that it is often directly proportional to any microbiological action which has taken place. Some sugars may be lost by thermophilic bacterial action during the diffusion process or other stages of production and lactic acid is formed. Bacterial action at lower temperature or thermal decomposition of sucrose during processing can result in the formation of propionic, butyric and valeric acids. The presence of these acids, particularly butyric and valeric, in syrups can cause inhibition of microbiological processes such as the production of baker's yeast, and therefore the analysis of these acids is of concern to the bread and flour confectionery industry, who use these raw materials. Oldfield, Parslow and Shore[16] measured lactic acid in beet molasses and processed juice by oxidation to acetaldehyde and determined it by GC. The colorimetric determination of lactic acid using semicarbazide is accurate but the necessary clean-up procedure on an anion exchange resin makes the analysis lengthy. Paper chromatographic methods are quick but not sufficiently quantitative, and Oldfield *et al.*[16] produced a method which was accurate and fairly rapid. Liquors were diluted so that the sucrose content was approximately 5%, HIO_4 was added and the mixture was transferred to the heated injection port of the chromatograph. The temperature of the injection port was adjusted such that sufficient water was evaporated from the sample to give a sufficiently concentrated HIO_4 solution to oxidise the lactic acid to acetaldehyde, which was swept on to the GC column by the carrier gas. A plot of peak height against lactic acid concentration was linear in the range 0–10 μg. Molasses samples from different UK factories contained 0.80–2.42% lactic acid. Lactic acid can be determined in raw beet juice provided an ion exchange resin clean-up procedure is used prior to the oxidation step.

Kiely and O'Drisceoil[17],[18] claim that their method for the determination of lactic acid in beet molasses and juice is more rapid and accurate than that of other workers, including Oldfield *et al.*[16]

A methanolic solution of molasses at pH 1 was filtered to remove sugar and the filtrate was methylated, the methyl lactate being determined by GC. The lactic acid concentration of raw juice from four Irish factories was in the range 0.004–0.020% in 1968–1969 and 0.023–0.028% in 1969–1970. The same factories produced molasses which contained 1.38–2.21% lactic acid in 1968–1969 and 0.98–1.49% in 1969–1970. Acetic, propionic and butyric acids were determined on the molasses as the free acids after removal of sugar by filtration from methanolic solution. In 1968–1969 acetic, propionic and butyric acids were found in molasses from the same four Irish factories at concentrations of 2.60–3.23%, 0.01–0.02% and 0.24–0.55%, respectively. Products from the following year showed significantly lesser concentrations of all three acids, and taken together with the lactic acid figures pointed to microbiologically cleaner products in 1969–1970 than in the previous year.

D. Glucose syrup and glucose caramel

Glucose syrup is known also as liquid glucose, corn syrup and starch syrup and is produced by the partial hydrolysis of starch to yield an aqueous solution of glucose containing maltose and dextrins.

Alexander and Garbutt[19] analysed glucose syrup for glucose via the TMS derivative and used sorbitol as internal standard to overcome some difficulties encountered in reproducibility of peak areas due to changes in sensitivity caused by deposition of eluting substances on to the FID. This GC method was considerably more rapid than paper chromatography and had the advantage of giving an absolute glucose content, whereas the various 'dextrose equivalent' methods gave an empirical figure.

It is often necessary to determine the concentrations of other sugars as well as glucose present in glucose syrup. Brobst and Lott[20] described an improved method for trimethylsilylating the carbohydrates of glucose syrup so that the presence of small amounts of water did not hinder the derivative formation. A typical analysis of glucose syrup showed 20.2–21.0% glucose, 14.4–16.3% maltose, 15.5–16.8% maltotriose and 9.8–10.9% maltotetraose. Hard candy contains liquid glucose and a large proportion of sucrose. The total sugar analysis of this confectionery was achieved by changing the GC conditions to accommodate the sucrose concentrations, which, in four samples of hard candy, ranged from 49 to 56%.

In an improved version of this method, Brobst and Lott[21] used 60/80 mesh glass beads to support 0.25% SE-52 and a subsequently lower operating temperature programme. *Figure 15.2* is a chromatogram of the TMS derivatives of the sugars from a glucose syrup specifically produced for the brewing industry.

Sennello[22] had difficulty in fully trimethylsilylating the carbohydrates of a glucose syrup containing an above-average concentration of fructose. He had used a 21% w/v solution of TSIM, which had the advantage of speed, did not liberate large amounts of heat and minimised anomerisation. After experimenting with various silylating reagents, he used a 1:1 mixture in pyridine of TSIM: TMCS, which was satisfactory in producing a single fructose derivative and made possible its quantitative determination.

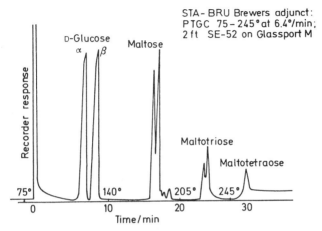

Figure 15.2 Chromatogram of corn syrup (STA-BRU Brewers Adjunct) on 2-ft column, SE-52 on glass beads. Separation by programmed-temperature gas chromatography (PTGC). (After Brobst and Lott[21]). For GC details, see Table 15.1

The problems inherent in quantifying overlapping peaks which occur with anomerisation of the four major sugars of glucose syrup have been overcome by Cayle, Viebrock and Schiaffino[23]. They reduced the sugars to the corresponding alditols with sodium borohydride, prior to formation of the TMS derivatives. Chromatograms demonstrated the loss of anomerisation for glucose and maltose, but a possible drawback to the method was that fructose, like glucose, was reduced to glucitol, so that a glucose determination would have to be carried out enzymically and the fructose concentration obtained by difference.

It has been suggested by Kheiri and Birch[24] that the laevoglucosan content of glucose syrup is a better guide to characterising the syrup than the 'dextrose equivalent' figure. There is a wide range of enzyme-converted and acid-converted syrups commercially available, and it was considered that not even the glucose to maltose ratio or even a complete sugar analysis could serve as a proper index of sample history. Laevoglucosan was determined by GC of the TMS derivative using *p*-terphenyl as internal standard. The laevoglucosan concentration of the syrup was found to be directly proportional to the degree of acid conversion of the starch, and eight out of nine syrups were correctly identified with respect to the degree of the acid and enzyme treatment they had received. These samples contained laevoglucosan in the range 0.015–0.121%.

Although there is considerable interest in the caramelisation of sugars, little investigation has been carried out into the volatiles responsible for caramel flavour. Sugisawa[25] investigated volatiles produced when glucose was heated

261

over a period of time at 150°C. He collected the volatiles on activated charcoal which was subsequently refluxed in boiling water to remove the neutral compounds. These compounds were trapped at −70°C and chromatographed on three different stationary phases. The acidic volatiles were obtained by steam distillation followed by extraction into diethyl ether and were chromatographed as their methyl esters. *Table 15.2* lists the 22 volatiles obtained from glucose caramel.

Table 15.2 Volatiles identified in glucose caramel (after Sugisawa[25])

Acids:	formic
	acetic
	propionic
	isobutyric
	n-butyric
	isovaleric
	n-valeric
Aldehydes:	acetaldehyde
	acrylaldehyde
	propionaldehyde
	isobutyraldehyde
	n-butyraldehyde
	isovaleraldehyde
	n-valeraldehyde
Ketones:	acetone
	methyl ethyl ketone
	methyl propyl ketone
	diethyl ketone
Alcohols:	methanol
	ethanol
Others:	furan
	2-methyl furan

E. Maple syrup

Maple syrup is a flavoured sugar syrup produced by evaporation of the sap of the sugar maple tree. The colourless, flavourless sap becomes coloured and flavoured after heating in air to produce the syrup, the final product being mainly composed of sucrose with smaller amounts of other sugars.

The main GC interest in maple syrup has concerned its flavour, and the Eastern Regional Research Laboratory of the United States Department of Agriculture, Philadelphia, has produced most of the literature on this subject. In 1963 and 1964 Underwood and Filipic[26,27] and Filipic and Underwood[28] made studies of the flavour contributors of maple syrup using a chloroform extract, and showed that vanillin, syringaldehyde and dihydroconiferyl alcohol were major constituents. In 1965 Underwood and Filipic[29] determined the combined syringaldehyde and dihydroconiferyl alcohol using dichloromethane in place of chloroform as extracting solvent, and found 5 and 8 p.p.m. of these combined compounds in two samples of syrup. Besides the above-mentioned flavour compounds, 1-hydroxypropan-2-one was also found as a major constituent by Filipic, Underwood and Willits[30]. Chloroform was the syrup extractant and in turn the chloroform solutions were extracted with diethyl ether. In addition to the major constituents, a small amount of methyl cyclopent-2-en-2-ol-1-one was found and was considered to be an important flavour contributor of the syrup. In a similar analysis Underwood, Filipic and Bell[31] identified 1-hydroxypropan-2-one, isomaltol, 2-hydroxy-3-methylcyclopent-2-en-1-one, furan-2-one, 5-hydroxymethyl-2-furaldehyde, vanillin, syringaldehyde and dihydroconiferyl alcohol in maple syrup.

A cyclic polyol, quebrachitol (1-O-methyl-L- inositol), was found in maple sap as a major constituent by Stinson *et al.*[32] Some of this compound was found in the final syrup but had no influence on colour or flavour.

Johnson and Fernandez-Flores[14] used GC of the TMS derivatives to separate the organic acids of various syrups including maple. The major acid was malic with some succinic acid and small concentrations of fumaric and syringic acids.

F. Honey

Honey is the saccharine substance produced by bees from flower nectar, and contains glucose, fructose, sucrose and dextrins.

As with maple syrup, the attention given to honey by the gas chromatographer has been mainly concerned with flavour constituents and not with sugar composition, although Hadorn, Zuercher and Strack[33] have determined the sugars in honey via TMS derivatives.

Doerrscheidt and Friedrich[34] used two methods to determine volatiles in honey. They used a concentration technique prior to GC on Carbowax 1500 with TC detection, and direct injection when utilising an FID. They isolated 31 volatile compounds and identified some of these, mainly alcohols and the methyl esters of formic, acetic, propionic, butyric, valeric and caproic acids.

The volatile carbonyl compounds of rape, clover and thyme honeys were isolated by ten Hoopen[35]. The aqueous honey extract was saturated with salt and the mixture either steam-distilled or vacuum-diffused into 2,4-dinitrophenylhydrazine solution. The carbonyls were liberated from the hydrazones with α-oxoglutaric acid in a capillary column attached directly to the GC sample inlet. Formaldehyde, acetaldehyde, isobutyraldehyde, acetone and biacetyl were found in the honey samples.

The volatile aromatic compounds of honey were isolated by Cremer and Riedmann[36] using a vapour-current distillation procedure and 120 components were found. These components included alcohols, ketones, aldehydes and esters, and were thought to originate not only from the flowers but also from the fermentation process.

Smith and McCaughey[37] extracted trace amounts of honey lipids with Skellysolve B and diethyl ether, and transesterified the extract using sodium methoxide. The methyl esters were purified on a silicic acid column prior to GC. The main acids present in honey lipids were found to be oleic and palmitic with small amounts of lauric, myristoleic, stearic and linoleic acids. The GC analytical findings were supported by IR spectroscopy.

Sugar Products and Syrups

References

1. Greenwood, C. T., Knox, J. H. and Milne, E., *Chem. & Ind.*, 1878 (1961)
2. Kircher, H. W., *Anal. Chem.*, **32**, 1103 (1960)
3. Kircher, H. W., *Tappi*, **45**, 143 (1962)
4. Hause, J. A., Hubicki, J. A. and Hazen, G. G., *Anal. Chem.*, **34**, 1567 (1962)
5. Tamura, Z. and Imanari, T., *Chem. Pharm. Bull., Japan*, **15**, 246 (1967)
6. Luke, M. A., *J. Ass. Off. Anal. Chem.*, **54**, 937 (1971)
7. Sweeley, C. C., Bentley, R., Makita, M. and Wells, W.W., *J. Amer. Chem. Soc.*, **85**, 2497 (1963)
8. Mueller, B. and Goeke, G., *Dt. LebensmittRdsch.*, **68**, 222 (1972)
9. Karkkainen, J. and Vihko, R., *Carbohyd. Res.*, **10**, 113 (1969)
10. Vidaurreta, L. E., Fournier, L.B. and Burks, M.L., *Anal. Chim. Acta*, **52**, 507 (1970)
11. Walker, H. G., *Int. Sug. J.*, **67**, 237 (1965)
12. Mahoney, V. C. and Lucas, P. C., *Int. Sug. J.*, **73**, 291 (1971)
13. Johnson, A. R., Corliss, R. L. and Fernandez-Flores, E., *J. Ass. Off. Anal. Chem.*, **54**, 61 (1971)
14. Johnson, A. R. and Fernandez-Flores, E., *J. Ass. Off. Anal. Chem.*, **52**, 559 (1969)
15. Mehltretter, C. L. and Otten, J. G., *Int. Sug. J.*, **73**, 235 (1971)
16. Oldfield, J. F. T., Parslow, R. and Shore, M., *Int. Sug. J.*, **72**, 35 (1970)
17. Kiely, M. and O'Drisceoil, P., *Int. Sug. J.*, **73**, 135 (1971)
18. Kiely, M. and O'Drisceoil, P., *Int. Sug. J.*, **73**, 196 (1971)
19. Alexander, R. J. and Garbutt, J. T., *Anal. Chem.*, **37**, 303 (1965)
20. Brobst, K. M. and Lott, C. E., *Cereal Chem.*, **43**, 35 (1966)
21. Brobst, K. M. and Lott, C. E., *Proc. Amer. Soc. Brew. Chem.*, 71 (1966)
22. Sennello, L. T., *J. Chromatog.*, **56**, 121 (1971)
23. Cayle, T., Viebrock, F. and Schiaffino, J., *Cereal Chem.*, **45**, 154 (1968)
24. Kheiri, M. S. A. and Birch, G. G., *Cereal Chem.*, **46**, 400 (1969)
25. Sugisawa, H., *J. Food Sci.*, **31**, 381 (1966)
26. Underwood, J. C. and Filipic, V. J., *J. Ass. Off. Agric. Chem.*, **46**, 334 (1963)
27. Underwood, J. C. and Filipic, V. J., *J. Food Sci.*, **29**, 814 (1964)
28. Filipic, V. J. and Underwood, J. C., *J. Food Sci.*, **29**, 464 (1964)
29. Underwood, J. C. and Filipic, V. J., *J. Ass. Off. Agric. Chem.*, **48**, 689 (1965)
30. Filipic, V. J., Underwood, J. C. and Willits, C. O., *J. Food Sci.*, **30**, 1008 (1965)
31. Underwood, J. C., Filipie, V. J. and Bell, R. A., *J. Ass. Off. Anal. Chem.*, **52**, 717 (1969)
32. Stinson, E. E., Dooley, C. J., Purcell, J. M. and Ard, J. S., *J. Agric. Food Chem.*, **15**, 394 (1967)
33. Hadorn, H., Zuercher, K. and Strack, C., *Mitt. Geb. Lebensmitt. Hyg.*, **65**, 198 (1974)
34. Doerrscheidt, W. and Friedrich, K., *J. Chromatog.*, **7**, 13 (1962)
35. ten Hoopen, H. J. G., *Z. Lebensmitt. Forsch.*, **119**, 478 (1963)
36. Cremer, E. and Riedmann, M., *Z. Anal. Chem.*, **212**, 31 (1965)
37. Smith, M. R. and McCaughey, W. F., *J. Food Sci.*, **31**, 902 (1966)

16
Cereals and their Products

A. Introduction

Bread and cereal products based on wheat, maize, rye, oats and rice, together with other starch-containing foods, such as potato and soya, provide the basis of the carbohydrate intake for man and animals. Some of these foods also contribute significantly to the protein in the human diet.

Wheat is the most important of the cereals, and bread made from wheat flour provides vitamins of the B group plus iron and calcium. Macroanalytical techniques have proved to be satisfactory in the estimation of the total starch, protein and fat of flour and bread, and the composition of the protein and the fat, particularly the latter, has become facilitated by the use of GC techniques. Bread contains between 1 and 4% of fat, and its composition has a direct bearing on the type and quality of the final product.

The amino acid composition of cereal protein, particularly wheat and maize, has been achieved using GC, as has the determination of flour sugars.

GC has been used as the main technique in the analysis of food flavours, and although bread does not have a strong flavour, gas chromatographers have deduced some of the compounds responsible for it.

Additives and contaminants of cereal products have been covered in the relevant chapters in this book. These include the analysis of propionic and sorbic acids in bread (Chapter 17), emulsifiers and stabilisers in bread (Chapter 18), antioxidants in breakfast cereals (Chapter 17), sorbitol in biscuit (Chapter 19) and fumigants in flour (Chapter 20).

The GC conditions of some of the more important references in this present chapter are given in *Table 16.1*.

B. Wheat flour, dough and bread

The composition of the free and bound lipids of wheat have a profound effect on the ageing of flour and, hence, on the quality of bread made from such flour. Great advances in this field of research have been made by the use of GC, frequently after silicic acid column chromatographic separation of lipid constituents.

Table 16.1 GC conditions used in the analysis of cereals and their products

Class of compound	Ref.	Column dimensions	Stationary phase	Support material	Carrier gas and conditions	Temperature or temperature programme/°C	Detector
Methyl esters of fatty acids	6	5ft x 0.25in O.D.	25% DEGS plus 2% H_2PO_4	80/100 AW Chromosorb W	He, 90ml/min	200	FI
Methyl esters of fatty acids	10	2m x 3mm	12% EGA	Chromosorb W	N_2, –	195	FI
Methyl esters of fatty acids	15,16	2m x 4.6mm	15% BDS	60/100 Celite	He, 103ml/min	187	–
Acetic acid	17	10ft x 0.125in O.D.	10% LAC-1-R (296)	80/100 Gas Chrom Q	N_2, 50ml/min	100	FI
TFA derivatives of butyl esters of amino acids	27	1.5m x 4mm I.D.	0.65% EGA	80/100 AW Chromosorb W	N_2, –	70–230 at 6/min	FI
		1.5m x 4mm I.D.	2% OV-17 plus 1% OV-210	100/120 Supelcoport	N_2, –	70–230 at 6/min	FI
TFA derivatives of methyl esters of amino acids	28	4.5ft x 0.125in O.D.	1.1% NPGS	80/100 AW, DMCS Chromosorb G	–	70–230	FI
		7.5ft x 0.125in O.D.	3.3% OV-17	80/100 AW, DMCS Chromosorb G	–	70–230, cooled to 160, programmed to 230	FI
TMS derivatives of sugars	19	15ft x 0.125in O.D.	1% SE-30	100/120 Gas Chrom Q	He, 50ml/min	170 for 38 min, 170–270 at 4/min	FI
Carbonyls	23	23ft x 0.125in O.D.	1% TCEP	AW 60/80 brick dust	N_2, 70ml/min	50 or 100 or 150	FI

266

As early as 1958, Coppock, Fisher and Ritchie[1] examined wheat flour oil and used GC in the analysis of the composite fatty acids of the acetone-insoluble fraction of the oil. They showed the presence of unsaturated C_6 and C_7 acids which were thought possibly to have an influence on bread aroma and flavour. GC of the methyl esters of the fatty acids was used in the analysis, and the saturated acids C_8-C_{13}, C_{16}, C_{17} and C_{18}, together with undecenoic, oleic and linoleic were also found. The C_{18} acid fraction was composed mainly of oleic and linoleic acids, and the C_{16} acid fraction was almost entirely palmitic acid. In addition to these acids, unsaturated C_{19} acid was found in flour oil for the first time. It was thought that certain ingredients of flour oil contributed to crumb firmness and loaf volume in bread and that the lipids might be those ingredients.

Muntoni, Tiscornia and de Giuli[2] studied the fat composition of flour and found after hydrolysis that the ratio of linoleic to oleic acid was higher in soft than in hard wheat. The fat was extracted by diethyl ether and, after hydrolysis, the fatty acids were converted into their methyl esters for GC analysis on an EGS stationary phase.

A fractionation of wheat flour lipids using different solvents for extraction was carried out by Fisher and Broughton[3] in order to study the fatty acid composition of each fraction. Wholemeal flour was extracted first with methanol and then with 1:1 methanol:chloroform. The two extracts were combined and the mixture was extracted with petrol to give soluble and insoluble fractions. A portion of the petrol-soluble fraction was treated with acetone to yield an

Table 16.2 Fatty acid methyl ester composition (%) in three fractions of crude flour lipids (after Fisher and Broughton[3])

Fatty acid by C number	Petrol-soluble fraction	Petrol-insoluble fraction	Petrol-soluble– acetone- insoluble fraction
12:0	trace	0.3	0.1
14:0	0.1	0.1	0.1
15:0	0.1	0.1	0.2
16:0	17.3	22.1	22.3
17:0	–	0.2	0.2
18:0	0.9	0.7	0.8
18:1	14.3	9.4	9.6
18:2	62.8	63.6	62.9
18:3	3.0	2.5	2.6
20:0–22:0 including unsaturateds	0.4	1.1	1.3

acetone-insoluble fraction; *Table 16.2* shows the fatty acid compositions of these three fractions of crude flour lipids. The table shows that linoleic acid is the dominant acid in all fractions and has approximately the same concentration in the three fractions. The other major acids are palmitic and oleic.

In addition to silicic acid column chromatographic separation prior to GC analysis, cereal chemists have used counter-current distribution processes in order to obtain extracts of greater definition. Both these methods of separation were used by Nelson, Glass and Geddes[4] in the examination of the triglycerides of wheat germ, bran and endosperm. Initial extraction of wheat flour was made by using water-saturated n-butanol, and after the column chromatography, the counter-current separation was achieved with a solvent comprising 2:2:5 nitro-ethane:furfuraldehyde:Skellysolve F. After saponification of the final extract, the fatty acids were methylated and the resulting methyl esters chromatographed on a BDS polyester stationary phase. The average fatty acid composition of whole wheat triglycerides showed linoleic acid at 59% to be the dominant acid and both oleic and palmitic acids to be present at just over 16%.

In order to attempt to understand more fully the influence of lipids present in flour on bread quality, free and bound lipids were separated before silicic acid column chromatography by McKillican and Sims[5]. Wheat endosperm was extracted with hexane to give the free lipid fraction and the residue was further extracted with water-saturated n-butanol to give the bound lipid fraction. After the column chromatography, the lipid extracts were further cleaned up by preparative TLC and the fractions were used for the separate determinations of triglycerides and sterol esters. GC was used to analyse the fatty acids of the triglycerides, the fatty acids of the sterol esters and also the sterols themselves. Three different types of Canadian wheats—namely Hard Red Spring, Soft White Spring and Amber Durum—were compared and similar endosperm lipid patterns were obtained, the differences being in the relative concentrations of some of the triglycerides and sterol esters. Bound lipids of all three wheat varieties consisted of more palmitic acid and less oleic acid than free lipids, both classes having a predominance of linoleic acid. The major sterol ester in both free and bound lipids was β-sitosteryl palmitate.

Burkwall and Glass[6] used diethyl ether to extract the Selkirk variety of whole wheat, bran, germ and endosperm to obtain the free lipids, and re-extracted the residue with water-saturated n-butanol to obtain the bound lipids. The lipid fractions were hydrolysed and methylated with diazomethane and the resulting methyl esters analysed by GC. *Table 16.3* shows the fatty acid

Table 16.3 Fatty acid composition of wheat and its milled products as a percentage of the original wheat (after Burkwall and Glass[6])

Fatty acid by C number	Wheat	Flour	Bran	Shorts
16:0	0.30	0.16	0.13	0.03
18:0	0.02	0.01	0.01	nil
18:1	0.24	0.09	0.13	0.03
18:2	1.05	0.49	0.44	0.11
18:3	0.09	0.04	0.04	0.01
others	0.03	0.01	0.01	nil

distribution in the total lipids of whole wheat, flour, bran and shorts. Linoleic acid was the major acid found in whole wheat and its milled products, and palmitic and oleic acids were the only other acids found in significant concentrations. Traces of the odd-numbered carbon acids from C_{11} to C_{21} were found in all the wheat products. The distribution of fatty acids in flour differed from that in bran and again from that in shorts. In all products free lipids had a higher oleic and linolenic acid content and a lower palmitic and linoleic acid content when compared with bound lipids.

The behaviour of lipids in doughmaking and breadmaking is a practical extension of the academic work carried out by cereal chemists concerning the nature of flour lipids. Daniels *et al.*[7] examined the binding of lipid to protein in three different commercial dough-mixing processes. This lipid–protein interaction was reported as extensive during dough-mixing and subsequent baking, and an analysis of the relative distribution of component fatty acids, in both the free and bound lipids, was carried out via GC of the methyl esters. The binding of protein with added as well as natural lipids was evaluated. It was concluded that non-selective binding of triglyceride lipid plays a greater part in dough-mixing than was hitherto supposed.

TMS derivatives of partial glycerides and other hydroxy compounds of flour, dough ingredients and dough were made by Wood[8] in a study of the changes which occur during dough-mixing. He found that after mixing the dough in air, the concentration of free fatty acids fell as the diglyceride concentration increased. The action of lipoxidase was evident in the decrease of total C_{18} relative to total C_{16} glycerides.

Lecithin is an important ingredient of flour, since it imparts a softness to the bread crumb. McKillican[9] examined Thatcher wheat endosperm specifically for lecithins and lysolecithins. The flour was extracted with hexane to obtain free lipids and the residue was re-extracted with water-saturated n-butanol to obtain the bound lipids. BF_3–methanol was the methylating medium used in the transesterification of lecithins, lysolecithins and hydrolysis products to produce esters for GC analysis after a TLC clean-up procedure. Results showed that the acid in the α position in lecithins was primarily palmitic or linoleic, whereas the β position was taken up usually by linoleic acid. Lysolecithins were similar in fatty acid composition to lecithins, but contained rather more linoleic than palmitic acid in the α position and differed from lecithins in containing approximately 2% linolenic acid in that position.

In an investigation of lipid content, samples of flour or freeze-dried dough were extracted by Graveland[10] using various solvents in a percolation procedure. He tried hexane, diethyl ether, chloroform, 2:1 ethanol:diethyl ether and 2:1 chloroform: methanol as extracting solvents before choosing 10:10:1 benzene: ethanol:water as the most suitable extractant for free and bound lipids. Fourteen polar and 12 non-polar lipids were separated by silica gel TLC, and each lipid was analysed for fatty acid composition via the GC of its methyl esters. In accordance with the findings of other workers[3,4,6], he found that linoleic was the major acid present in both flour and dough lipids and that palmitic and oleic acids were present in significant concentration. The linoleic acid concentration of the components of flour lipids is affected in different ways according to the conditions under which dough-mixing is conducted. *Figure 16.1* is a graph comparing the oxidation of linoleic acid with mixing time in atmospheres of nitrogen, oxygen and air, the latter with and without lipoxidase, and shows virtually no oxidation in nitrogen but a rapid decrease in the linoleic acid concentration in the presence of oxygen or air.

Experiments have been carried out by Daniels[11] to ascertain at what level the addition to flour of chlorine dioxide or chlorine itself causes changes in lipid concentration. GC was used to determine the lipid fatty acids, and it was found that the addition of 120 p.p.m. chlorine dioxide plus storage up to 15 weeks showed no alteration in lipid content. Treatment of flour with larger doses of chlorine dioxide and storage for longer periods than 15 weeks, at the 120 p.p.m. level, caused changes in the lipid composition. The normal usage of chlorine dioxide is in the range 15–30 p.p.m., and therefore the normal addition to flour of this improver and bleacher should not alter the lipid composition. In the experimental work, Daniels percolated petrol and then acetone through the flour and distilled off these solvents under vacuum and dried the resulting flour oil over P_2O_5. The oil was hydrolysed, the resulting aqueous phase acidified and the fatty acids extracted into petrol. Diazomethane was used to form the methyl esters for GC analysis.

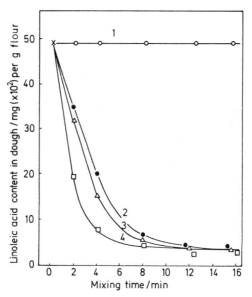

Figure 16.1 Rates of oxidation of linoleic acid in dough: 1, mixing in nitrogen; 2, mixing in air; 3, mixing in air with lipoxidase; 4, mixing in oxygen (after Graveland[10])

Methyl esters of fatty acid components of the glycerides of fats added to the dough in breadmaking have been utilised in GC analysis. Mutoni, Tiscornia and de Giuli[12] made three breads containing 5% of olive oil, lard and butter, respectively, and compared the fatty acid ratios of the extracted and hydrolysed triglycerides with those in a control bread. The presence of these added fats was verified up to 6 months after their additions to the breads. A similar study was carried out by Olivari and Benassi[13,14], who analysed flour made from five different wheat varieties with and without the addition of olive oil and lard. The ratio of saturated to unsaturated fatty acids was established as a criterion for the verification of the addition of olive oil and lard to flour. Baking diminished the linoleic and linolenic acid content and slightly increased the content of oleic acid and saturated acids.

Franciosi and Giovannini[15,16] studied the effects of the addition of egg to flour on the ratios of fatty acids produced after diethyl ether extraction of alimentary pasta. They also compared the differences in the same fatty acid ratios of pasta made entirely of hard wheat and of pasta made from mixed hard and soft wheat. GC of the methyl esters was used for analysis, and the ratios of various combinations of palmitic, oleic and linoleic acids were used to verify egg addition to such pasta and to authenticate hard wheat flour pasta. *Table 16.4* shows the range of the fatty acid ratios for the various pastas analysed. The addition of even a small concentration of egg to pasta reduces the linoleic to oleic acid ratio to below 2.0, which is outside the normal range for pasta without egg. The other ratio of value in determining the presence of egg is that of the palmitic plus oleic to linoleic acid.

Table 16.4 Ratios of fatty acids in different types of pasta (after Franciosi and Giovannini [15,16])

Ratio of acids by C number	Pasta without egg	Pasta with egg	Hard wheat pasta	Mixed wheat pasta
$\frac{16:0}{18:1}$	0.93−1.23	0.55−1.15	0.92−1.16	1.20−1.99
$\frac{18:2}{18:1}$	2.09−3.48	0.20−1.71	1.27−3.16	2.59−4.02
$\frac{18:2}{16:0}$	2.05−3.30	0.27−2.61	−	−
$\frac{16:0 + 18:1}{18:2}$	0.64−0.96	0.96−8.63	−	−
$\frac{16:0 + 18:2}{18:1}$	−	−	2.19−4.30	4.19−5.83

In the distinction of the pasta made solely from hard wheat, these workers suggest that the palmitic to oleic acid ratio should not exceed 1.20, the linoleic to oleic acid ratio should not exceed 3.30 and the palmitic plus linoleic to oleic acid ratio should not exceed 4.50.

Although the bulk of the uses of GC in the analysis of bread and flour concerns natural lipids and added oils or fats, analyses of acetic acid, sugars and carotenoids have been made.

Hunter, Walden and Kline[17] determined the acetic acid of sourdough French bread after extraction with acetone, removal of non-acidic substances with dichloromethane, acidification and centrifugation of the extract. Sourdough French bread, which is peculiar to Northern California, is made from flour, water, salt and a starter sponge which provides the micro-organisms necessary for fermentation and souring. This bread contained 5−10 times the acetic acid content of conventional bread, the acidity of which normally comes from carbon dioxide and lactic acid.

Diemair and Schams[18] examined several different foods for their lower fatty acid contents, and included rye bread in the study, which is of academic interest only.

The GC determination of sugars in wheat, flour, bread and wheat flakes was carried out by Mason and Slover[19]. They converted the reducing sugars to their oximes and then silylated the mixture of these oximes and non-reducing sugars. All hydroxyl groups were silylated and the resulting TMS derivatives were chromatographed on a 1% SE-30 stationary phase. Wheat and flour contained sucrose and raffinose as the major components with smaller amounts of glucose and fructose, whereas bread contained glucose and lactose as the major components and smaller amounts of fructose, glucose and maltose. Wheat flakes cereal contained 7% sucrose as the major sugar with smaller amounts of fructose, glucose, maltose and traces of raffinose. *Figure 16.2* is a chromatogram showing sugar derivatives of wheat flakes with inositol as the internal standard.

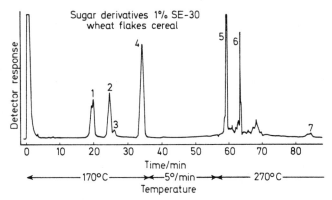

Figure 16.2 Chromatogram of oxime TMS ethers of reducing sugars and TMS ethers of non-reducing sugars from wheat flakes cereal. The peak identities are: 1, fructose; 2, glucose; 3, glucose; 4, inositol; 5, sucrose; 6, maltose; 7, raffinose. (After Mason and Slover[19]). For full GC details, see Table 16.1

The wheat flour carotenoids were extracted from Mindum and Thatcher variety wheats and TLC was used to separate the carotenoids in an academic study by Lepage and Sims[20]. GC was used in the identification of the fatty acids, which were palmitic, stearic, oleic, linoleic and linolenic acids. In Thatcher flour, esters accounted for 78% of total carotenoids, whereas free lutein accounted for 85% in Mindum flour. No β-carotene was found in any extract.

Although bread and flour would not be commonly considered as highly flavoured food products, nevertheless some interest has been shown in the flavour and aroma constituents of bread and bread pre-ferment liquid. GC, IR spectroscopy and MS were used in the identification of 27 diethyl ether-soluble neutral components of bread pre-ferment liquid by Smith and Coffman[21]. Most of these compounds are well-known products of yeast fermentation processes and were dominated by ethanol and isomers of 2,3-butanediol. Compounds not previously reported as fermentation products of bread were ethyl formate, γ-butyrolactone, 3-hydroxypropyl acetate and 2-phenylethanol.

Ethanol was also the largest component of bread prepared by a pre-ferment method as found by Wick, Figueiredo and Wallace[22]. Flavour distillates were prepared and many volatiles were identified or tentatively identified by GC and TLC methods, including several alcohols and carbonyls. Addition of 1000 p.p.m. proline to the dough gave improved bread aroma, and although the composition of the volatile fraction was unchanged, the total odour concentration increased from between 3 and 5 p.p.m. to between 8 and 10 p.p.m. The major components were ethanol as the largest, plus propanol, isobutanol, isopentanol, acetoin, furfuraldehyde and acetic acid.

Hunter and Walden[23] used semicarbazide to form the semicarbazones of carbonyl volatiles in a preliminary separation stage prior to GC of the carbonyl constituents of bread. The semicarbazones were formed in a dichloromethane solvent medium and the original carbonyls were subsequently regenerated by phosphoric acid. Tentative identifications of various aldehydes and ketones were made, and a typical chromatogram of bread aroma concentrate is shown in *Figure 16.3*. Kevei[24] has more recently studied the changes in aroma composition during breadmaking, using a GC method.

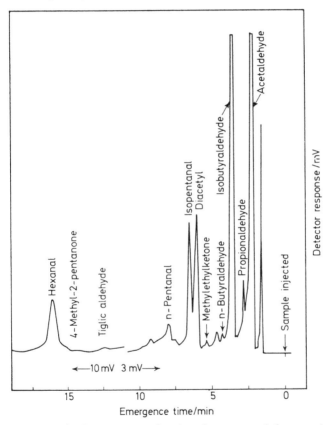

Figure 16.3 Chromatogram of carbonyls regenerated from semi-carbazones obtained from a bread aroma concentrate (after Hunter and Walden[23]). For GC details, see Table 16.1

C. Other cereals and pulses

Very little GC analysis has been applied to studies concerning cereals other than wheat. The oil content and amino acid composition of maize has been reported and there has been a small interest in the composition of soyabean meal.

Black, Spyres and Brekke[25,26] extracted the oil from dry-milled maize meal and simultaneously transesterified the fatty acids of the glycerides to yield methyl esters for GC. Benzene was more effective than petrol in extracting a larger amount of oil. The oil content of a sample containing bound or oxidised oil can be determined from the methyl palmitate content, and this method is especially suitable for maize samples containing less than 1% of oil. The oil of wheat, soyabean, mustard and rape seed meals can also be determined by this method.

The amino acid contents of maize and soyabean meal have been analysed using GC of the butyl esters of the *N*-TFA derivatives by Zumwalt, Kuo and Gehrke[27]. The proteins of both foods were hydrolysed with 6N hydrochloric acid and the resulting amino acids were separated on a cation exchange resin. After the derivative formation, the esters from the maize were chromatographed on an EGA stationary phase and those of the soyabean meal on an OV-17 plus OV-210 stationary phase. *Table 16.5* shows the amino acid composition of whole maize and soyabean meal.

Table 16.5 Amino acid (%) of whole maize and soyabean meal (after Zumwalt *et al.*[27])

Amino acid	Maize	Soyabean meal
Alanine	0.65	2.17
Valine	0.43	2.33
Glycine	0.33	2.03
Isoleucine	0.29	2.11
Leucine	0.97	3.45
Proline	0.90	2.86
Threonine	0.33	1.87
Serine	0.48	2.79
Methionine	0.19	0.48
Hydroxyproline	0.03	0.10
Phenylalanine	0.40	2.34
Aspartic acid	0.52	5.23
Glutamic acid	1.40	8.05
Tyrosine	0.19	1.34
Ornithine	trace	trace
Lysine	0.26	2.96
Arginine	0.36	3.18
Cystine	0.04	0.26
Histidine	0.35	1.58
Tryptophan (destroyed in the analysis)	–	–

The methyl esters of the *N*-TFA derivatives of amino acids were preferred to the butyl esters by Shearer and Warner[28] in their analyses of maize and oats. The cereal protein was hydrolysed with 6N hydrochloric acid and, after derivative formation, two stationary phases were used to separate 20 amino acids. Neither NPGS nor OV-17 phases on their own completely separated all the amino acids and therefore both were used. Before the acid hydrolysis, oats were initially extracted with petrol to remove some of the interfering substances. The major amino acid found in maize was glutamic acid, at 19.6 g per 100 g crude protein, with approximately half as much leucine. Proline, phenylalanine and aspartic acid were next highest in concentration at approximately

6–7 g per 100 g crude protein. Glutamic acid was the major amino acid of oats at 20.8 g per 100 g crude protein, followed by valine, arginine and leucine at concentrations of 16.1, 8.7 and 7.8 g per 100 g crude protein, respectively. The extra clean-up afforded by the ion exchange procedure as given by Zumwalt *et al.*[27] yields a much cleaner extract for subsequent derivative formation and GC, and is recommended for inclusion in any amino acid analysis of cereals.

D. Biscuit

GC has been used for the detection of fats other than butter fat in butter biscuits by Hahn and Keding[29]. The sample was extracted with a 1:1 mixture of diethyl ether and petrol, and the concentrated extract refluxed with methanolic hydrochloric acid to produce methyl esters for GC. Lauric, myristic, palmitic, stearic, oleic, linoleic and palmitoleic acids were determined quantitatively, and adulteration with fat other than butter fat was indicated when either the lauric to palmitic acid ratio or the linoleic to palmitic acid ratio exceeded 0.15.

References

1. Coppock, J. B. M., Fisher, N. and Ritchie, M. L., *J. Sci. Food Agric.,* **9,** 498 (1958)
2. Muntoni, F., Tiscornia, E. and de Giuli, G., *Riv. Ital. Sostanze Grasse,* **41,** 154 (1964)
3. Fisher, N. and Broughton, M. E., *Chem. & Ind.,* 869 (1960)
4. Nelson, J. H., Glass, R. L. and Geddes, W. F., *Cereal Chem.,* **40,** 343 (1963)
5. McKillican, M. E. and Sims, R. P. A., *J. Amer. Oil Chem. Soc.,* **41,** 340 (1964)
6. Burkwall, M. P. and Glass, R. L., *Cereal Chem.,* **42,** 236 (1965)
7. Daniels, N. W. R., Richmond, J. W., Eggitt, P. W. R. and Coppock, J. B. M., *J. Sci. Food Agric.,* **17,** 20 (1966)
8. Wood, P. S., *Column,* **3,** 7 (1969)
9. McKillican, M. E., *J. Amer. Oil Chem. Soc.,* **44,** 200 (1967)
10. Graveland, A., *J. Amer. Oil Chem. Soc.,* **45,** 834 (1968)
11. Daniels, D. G. H., *J. Sci. Food Agric.,* **11,** 664 (1960)
12. Muntoni, F., Tiscornia, E. and de Giuli, G., *Boll. Lab. Chim. Prov.,* **15,** 427 (1964)
13. Olivari, L. and Benassi, R., *Boll. Lab. Chim. Prov.,* **16,** 212 (1965)
14. Olivari, L. and Benassi, R., *Boll. Lab. Chim. Prov.,* **16,** 377 (1965)
15. Franciosi, A. and Giovannini, G., *Boll. Lab. Chim. Prov.,* **15,** 131 (1964)
16. Franciosi, A. and Giovannini, G., *Boll. Lab. Chim. Prov.,* **15,** 336 (1964)
17. Hunter, I. R., Walden, M. K. and Kline, L., *Cereal Chem.,* **47,** 189 (1970)
18. Diemair, W. and Schams, F., *Z. Lebensmitt. Forsch.,* **112,** 457 (1960)
19. Mason, B. S. and Slover, H. T., *J. Agric. Food Chem.,* **19,** 551 (1971)
20. Lepage, M. and Sims, R. P. A., *Cereal Chem.,* **45,** 600 (1968)
21. Smith, D. E. and Coffman, J. R., *Anal. Chem.,* **32,** 1733 (1960)
22. Wick, E. L., Figueiredo, M. de and Wallace, D. H., *Cereal Chem.,* **41,** 300 (1964)
23. Hunter, I. R. and Walden, M. K., *J. Gas Chromat.,* **4,** 246 (1966)
24. Kevei, E., *Nahrung,* **18,** 269 (1974)
25. Black, L. T., Spyres, G. G. and Brekke, O. L., *Cereal Chem.,* **44,** 152 (1967)
26. Black, L. T., Spyres, G. G. and Brekke, O. L., *Cereal Chem.,* **46,** 63 (1969)
27. Zumwalt, R. W., Kuo, K. and Gehrke, C. W., *J. Chromatog.,* **55,** 267 (1971)
28. Shearer, D. A. and Warner, R. M., *Int. J. Environ. Anal. Chem.,* **1,** 11 (1971)
29. Hahn, H. and Keding, H., *Dt. Lebensmitt.-Rdsch.,* **65,** 320 (1969)

Part 4

Food Additives

17

Preservatives and Antioxidants

A. *Introduction*

Most countries of the world allow the use of preservatives and antioxidants in certain foods and have legislated accordingly. In the UK *The Preservatives in Food Regulations 1974*[1] define a preservative as 'any substance which is capable of inhibiting, retarding or arresting the growth of micro-organisms or any deterioration of food due to micro-organisms or of masking the evidence of any such deterioration . . .'. The *Antioxidant in Food Regulations 1974*[2] define an antioxidant as 'any substance which is capable of delaying, retarding or preventing the development in food of rancidity or other flavour deterioration due to oxidation . . .'. In the UK the legislation has been framed to allow certain preservatives and antioxidants to be present in specified foods up to a maximum concentration given in mg/kg.

Before the advent of GC, most analytical methods for these two classes of additives, with the notable exception of the distillation and volumetric methods for the determination of sulphur dioxide, were based on colorimetry after extraction of the additives from the food. Such methods were, and to some extent are, the generally accepted ones. GC has three main advantages: sensitivity in the quantitative determination of individual preservatives and antioxidants; versatility in the quantitative determination of more than one additive in the same analysis; and specificity in the differentiation of very similar compounds which otherwise behave identically in colorimetric reactions.

Many GC methods for preservatives and antioxidants have superseded earlier chemical methods and are routine in many food analytical laboratories. The GC conditions of some of the more important publications cited are listed in *Table 17.1*.

B. *Preservatives*

GC has been used in the analysis of the following preservatives: benzoic acid and its hydroxy and halogenated derivatives; dehydroacetic acid and acetic acid derivatives; propionic acid; sorbic acid; salicylic acid; diethyl pyrocarbonate; diphenyl and *o*-phenyl phenol; ethylene oxide; nitrous oxide; and thiabendazole.

Table 17.1 GC conditions used in the analysis of preservatives and antioxidants

Class of compound	Ref.	Column dimensions	Stationary phase	Support material	Carrier gas and conditions	Temperature or temperature programme/°C	Detector
Benzoic acid, sorbic acid, hydroxybenzoate esters	15	0.5m x 6mm O.D.	100/120 Porapak Q	100/120 Porapak Q	N_2, 80 ml/min	240	FI
Sorbic acid	21	6ft x 4mm I.D.	7.5% EGA + 2% H_3PO_4	90/100 Anakrom ABS	N_2, 115ml/min	165	FI
Benzoic, sorbic and propionic acids	14	2m x 2mm I.D.	5% Carbowax 20M – terephthalic acid	60/80 AW, DMCS Chromosorb W	N_2, 65ml/min	100–210 at 5/min	FI
TMS derivatives of benzoic and sorbic acids and hydroxybenzoate esters	11	3m x –	3% SE-30	100/120 Aeropak	N_2, 30ml/min	90–290 at 8/min	FI
Diethyl carbonate	24	6ft x 0.125in O.D.	15% trimethylol-propantripelargonate	60/100 Celite 545	N_2, 35ml/min	80	FI
n-Heptanol	16	4ft x 0.25in	25% Carbowax 20M	80/100 Chromosorb P	He, 60ml/min	50–150 at 4/min	FI
Diphenyl, o-phenyl phenol	34	6ft x –	20% silicone oil	100/120 Celite	Ar, 20ml/min	160	AI
BHA, BHT	57	6ft x 4.5mm I.D.	10% DO-200	80/100 Gas Chrom Q	He, 50ml/min	160	FI
BHA, BHT	57	4ft x 4.5mm I.D.	10% Carbowax 20M	80/100 Gas Chrom Q	He, 50ml/min	190	FI
BHA, BHT, NDGA, DLTDP, the TMS derivatives of PG and THBP and the dimethyl ester of TDPA	59	16ft x 3mm	3% GE-XE-60	60/80 Gas Chrom Q	N_2, 125ml/min	100–150 at 10/min 150–250 at 5/min and other programmes	FI

1. Benzoic acid and its hydroxy and halogenated derivatives

Undissociated benzoic acid has a retarding influence on the growth of yeasts and moulds, although it is in the form of the sodium or potassium salts that it is most often added to foods. These foods are therefore more acidic than basic in nature for this preservative to be effective.

4-Hydroxy benzoic acid has the same action as the parent acid and is used in the alkyl ester form, usually methyl or propyl. The antimicrobial action of these esters increases with the length of the alkyl chain from methyl to amyl, and it is in the ester form that this preservative is most effective.

Benzoic acid is normally extracted from foods by either steam distillation or diethyl ether extraction from acid solution. Colorimetric methods of analysis are standard, although sometimes interfering substances from the food can be troublesome and TLC or column chromatographic clean-up procedures are necessary prior to analysis. Colorimetric methods are also used in the determination of the esters of 4-hydroxy benzoic acid, after hydrolysis of the esters and extraction of the free acid from acid solution.

Similar extraction and colorimetric analysis is also applied to the analysis of 4-chlorobenzoic acid, which is used in some countries to inhibit mould and rope in bread.

The problems associated with clean and efficient extraction of benzoic acid and 4-hydroxybenzoate esters from foods still remain to some extent, although some co-extracted substances may be separated from the preservatives by GC. Nobe and Stanley[3] used an acid digestion technique in the analysis of sodium benzoate in meat and extracted the acid digest with diethyl ether. After the preliminary extraction procedure, the benzoic acid was re-extracted into chloroform and chromatographed on a 2,2-dimethylpropane-1,3-diol isophthalate stationary phase. Sodium benzoate was determined in meat down to 11 p.p.m. by this method.

Benzoic and soroic acids were extracted from wine by Wuerdig[4] after salt addition and acidification. 2:1 diethyl ether:pentane was used as extractant and the preservative acids plus *o*-xylene internal standard were methylated with diazomethane for GC analysis. Standard additions of both benzoic and sorbic acids to wine resulted in very good recoveries.

Benzoic acid in meat and meat products was determined using GC by Clarke, Humphreys and Stoilis[5]. Although the method was primarily intended for the determination of benzoic, *o*- and *p*-chlorobenzoic and sorbic acids and also 4-hydroxybenzoates in meat sold as pet food, it could be used for the determination of these preservatives in meat products sold for human consumption. A saline slurry of the meat was acidified with phosphoric acid and steam distilled, the distillate being acidified and extracted with diethyl ether. After evaporation, dimethyl phthalate, acting as internal standard, was added in methanol and the methanolic solution used for GC. The method will determine benzoic acid in the range 0.1–0.5%.

Eight preservatives and 13 antioxidants were analysed by Nishimoto and Uyeta[6] using a total of seven stationary phases to effect the necessary separations. The preservatives were extracted by conventional procedures, and benzoic acid and various alkyl 4-hydroxybenzoates were included in the analysis.

Goddijn, Praag and Hardon[7] extracted benzoic acid and its derivatives, salicylic acid and sorbic acid, from foods such as fruit juices, syrups and preserves, and determined these preservatives after methylation with diazomethane. Methyl esters of benzoic, 2-chlorobenzoic, 4-chlorobenzoic, 4-bromobenzoic and sorbic acids were formed completely, whereas only partial methylation of the hydroxyl group of salicyclic acid occurred. Prolonged methylation ensured the conversion of the hydroxyl group as well as the carboxyl group of 4-hydroxy benzoic acid.

A preliminary GC examination of the concentrated preservative extract before methylation was necessary in order to detect the presence of 4-hydroxy benzoic acid and, hence, to perform the prolonged methylation. 5% SE-52 and 20% PEG were the stationary phases for the two columns necessary to separate all the preservatives. Many of these same preservatives were separated on a 15% DEGS plus 3% H_3PO_4 stationary phase by Vogel and Deshusses[8]. Those preservatives separated were benzoic, 2-chlorobenzoic, 4-chlorobenzoic, dehydroacetic, sorbic and salicylic acids, and the methyl, ethyl, propyl and butyl esters of 4-hydroxybenzoic acid. In similar fashion Groebel[9] separated benzoic, 4-chlorobenzoic, sorbic and salicylic acids and esters of 4-hydroxybenzoic acid using extracts prepared from foods such as margarine, mayonnaise, marinated fish and meat and fish pastes. The homogenised sample was acidified and diethyl ether-extracted and the ether solution was then re-extracted into sodium hydroxide solution. This solution was acidified and re-extracted into diethyl ether. After evaporation the acidic residue was methylated and chromatographed on a 20% SE-30 stationary phase.

A detailed scheme for the GC analysis of 22 preservatives and antioxidants in beer was demonstrated by Silbereisen and Wagner[10], who used a capillary column method with a PEGA stationary phase. Benzoic, 2-chlorobenzoic, 4-chlorobenzoic, dehydroacetic, chloroacetic, sorbic and salicylic acids plus 4-hydroxybenzoate and bromoacetic acid esters were among the preservatives separated as their methyl esters.

Benzoic acid, sorbic acid, and the methyl, ethyl and propyl esters of 4-hydroxybenzoic acid were separated and analysed in marmalade, margarine, mustard, mayonnaise, canned fish, beer and wine by Gossele[11] using the TMS derivatives of these acids. Sand was mixed with the sample undergoing analysis to facilitate penetration of the diethyl ether which was the extractant, following acidification of the sample. After clean-up of the ether extract with sodium hydroxide solution, the preservatives were re-extracted into chloroform, which was evaporated prior to derivative formation and GC using methyl gallate as internal standard. There was no need to use the sand treatment for the analysis of beer and wine, which were acidified and extracted with diethyl ether. Recoveries were greater than 85% for all the preservatives except ethyl and propyl 4-hydroxybenzoates in mustard, for which recoveries were 76% and 57%, respectively.

TMS derivatives have also been used by Takemura[12] and by Larsson and Fuchs[13].

A very simple GC method for the determination of benzoic acid in margarine was used by Graveland[14]. He suspended the margarine in diethyl ether which contained phosphoric acid and also valeric acid as internal standard and separated the supernatant after centrifugation. GC was performed on a stationary phase consisting of a mixture of Carbowax 20M and terephthalic acid, and as little as 5 ng of benzoic acid could be detected.

The detection and determination of benzoic acid, methyl, ethyl, n-propyl esters of *p*-hydroxybenzoic acid and sorbic acid in a variety of foodstuffs have been achieved by the simple GSC method of Fogden, Fryer and Urry[15]. The feature of their method is evaporation to dryness of the extract in the presence of tris(hydroxymethyl) methylamine at 60–70°C, in order to minimise loss of benzoic and sorbic acids. After evaporation of the extract, 2-phenoxyethanol is added in aqueous methanol as internal standard prior to chromatography on Porapak Q. This method is recommended for its simplicity, versatility and good recoveries of added preservatives.

There is a potential use of 8–12 p.p.m. of n-heptyl-4-hydroxybenzoate in beer, according to Schulman, Burris and Li[16]. Having added this preservative to beer, they distilled the alkaline sample and extracted the distillate with diethyl ether. This solvent removed n-heptanol from the distillate, which was chromatographed on a 25% Carbowax 20M stationary phase. There appears to be no evidence of the use of this preservative in beer but the analytical implication is of interest, since if 4-hydroxybenzoates with longer carbon chain lengths than heptyl were ever used, the GC analysis of the alcohol moiety could be advantageous.

2. *Dehydroacetic acid and acetic acid derivatives*

Dehydroacetic acid, chloroacetic acid and bromoacetic acid esters have been used as preservatives in food. Dehydroacetic acid is a lactone which has been used to control mould growth in cheese, but because there is a doubt regarding its safety in use, it is not a permitted preservative in the UK or in the USA. It can be determined by colorimetry after its diethyl ether extraction from acid solution, and this principle has been applied to its analysis in food. In this chapter, under Benzoic Acid and its Hydroxy and Halogenated Derivatives, methods have been cited which include the analysis of dehydroacetic acid in food by GC[6,8,10]. In the latter method, dehydroacetic acid was separated from salicylic acid and octyl gallate on a 5% OV-17 stationary phase. In the same method a capillary column coated with PEGA was used to separate chloroacetic acid and bromoacetic acid esters.

3. *Propionic acid*

Propionic acid and its sodium or calcium salt is used primarily to inhibit microbial action in flour which can lead to mould and ropiness in bread or flour confectionery.

Methods of direct solvent extraction have not proved successful and propionic acid has been best isolated by steam distillation. Since other lower aliphatic acids which may be present in the food are also steam-distilled, subsequent analysis has been difficult, depending on silicic acid column chromatographic separation.

Shelley, Salwin and Horwitz[17], in their determination of lower aliphatic acids in food, steam-distilled propionic acid into an alkaline solution and liberated the free acid with dichloroacetic acid in acetone. GC of propionic acid was carried out with methyl heptanoate as internal standard.

Propionic and sorbic acids were extracted together from bread using a steam distillation procedure by Walker, Green and Fenn[18]. The bread was crumbled or minced before steam distillation from acid solution into sodium hydroxide solution, which was subsequently acidified and diethyl ether-extracted to give the two acid preservatives. Propionic acid was chromatographed on a PEGS stationary phase and recoveries of the acid added to bread were greater than 95%.

Propionic acid has been used in cheese, and Grosjean and Fouassin[19] steam-distilled the preservative from samples of both cheese and bread and, after collecting some distillate, added more water to the distillation flask to collect a further distillate. The combined distillates were evaporated carefully to dryness and esterified with propanol containing crotonic acid as internal standard. A 20% Apiezon L stationary phase was used to separate propyl propionate and propyl crotonate.

Steam distillation followed by GSC analysis of propionic acid in saltine crackers has been successfully applied by Jones[20].

Graveland[14] described a simple and effective GC method for the determination of both propionic and sorbic acids in rye bread, which avoids steam distillation. The sample of bread was disintegrated at 0°C for 5 min in a solution of diethyl ether containing phosphoric acid and also valeric acid as internal standard. The extract was injected directly on to a column containing a 5%

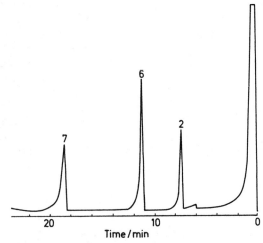

Figure 17.1 Gas chromatogram of two preservatives found in rye bread: 2, propionic acid; 6, valeric acid (internal standard); 7, sorbic acid (after Graveland[14]). For GC details, see Table 17.1

Carbowax 20M–terephthalic acid stationary phase; *Figure 17.1* shows a chromatogram of rye bread and the separation of propionic, valeric and sorbic acids. Recoveries of propionic and sorbic acids added to bread were 98–101% and the method detected down to 5 ng of both acids. Because of its rapidity and reproducibility and since it removes the inconvenience of steam distillation, it is recommended for use in the determination of both propionic and sorbic acids in bread and flour confectionery.

284

4. Sorbic acid

The sodium, potassium or calcium salts of sorbic acid can be added to food as mould and yeast inhibitors. As with benzoic acid, it is the free acid which is the effective inhibitor and therefore sorbic acid may be found in foods which have an acid reaction, such as cheese, wine, beer, fruit juice, fruit pulp, preserves and mayonnaise. It is also used, or has a potential use, in flour confectionery, bread, marzipan, prunes, margarine, marinated fish, canned fish, meat and fish pastes and mustard.

The most general method of extraction of sorbic acid from a food is by steam distillation into an alkaline solution and extraction of the free acid with diethyl ether after acidification of the distillate. Sorbic acid can be determined subsequently by UV spectrophotometry or by colorimetry following oxidation.

Many applications of GC methods to sorbic acid determination combine benzoic acid, and references have been made to these methods in this chapter under Benzoic Acid and its Hydroxy and Halogenated Derivatives[7-11, 13, 15].

Nishimoto and Uyeta[6] included sorbic acid in their composite analysis of eight preservatives and thirteen antioxidants after their extraction by conventional procedures.

By virtue of the possible addition of sorbic acid to flour for breadmaking, Graveland[14] used a simple GC method for its determination in rye bread, which has been described in the section of this chapter on Propionic Acid (see also *Figure 17.1*). Recovery of added sorbic acid to rye bread was 98–101%, and since the method is very rapid and simple, it offers great saving in time and convenience when compared with previous methods for the determination of both sorbic acid and propionic acid in bread and flour confectionery.

For analysts who still prefer to use steam distillation to remove sorbic acid from bread or flour confectionery, Walker *et al.*[18] described a method which has been given in this chapter in the section on Propionic Acid. Sorbic acid was chromatographed on an EGA stationary phase and recovery was in excess of 95% for the added acid.

One distinct advantage which sorbic acid has as a mould and yeast inhibitor is that it does not prevent bacterial growth which is essential in the production of certain foods, e.g. cheese. The preservative is applied either by dipping the cut cheese or by using it as a dusting powder on the lining of the wrapper. La Croix and Wong[21] found that conventional colorimetric methods for the determination of sorbic acid in Cheddar cheese were subject to interference and gave high blank values. They finally shredded the cheese, mixed it with Celite, transferred the mixture to a chromatographic column and eluted with acetonitrile over a period of 3 h. The eluate was evaporated and used for GC on an EGA–phosphoric acid column which effectively separated any co-extracted benzoic acid. A standard curve was prepared in the range 4–12 μg sorbic acid and the sample dilution was arranged to fit that range. These authors reiterated the importance of completely homogenising the cheese, since the method of application of the preservative means that it may not be evenly spread through the sample.

Reference has been made in this chapter under Benzoic Acid and its Hydroxy and Halogenated Derivatives to the analysis of sorbic acid in wine and beer[4, 10, 11]. As with the extraction of other preservative acids from wine and beer, the simplest procedure is extraction of the acidified sample with diethyl ether. di Stefano, Vercillo and Boniforti[22] used this extraction procedure and made the ethyl ester for GC purposes, using ethyl nonanoate as internal standard. After the analysis of 13 samples of wine, these workers found that although some recoveries of added sorbic acid were virtually complete, others were unsatisfactory and needed a more rapid extraction to achieve satisfactory recovery. This problem, coupled with that of the length of time necessary to produce the ethyl sorbate using ethanol and sulphuric acid, makes it one of the less practical methods described in the literature concerning the GC analysis of preservatives.

Reference has been made in this chapter in the section on Benzoic Acid and its Hydroxy and Halogenated Derivatives to the analysis of sorbic acid in fruit juices, syrups and preserves[7]. A similar extraction procedure and ethyl sorbate formation to that already mentioned[22] was applied by the same authors, Boniforti, di Stefano and Vercillo[23], to the analysis of sorbic acid in commercial fruit juices and pulps.

Reference has been made in the same section to the GC determination of sorbic acid in margarine, mustard, mayonnaise and canned fish[11], and also in margarine, mayonnaise, marinated fish and meat and fish pastes[9].

5. Salicylic acid

The use of salicylic acid as a preservative was once fairly widespread but it is not permitted in food at the present time in many countries, including the UK.

It has been included in GC methods of analysis where workers have evaluated most of the preservatives and antioxidants that have ever been tried. Such methods have been cited in this chapter in the section on Benzoic Acid and its Hydroxy and Halogenated Derivatives[7-10, 12].

6. Diethyl pyrocarbonate

Diethyl pyrocarbonate (or diethyl dicarbonate or diethyl oxidiformate) is used as an additive in wine, beer and non-alcoholic beverages to prevent unwanted fermentation. This preservative has been developed by the brewing and soft drinks industries in order to replace sulphur dioxide and benzoic acid, which can impart undesirable flavours to the products. Diethyl pyrocarbonate has the advantage that it breaks down finally to ethanol and carbon dioxide, neither of which has a deleterious effect on the organoleptic qualities of the beverages.

Most of the GC methods for the detection or determination of diethyl pyrocarbonate depend on its hydrolysis to ethanol and carbon dioxide and further reaction of ethanol with diethyl pyrocarbonate to form diethyl carbonate plus water:

$$C_2H_5O-\overset{\overset{\text{O}}{\|}}{C}-O-\overset{\overset{\text{O}}{\|}}{C}-OC_2H_5 + H_2O \rightarrow 2C_2H_5OH + 2CO_2$$

$$C_2H_5O-\overset{\overset{\displaystyle O}{\|}}{C}-O-\overset{\overset{\displaystyle O}{\|}}{C}-OC_2H_5 + 2C_2H_5OH \rightarrow 2C_2H_5O-\overset{\overset{\displaystyle O}{\|}}{C}-OC_2H_5 + H_2O$$

The ratio between these two reactions depends on the concentration of ethanol in the sample, and Wunderlich[24] calculated the diethyl carbonate concentration according to the formula:

$$\text{mg diethyl carbonate} = \frac{0.55 \times \text{mg diethyl pyrocarbonate} \times \% \text{ ethanol}}{100}$$

Only a small percentage of diethyl carbonate is normally formed and it is proportional to the alcoholic strength of the sample. Wunderlich[24] reported the findings of 12 collaborators for a method which used carbon disulphide to extract diethyl carbonate from wine, centrifugation and evaporation of the extract and GLC on a 15% trimethylol propantripelargonate stationary phase. The method is simple and gives satisfactory results in the range 0–10 p.p.m. of diethyl carbonate in wine. The original concentration of diethyl pyrocarbonate is calculated from the above-mentioned formula. It has been adopted as an official first action method by the FDA in the USA and is recommended as rapid and reasonably accurate.

Several other workers have used GC methods to estimate the diethyl pyrocarbonate content of wine[25-30] and also of alcoholic beverages[31], beer[10, 32] and fruit juice[33].

7. Diphenyl and o-phenyl phenol

Diphenyl is normally incorporated in the wrapping papers of oranges and other citrus fruits, to prevent mould growth. o-Phenyl phenol or its sodium salt is used as a dipping solution for these fruits and these act as bacteriocides and fungicides.

Steam distillation followed by UV spectrophotometry forms the basis for the non-chromatographic determination of diphenyl and o-phenyl phenol, and while reasonably satisfactory, some interference can occur from natural substances. Thomas[34] steam-distilled concentrated orange juice and extracted the distillate with chloroform prior to GLC on a silicone oil stationary phase. He found that the technique was suitable for the determination of diphenyl but obtained a peak which interfered with that of o-phenyl phenol in one sample of Spanish juice. Otherwise the method was sensitive down to 1 p.p.m. for both compounds.

Several other workers have used GC methods to detect or determine diphenyl and o-phenyl phenol in citrus fruit[35-43].

8. Ethylene oxide

Bread has been preserved by its storage in air-tight plastics bags in a mixture of ethylene oxide and carbon dioxide. Buquet and Manchon[44] used a GC

287

method to determine the concentrations of ethylene oxide, ethylene glycol and 2-chloroethanol in bread which remained after such a treatment. Approximately 300 p.p.m. each of ethylene oxide and 2-chloroethanol were found in an acetone extract of the bread.

9. Nitrous oxide

In experiments carried out to discover the reaction products of nitrite preservation of meat products, Moehler and Ebert[45] determined the nitrous oxide concentrations by a GC method. Chopped lean beef was treated with salt, nitrite and cysteine and heated for various periods at various temperatures in air-tight containers, the resulting meat being minced and suspended in hot water. The liberated nitrous oxide was swept by a helium stream on to a Porapak Q column for quantitative evaluation.

10. Thiabendazole

Thiabendazole, 2-(4-thiazolyl)benzimidazole, is used to control the rotting of fruits and vegetables, e.g. oranges and bananas in storage.

Mestres, Campo and Tourte[46] employed a GC method for the determination of thiabendazole in whole citrus fruit and in citrus and banana pulp. Samples were made alkaline with ammonia solution and macerated with ethyl acetate and filtered. The filtrate was cleaned up by extraction into 0.1N hydrochloric acid, then making this aqueous phase alkaline and re-extracting back into ethyl acetate. The concentrated extract was chromatographed on a 10% DC-200 stationary phase, using FPD set in the mode for determining sulphur. The limit of detection was of the order of 0.1 μg.

Thiabendazole residues were extracted from the skins of citrus fruits and bananas by Hey[47] using dichloromethane to reflux the samples. GC was performed on a 5% SE-30 stationary phase, using a nitrogen-sensitive detector.

C. Antioxidants

GC has been used in the analysis of the following food antioxidants: butylated hydroxyanisole (BHA) and butylated hydroxytoluene (BHT); gallate esters; others—nordihydroguaiaretic acid (NDGA), dilauryl 3,3'-thiodipropionate (DLTDP), 3,3'-thiodipropionic acid (TDPA), 4-hydroxymethyl-2,6-di-tert-butylphenol (Ionox-100), 2,4,5-trihydroxybutyrophenone (THBP), mono-tert-butylhydroquinone (TBHQ), ethyl protocatechuate and guaiac resin.

1. BHA and BHT

Conventional methods for the determination of BHA and BHT have relied on steam distillation or solvent extraction methods followed by colorimetric analysis. The addition of both these antioxidants is primarily to prevent oxidation of fatty materials, and the foods to which they are added, namely fats, oils, vitamin oil concentrates and glyceride emulsifiers, contain many substances which interfere with most analytical methods. Provided the fatty materials can be mainly removed, GC gives efficient separation of mixed antioxidant additives, such as BHA with BHT, and it is very sensitive.

These two antioxidants are used in some countries, e.g. the USA, for preventing flavour deterioration of breakfast cereals and potato granules.

GC methods have been used to determine BHA and BHT in dehydrated potato granules[48, 49], breakfast cereals[50-56] and fats and oils[57-61].

Two stationary phases were used by Hartman and Rose[57] to reverse the elution order of BHA and BHT to verify their presence in vegetable oils. *Figure 17.2* shows two chromatograms on 10% DC-200 and 10% Carbowax 20M in-

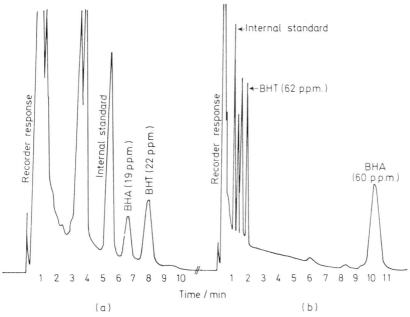

Figure 17.2. (a) A 6 ft × 4.5 mm I.D. aluminium column packed with 10% DC-200 on 80–100 mesh Gas Chrom Q Column, 160°C; detector, 250°C; injector port, 250°C; flow rate, 50 ml/min helium; sensitivity, 4 × 10⁻¹³. A full scale deflection; sample, cotton seed oil fortified with BHA and BHT; sample size injected, 6.0 μl containing 0.100 g oil/ml. (b) A 4 ft × 4.5 mm I.D. aluminium column packed with 10% Carbowax 20 M on 80–100 mesh Gas Chrom Q. Column, 190°C; remaining conditions same as (a); sample, cottonseed oil fortified with BHA and BHT; sample size injected, 5.0 μl containing 0.100 g oil/ml. (After Hartman and Rose[57])

dicating the reversed elution order of the two antioxidants with methyl undecanoate as the internal standard. These workers dissolved the vegetable oil in carbon disulphide plus internal standard and submitted the mixture directly to the chromatograph. To overcome the inevitable contamination of the column which would ensue from deposition of the oil, a stainless steel sleeve containing a 0.25 in siliconised glass wool plug was fitted inside the inlet port of the standard aluminium column to collect the oil, this plug being changed daily. The method gave recoveries of 97–104% of added BHA and BHT to soyabean, cotton seed, corn and peanut oils, and each analysis took only 20 min to perform, which is considerably shorter than the time taken by other GC methods in this field. Schwien, Miller and Conroy[58], for instance, found it necessary to remove fatty materials from fats and oils by alumina column chromatographic clean-up prior to GC analysis of BHA and BHT. The fat or oil was mixed with Celite and the mixture was extracted with acetonitrile, and this in turn was extracted with 1:1 diethyl ether:petrol. The extract was evaporated to low bulk before clean-up on the alumina column, and the antioxidants were eluted with 75% aqueous methanol and the eluate re-extracted with the mixed ether–petrol solvent. Antioxidant-free samples of tallow, lard and corn oil were spiked at 67 and 17 p.p.m. levels of antioxidants, and recoveries were 80–95% for BHA and 80–115% for BHT.

Several antioxidants, including BHA and BHT, were determined in lard by McCaulley *et al.*[59] They appreciated that conventional extraction procedures without a clean-up step produced fatty materials which would interfere with the GC determination of antioxidants, and devised an alternative method whereby the compounds were vacuum-sublimed on to a surface which was cooled with liquid nitrogen. The antioxidant residue was dissolved in solvent and the GLC performed on a 3% GE-XE-60 stationary phase. This phase was selected for its satisfactory separation of eight antioxidants, some of which could be used in combination in certain foods. Very often, mixtures of antioxidants are used in foods because they appear to have a synergistic effect on each other and therefore the total concentration of antioxidants can be minimised. BHA, BHT and gallates have been used in combination at low concentrations in lard in place of much larger concentrations of either BHA or BHT.

Besides BHA and BHT, other antioxidants determined were propyl gallate (PG), 2,4,5-trihydroxybutyrophenone (THBP), 3,3′-thiodipropionic acid (TDPA) and 4-hydroxymethyl-2,6-di-tert-butylphenol (Ionox-100).

The method was able to quantitatively determine BHA or BHT at the 10 p.p.m. level and the other four antioxidants at the 25 p.p.m. level. The method could also be used to detect the presence of the antioxidants nordihydroguaiaretic acid (NDGA) and dilauryl 3,3′-thiodipropionate (DLTDP).

Antioxidants which may be used in vegetable oils include BHA, BHT, mono-tert-butylhydroquinone (TBHQ), PG and NDGA. Stoddard[61] extracted a hexane solution of vegetable oil with acetonitrile and 80% aqueous ethanol and combined the extracts. After evaporation of the combined extracts, the TMS derivatives were formed for GLC on a 3% JXR stationary phase. *Figure 17.3* is a chromatogram of the TMS derivatives of the five antioxidants. For the determination of BHT alone, a Florisil clean-up procedure was used and no derivative formation was necessary. Although of antioxidants added to peanut oil recoveries of the other four antioxidants were greater than 87%, the recovery of 3-BHA was only 68% and of 2-BHA only 76%. One of the drawbacks of this method is the time-consuming extraction, which comprises eight acetonitrile extractions and six of ethanol.

In many countries BHA and BHT are permitted in concentrated vitamin oils, such as vitamin A oil derived from sea animals, e.g. halibut liver oil, cod liver oil. Kamio and Shima[62] determined BHT in such an oil by GC and found the method satisfactory in the range 0.5–5.0% BHT.

Vitamin A oil preparations containing levels of BHA and BHT of 2500 and 5000 p.p.m. were successfully analysed by Choy, Quattrone and Alicino[63]. The oil was dissolved in a low-boiling hydrocarbon and the solution was concentrated and used for direct injection. The same simple method was used for the determination of BHA and BHT in multiple-vitamin preparations containing 250 p.p.m. of each antioxidant.

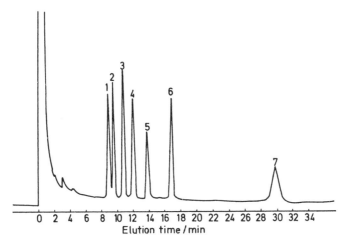

Figure 17.3 Tracing of gas chromatogram of trimethylsilyl derivatives of antioxidants and methyl pentadecanoate. Each peak is equivalent to 1 μg (range 10, attenuation 16): 1, 2-BHA; 2, 3-BHA; 3, TBHQ; 4, BHT; 5, C_{15}; 6, PG; 7, NDGA. (After Stoddard[61]). For GC conditions, see Table 17.1

Menon and Kulkarni[64] found that the established colorimetric procedures for the analysis of BHA and BHT, particularly for the latter, were laborious and non-specific, when applied to vitamin A concentrates. They obtained satisfactory analytical results by dissolving the concentrate in petrol and directly injecting the solution on to a 10% Apiezon M stationary phase. In this way a determination took only 0.25 h to complete, and recoveries of added BHA and BHT were in the range 96–104%.

2. Gallate esters

Gallate esters have been conventionally analysed by colorimetric methods. These suffer the disadvantage of not being able to distinguish one alkyl gallate from another, whereas GC methods can both separate and estimate these very similar antioxidants.

In their analysis of eight preservatives and thirteen antioxidants, Nishimoto and Uyeta[6] included alkyl gallates, separating them as their TMS derivatives on any one of three stationary phases. TMS derivatives were also used by Wachs and Gassmann[65] to separate ethyl, propyl, butyl, octyl and decyl gallates from BHA, a procedure which could be made quantitative.

Octyl gallate in beer was determined by Silbereisen and Wagner[10] and propyl gallate in lard was determined by McCaulley *et al.*[59]

Figure 17.3 includes propyl gallate in a chromatogram of antioxidants added to vegetable oil and determined by Stoddard[61]. Using the TMS derivative for GLC on a 3% JXR stationary phase, 93% of propyl gallate added to peanut oil was recovered by this method, which was described in the section on BHA and BHT in this chapter.

3. Others

Nordihydroguaiaretic acid (NDGA) NDGA is a permitted antioxidant in some countries, including the USA, but is not permitted in the UK. It has been added to edible oils and fats, and its determination is difficult because of its interaction with fat.

McCaulley *et al.*[59] attempted to quantitatively determine NDGA in lard by the method described in this chapter in the section on BHA and BHT but they were only able to measure it qualitatively. Further work on this method was conducted and certain modifications made to the original, including the simplification of the sublimation procedure. Unfortunately, this method, published by Fazio, Howard and Sandoval[66], did not result in any improvement for the determination of NDGA.

However, the later work of Stoddard[61], already described in this chapter produced a long but successful method for the determination of NDGA by GC; and 100% recovery of the antioxidant added to peanut oil was recorded and its separation from some other antioxidants was included in the chromatogram in *Figure 17.3*.

Dilauryl 3,3'-thiodipropionate (DLTDP) DLTDP is a permitted antioxidant in the USA and is used in edible oils, fats and polyolefins.

Attempts were made by McCaulley *et al.*[59] to quantitatively determine DLTDP in lard, but only a qualitative measurement was possible. Further work was carried out to modify the method, particularly in the sublimation clean-up technique involved in the extraction procedure, and Fazio *et al.*[66] produced a GC method capable of determining down to 10–25 p.p.m. of DLTDP and giving recoveries of 93–100% of this antioxidant added to lard.

3,3'-Thiodipropionic acid (TDPA) TDPA is a permitted antioxidant in the USA and is used in edible oils and fats.

McCaulley *et al.*[59] included it in their GC analysis of six antioxidants in lard.

4-Hydroxymethyl-2,6-di-tert-butylphenol (Ionox-100) Ionox-100 has been proposed as a possible antioxidant for use in edible oils and fats.

McCaulley *et al.*[59] included it in their GC analysis of six antioxidants in lard. It can be quantitatively determined down to 25 p.p.m. Fazio *et al.*[66] simplified that method and Ionox-100 was included in the analysis.

2,4,5-Trihydroxybutyrophenone (THBP) THBP is a permitted antioxidant in the USA and is used in edible oils and fats. McCaulley *et al.*[59] included it in their GC analysis of six antioxidants in lard. It can be quantitatively determined down to 25 p.p.m.

Mono-tert-butylhydroquinone (TBHQ) TBHQ is an antioxidant which is effective in preventing oxidation of vegetable oils. Stoddard[61] included this compound in the analysis of antioxidants in vegetable oils by the GC of their TMS derivatives. Recovery experiments carried out using TBHQ added to peanut oil gave 87% recovery.

Ethyl protocatechuate and guaiac resin Both these antioxidants were included in the general GC method for the analysis of food additives as described by Nishimoto and Uyeta[6] in this chapter in the section on Benzoic Acid and its Hydroxy and Halogenated Derivatives.

References

1. *The Preservatives in Food Regulations, 1974*, Statutory Instrument No. 1119, HMSO (1974)
2. *The Antioxidant in Food Regulations, 1974*, Statutory Instrument No. 1120, HMSO (1974)
3. Nobe, B. and Stanley, R. L., *J. Ass. Off. Agric. Chem.*, **48**, 791 (1965)
4. Wuerdig, G., *Dt. Lebensmitt.-Rdsch.*, **62**, 147 (1966)
5. Clarke, E. G. C., Humphreys, D. J. and Stoilis, E., *Analyst*, **97**, 433 (1972)
6. Nishimoto, T. and Uyeta, M., *J. Food Hyg. Soc., Japan*, **5**, 287 (1964)
7. Goddijn, J. P., van Praag, M. and Hardon, H. J., *Z. Lebensmitt. Forsch.*, **123**, 300 (1963)
8. Vogel, J. and Deshusses, J., *Mitt. Geb. Lebensmitt. Hyg.*, **56**, 35 (1965)
9. Groebel, W., *Dt. Lebensmitt.-Rdsch.*, **61**, 209 (1965)
10. Silbereisen, K. and Wagner, B., *Mschr. Brau.*, **23**, 57 (1970)
11. Gossele, J. A. W., *J. Chromatog.*, **63**, 429 (1971)
12. Takemura, I., *Bunseki Kagaku*, **20**, 61 (1971)
13. Larsson, B. and Fuchs, G., *Swed. J. Agric. Res.*, **4**, 109 (1974)
14. Graveland, A., *J. Ass. Off. Anal. Chem.*, **55**, 1024 (1972)
15. Fogden, E., Fryer, M. and Urry, S., *J. Ass. Publ. Anal.*, **12**, 93 (1974)
16. Schulman, H. L., Burris, G. H. and Li, F. K. C., *Proc. Amer. Soc. Brew. Chem.*, 67 (1965)
17. Shelley, R. N., Salwin, H. and Horwitz, W., *J. Ass. Off. Agric. Chem.*, **46**, 486 (1963)
18. Walker, G. H., Green, M. S. and Fenn, C. E., *J. Ass. Publ. Anal.*, **2**, 2 (1964)
19. Grosjean, M. H. and Fouassin, A., *Rev. Ferment. Ind. Aliment.*, **22**, 211 (1967)
20. Jones, F. B., *J. Ass. Off. Anal. Chem.*, **56**, 1415 (1973)
21. La Croix, D. E. and Wong, N. P., *J. Ass. Off. Anal. Chem.*, **54**, 361 (1971)
22. di Stefano, F., Vercillo, A. and Boniforti, L., *Boll. Lab. Chim. Prov.*, **15**, 523 (1964)
23. Boniforti, L., di Stefano, F. and Vercillo, A., *Boll. Lab. Chim. Prov.*, **12**, 505 (1961)
24. Wunderlich, H., *J. Ass. Off. Anal. Chem.*, **55**, 557 (1972)
25. Kielhoefer, E. and Wuerdig, G., *Dt. Lebensmitt.-Rdsch.*, **59**, 197 (1963)
26. Kielhoefer, E. and Wuerdig, G., *Dt. Lebensmitt.-Rdsch.*, **59**, 224 (1963)
27. Prillinger, F. and Horwatitsch, M., *Mitt. Rebe Wein, Obstb. Fruchteverwert*, **14**, 251 (1964)
28. Reinhard, C., *Dt. Lebensmitt.-Rdsch.*, **63**, 151 (1967)
29. Bandion, F., *Mitt. Rebe Wein, Obstb. Fruchteverwert*, **19**, 37 (1969)
30. Brandenburg, G. and Rohleder, K., *Dt. Lebensmitt.-Rdsch.*, **64**, 71 (1968)
31. Kunitake, N., *Bull. Brew. Sci.*, **15**, 1 (1969)

32. Cuzner, J., Bayne, P. D. and Rehberger, A. J., *Proc. Amer. Soc. Brew. Chem.,* 116 (1971)
33. Bandion, F. and Kain, W., *Fruchtsaft-Ind.,* **14,** 50 (1969)
34. Thomas, R., *Analyst,* **85,** 551 (1960)
35. Vogel, J. and Deshusses, J., *Mitt. Geb. Lebensmitt. Hyg.,* **56,** 185 (1965)
36. Vogel, J. and Deshusses, J., *Mitt. Geb. Lebensmitt. Hyg.,* **54,** 330 (1963)
37. Canuti, A., *Boll. Lab. Chim. Prov.,* **16,** 660 (1965)
38. Mestres, R. and Chave, C., *Trav. Soc. Pharm. Montpellier,* **24,** 272 (1965)
39. Mestres, R., Chave, C. and Dudieuzere-Priu, M., *Ann. Falsif. Exp. Chim.,* **60,** 73 (1967)
40. Mestres, R., Dudieuzere-Priu, M., Gaillard, J. C. and Tourte, J., *Ann. Falsif. Exp. Chim.,* **60,** 331 (1967)
41. Beernaert, H., *J. Chromatog.,* **77,** 331 (1973)
42. Hahn, H. and Thier, H-P., *Lebensmittelchem. Gerichtl. Chem.,* **26,** 185 (1972)
43. Morries, P., *J. Ass. Publ. Anal.,* **11,** 44 (1973)
44. Buquet, A. and Manchon, P., *Chim. Anal.,* **52,** 978 (1970)
45. Moehler, K. and Ebert, H., *Z. Lebensmitt. Forsch.,* **147,** 251 (1971)
46. Mestres, R., Campo, M. and Tourte, J., *Ann. Falsif. Exp. Chim.,* **63,** 160 (1970)
47. Hey, H., *Z. Lebensmitt. Forsch.,* **149,** 79 (1972)
48. Buttery, R. G. and Stuckey, B. N., *J. Agric. Food Chem.,* **9,** 283 (1961)
49. Thomson, W. A. B., *J. Chromatog.,* **19,** 599 (1965)
50. Schwecke, W. M. and Nelson, J. H., *J. Agric. Food Chem.,* **12,** 86 (1964)
51. Anderson, R. H. and Nelson, J. P., *Food Technol.,* **17,** 95 (1963)
52. Takahashi, D. M., *J. Ass. Off. Agric. Chem.,* **47,** 367 (1964)
53. Takahashi, D. M., *J. Ass. Off. Anal. Chem.,* **49,** 704 (1966)
54. Takahashi, D. M., *J. Ass. Off. Anal. Chem.,* **50,** 880 (1967)
55. Takahashi, D. M., *J. Ass. Off. Anal. Chem.,* **51,** 943 (1968)
56. Takahashi, D. M., *J. Ass. Off. Anal. Chem.,* **53,** 39 (1970)
57. Hartman, K. T. and Rose, L. C., *J. Amer. Oil Chem. Soc.,* **47,** 7 (1970)
58. Schwien, W. G., Miller, B. J. and Conroy, H. W., *J. Ass. Off. Anal. Chem.,* **49,** 809 (1966)
59. McCaulley, D. F., Fazio, T., Howard, J. W., Di Ciurcio, F. M. and Ives, J., *J. Ass. Off. Anal. Chem.,* **50,** 243 (1967)
60. Wolff, J-P. and Audiau, F., *Bull. Soc. Chim. France,* 2662 (1964)
61. Stoddard, E. E., *J. Ass. Off. Anal. Chem.,* **55,** 1081 (1972)
62. Kamio, H. and Shima, T., *Bunseki Kagaku,* **11,** 731 (1962)
63. Choy, T. K., Quattrone, J. J. and Alicino, N. J., *J. Chromatog.,* **12,** 171 (1963)
64. Menon, P. S. and Kulkarni, V. S., *Indian J. Technol.,* **5,** 168 (1967)
65. Wachs, W. and Gassmann, L., *Dt. Lebensmitt.-Rdsch.,* **66,** 37 (1970)
66. Fazio, T., Howard, J. W. and Sandoval, A., *J. Ass. Off. Anal. Chem.,* **51,** 17 (1968)

18

Emulsifiers and Stabilisers

A. Introduction

Several categories of food contain natural substances which have emulsifying or stabilising functions. Examples of such substances are glycerides, gums, starches and pectins, and these have been isolated for use in foods in which they do not normally occur. Modern food processing, however, has demanded emulsifiers of greater versatility, and such compounds have been tailor-made to fulfil specific needs.

In the UK the *Emulsifiers and Stabilisers in Food Regulations 1962*[1] define an emulsifier and stabiliser as: '(a) in the case of an emulsifier, of aiding the formation of, and (b) in the case of a stabiliser, of maintaining the uniform dispersion of two or more immiscible substances . . .'.

Some countries, including the UK, have drawn up lists of permitted emulsifiers and stabilisers, most of which may be added to unspecified foods. The levels of addition are left to the food technologist to formulate in accordance with sound commercial practice, and the need for quantitative analytical methods for these additives is largely the concern of the food industry. Law enforcement analysts would normally be only concerned with the detection of emulsifiers or stabilisers which are not permitted in food.

As new emulsifiers and stabilisers have been synthesised, analytical methods based on column and TLC separations and determinations have evolved, and in recent years GC has played an increasing part in the analysis of these additives. GC methods have been applied to the following categories of emulsifiers and stabilisers: glycerides and polyglycerides, sorbitan esters of fatty acids and their polyoxyethylene derivatives; brominated vegetable oils; sucrose diacetate hexaisobutyrate; and triethyl citrate and triacetin.

The compounds which are most readily analysed by GC are glycerides, polyglycerides, sorbitan esters of fatty acids and brominated vegetable oils. *Table 18.1* includes the GC conditions of some of the more important methods applied to these as outlined in this chapter.

B. Glycerides and polyglycerides

Emulsifiers are often natural ingredients of a food and in such instances their presence must forever be permitted. Fatty materials contain glycerides and

Table 18.1 GC conditions used in the analysis of emulsifiers and stabilisers

Class of compound	Ref.	Column dimensions	Stationary phase	Support material	Carrier gas and conditions	Temperature or temperature programme/°C	Detector
Polyols	16	6ft x 0.25in O.D.	15% Carbowax 20M	silanised Chromosorb T	He, 38ml/min	120	TC
TMS derivatives of polyols	5,11 15	3ft x 0.125in O.D.	3% JXR	110/120 Gas Chrom Q	He, 33–37 ml/min	125–325 at 10/min	FI
TMS derivatives of monoglycerides	10	3ft x 0.25in	2% SE-20	Aeropak 30	N_2, 70ml/min	175–375 at 6/min	FI
Isosorbide	13	1.5ft x 7mm O.D.	15% Carbowax 20M	80/100 AW Chromosorb W	Ar, 115ml/min	195	AI
Methyl esters of brominated vegetable oils	19	3ft x 0.125in O.D.	3% JXR	80/90 Anakrom ABS	He, 40ml/min	150–270 at 10/min	FI
Triethyl citrate, triacetin	22	6ft x 0.25in	20% SE-30	30/60 Chromosorb W	He, 10 p.s.i. inlet pressure	200	TC
Sucrose diacetate, hexaisobutyrate	20	3ft x 0.125in	3% JXR	80/90 Anakrom ABS	He, 35ml/min	210–320 at 20/min	FI
Decyl acetate, decyl isobutyrate	21	5ft x 0.125in	3% SE-30	100/120 Varaport 30	He, 35ml/min	100–290 at 10/min	FI

polyglycerides and their composition is of commercial interest to the fats and oils industry. The complexity of glyceride mixtures has made difficult their analysis, and although mono-,di- and triglycerides have been separated successfully by silicic acid column chromatography, the identification of the individual members of the three groups was not possible in complex mixtures, prior to the advent of GC. Even so, it was some time before the problem of suitable derivatives was solved. In 1959 Huebner[2] separated monoglycerides as their acetyl derivatives and a year later McInnes, Tattrie and Kates[3] used allyl ethers and iso-propylidine derivatives, but these methods, although the best in their time, were not very satisfactory.

In 1965 Wood, Raju and Reister[4] applied the TMS derivative technique to glycerides, which was an already established approach in GC analysis of carbohydrates. By this means, the natural monoglycerides of corn and coconut oils were separated and determined either on a column consisting of 20% PEGS on Chromosorb W or by a capillary column coated with Apiezon L. Three mixtures comprising the 1- and 2-isomers of monopalmitin and mono-olein, of monodecanoin, monolaurin, monomyristin, monopalmitin and monostearin and of the 1-isomers of monostearin, mono-olein and monolinolein were separated by this method. The two columns were also utilised to separate the methyl esters of the fatty acid components of the glycerides.

Most of the subsequent GC analyses applied to glycerides have employed the TMS derivatives. Sahasrabudhe and Legari[5] described a procedure based on the separation of mono- and diglycerides by silicic acid column chromatography and the analysis of each fraction by GC using a 3% JXR stationary phase, having made the TMS derivatives. In this way, the mono- and diglycerides of myristic, palmitic, stearic and oleic acids were determined in commercial preparations and shortening samples, 96—101% recoveries being recorded for added glycerides. It was found that the monoglyceride analysis agreed with the fatty acid distribution within the samples, as determined by GC.

A similar method was used by Watts and Dils[6] in the analysis of diglyceride composition of egg lecithins.

General mono- and diglycerides plus free glycerol in commercial emulsifiers and shortenings and also in soyabean oil and margarine based on soyabean oil were determined by Blum and Koehler[7]. After preliminary separation, the TMS derivatives were prepared and GLC carried out on a 3% OV-1 stationary phase, using cholesteryl acetate as internal standard. This method was found to be unsuitable for the analysis of triglycerides.

Complete and partial glycerol esters can contain acetic, lactic, phosphoric, citric, succinic, tartaric or diacetyl tartaric acid moieties. Pentiti[8] determined the diacetyltartaric acid esters of mono- and diglycerides by GC after removing free tartaric acid, separating the mono- from the diglycerides by TLC and preparing the methyl esters of both TLC fractions.

Succinylated monoglycerides have been used in bread and cakes, and Ma and Morris[9] described a method of extraction of these emulsifiers with water and n-propanol, using hydrolysis to give succinic acid followed by GC of the dimethyl succinate after methylation.

Lactic acid glycerides have been used in shortenings, and Neckermann and Noznick[10] examined fats for these particular emulsifiers. They separated both lactic acid monoglycerides and other monoglycerides from di- and triglycerides by silicic acid column chromatography and prepared the TMS derivatives of the appropriate fraction for GC analysis. The method was applied successfully to fat samples containing as little as 6% of total monoglycerides plus lactic acid monoglycerides.

Partial polyglycerol esters are used as emulsifiers, stabilisers and defoamers and can be found in margarine, vegetable oils and flour confectionery. Ma and Morris[9] determined polyglycerol esters after formation of the methyl esters for GC, whereas Sahasrabudhe[11] used the TMS derivatives. He found that commercial polyglycerol esters were made up of glycerol, free fatty acids, mono- and diglycerides and mono fatty acid esters of diglycerol and triglycerol. Pre-

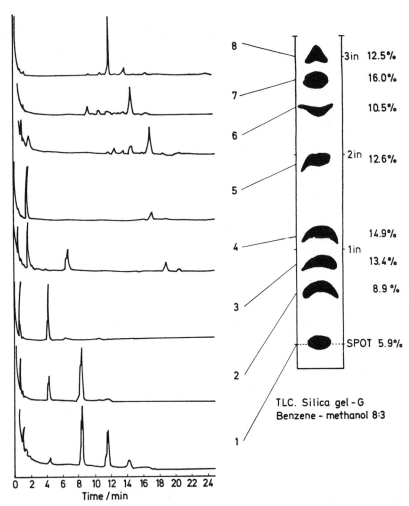

Figure 18.1 Diagrammatic representation of separation of polyglycerols and poly-glycerol esters by TLC and their identification by GLC: TLC fractions 1, 2 and 3, polyglycerols; 4 and 5, cyclic polyglycerols; 6, triglycerol monostearate; 7, diglycerol monostearate; 8, glycerol monostearate (after Sahasrabudhe[11]). For GC details, see Table 18.1

parative TLC on silica gel was used to separate eight fractions and each fraction was trimethylsilylated and analysed by GC. *Figure 18.1* is a diagrammatic representation of the TLC separation alongside the GC chromatograms of each fraction, showing the efficient separation of the polyglycerol fractions from those of the cyclic, tri-, di- and monoglycerides.

Propylene glycol esters are used in shortenings, cake mixes and ice creams, very often in conjunction with mono- and diglycerides. Sahasrabudhe and Legari[12] separated these mixtures of emulsifiers by silicic acid column chromatography and used the TMS derivatives of the three fractions in the GC analysis. The three fractions were comprised of triglycerides plus propylene glycol diesters, diglycerides plus propylene glycol monoesters and monoglycerides. In the fractions containing the propylene glycol esters, these were well separated from the glycerides on the 3% JXR stationary phase.

C. Sorbitan esters of fatty acids and their polyoxyethylene derivatives

The oil-soluble sorbitan esters of fatty acids and their water-soluble polyoxyethylene derivatives are used independently and also in admixture as emulsifiers and thickeners in a number of foods. These foods include bread and flour confectionery, sugar confectionery, ice cream and frozen dessert items, pickles and fruit juice. These additives are therefore widely used, and this is reflected in the relatively large amount of attention that has been devoted to their analysis. GC features prominently among the methods used for their analysis, which often include column chromatography and TLC.

In 1964 Wetterau, Olsanski and Smullin[13] developed a GC method for the determination of sorbitan monostearate in cake mixes and baked cakes. It was known that sorbitan monostearate, added in order to improve cake texture and volume, was incorporated at a concentration between 0.1 and 0.6% and that commercial sorbitan monostearate consisted of a mixture of polyol esters and anhydrides. One of these anhydrides, 1,4-3,6-dianhydro sorbitol, otherwise known as isosorbide, was found in the sorbitan monostearate at a concentration of 4—5%, and Wetterau *et al.*[13] analysed the cake preparations for this compound and related their answer to the actual sorbitan monostearate concentration. The sample was extracted with ethanol, leaving behind sugars and triglycerides, and the ethanol was replaced by n-heptane in an azeotropic distillation procedure. Interfering substances were removed by column chromatographic clean-up and the sorbitan monostearate fraction hydrolysed with KOH. After deionisation of the hydrolysate on an ion exchange resin, the polyol solution was analysed by GC and the isosorbide used for quantitative measurement.

At the time, despite its very time-consuming extraction procedure, this method was the only GC attempt at analysing these emulsifiers, and Murphy and Grisley[14] subsequently pointed out that sorbitan monostearate obtained from various sources has isosorbide contents varying between 1.2 and 6.1%. Therefore, unless the source was known, quantitative results by the Wetterau *et al.* method could be misleading. With that proviso, however, the method was used by Murphy and Grisley, who extended it to include the GC analysis of other sorbitan esters, the isosorbide content of which lay between 1.5 and 15% depending on the particular ester and its manufacturer. These workers also used the TMS derivative of isosorbide in its analysis and applied their methods to the determination of sorbitan esters in dessert toppings and ice cream powders.

Sahasrabudhe and Chadha[15] separated sorbitan fatty acid esters by column chromatography and TLC, and the TMS derivatives of the fractions were chromatographed. The TMS derivatives of the polyols were chromatographed on a 3% JXR stationary phase and the TMS derivatives of the fatty acids were chromatographed on a 6% BDS phase. *Figure 18.2* shows the separation of polyols and their esters in three sorbitan ester preparations.

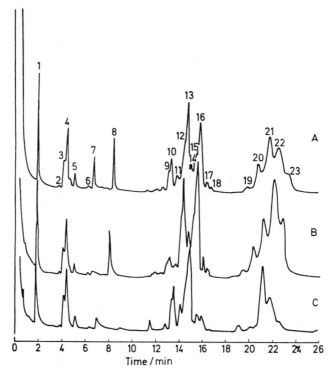

Figure 18.2 GLC separation of polyols and their esters. A, Span 60; B, Span 80; C, Span 40. 1: Isosorbide,; 2, 3 and 5: unconfirmed sorbitol anhydrides; 4: 1,4-sorbitan; 6: sorbitol; 7: palmitic acid; 8: stearic acid; 9 and 10: isosorbide monopalmitates; 11: unknown; 12: isosorbide monostearate; 13: 1,4-sorbitan monopalmitate; 14: unknown; 15: 1,4-sorbitan mono-oleate and sorbitol monopalmitate; 16: 1,4-sorbitan monostearate; 17: sorbitol monostearate; 18, 19: unidentified; 20, sorbitan dipalmitate; 23: unknown. (After mitostearate; 22: sorbitan distearate; 23: unknown. (After Sahasrabudhe and Chadha[15]). For GC details, see Table 18.1

Having found that one analysis of sorbitan monostearate in whipped cream, vegetable oil topping, etc., using the official 1965 FDA method of the US, could take 3 weeks to complete, Lundquist and Meloan[16] developed a method based on hydrolysis of the esters followed by GC of the resulting polyols. Ethanol and water were added to the sample and extraction carried out using diethyl ether and petrol. The evaporated extract was injected on to a 12 in column of 30 mesh soda-lime beads which was connected to an orthodox 15% Carbowax 20M on silanised Chromosorb T column. Hydrolysis was achieved on the reactive precolumn and the resulting polyols were separated on the analytical column. The method was applied successfully to the analysis of both sorbitan esters and their polyoxyethylene derivatives in whipped cream, an analysis taking 2.5 h to complete.

D. Brominated vegetable oils

Brominated vegetable oils are bromine addition products of vegetable oils such as sesame, olive, corn and cotton seed. They are used as dispersing agents for the flavouring oils added to soft drinks and also have the ability to stabilise the resulting cloudy appearance, which simulates the dispersion effect of natural oils present in fruit juices, particularly of the citrus type.

The use of these oils was prohibited in the UK as from 1970, and the levels allowed in Canada have been reduced from a maximum of 150 p.p.m. to a tenth of that value. Because of the necessity for determining quantitatively the brominated vegetable oil concentration in soft drinks, the Canadian Food and Drug Directorate was responsible for stimulating research for methods of analysis and GC was successfully applied. Conacher, Chadha and Sahasrabudhe[17] attempted to chromatograph brominated vegetable oils directly but found that there was a build-up of oily substances in the injection port. Methyl esters of the acids of the oils did not have that disadvantage and were used in the analysis of the brominated vegetable oils after their extraction. Methyl pentadecanoate internal standard in methanol was added to the soft drink, which was saturated with salt and extracted with diethyl ether. After evaporation of the ether, the methyl esters were formed using sodium and anhydrous methanol and these were chromatographed on a 3% JXR stationary phase. *Figure 18.3* compares chromatograms of the fatty acid esters of the four brominated vegetable oils and shows their separation from methyl esters of fatty acids and their bromo derivatives. These workers analysed six orange drinks for brominated vegetable oils and found that over half of the fatty acid esters of the oils were tetrabromo derivatives and the remainder of the bromine was present in the dibromo derivatives. By comparing the brominated vegetable oil content and free palmitic and stearic acids contents of standards with those of the samples, it was possible to deduce which oils had been used. Thus, one sample was found to contain brominated corn oil and another sample brominated sesame oil. Four samples gave indecisive results and the presence of no one brominated vegetable oil could be confirmed. *Table 18.2* gives the fatty acid composition of brominated olive, sesame, cotton seed and corn oils as used by Conacher, Meranger and Leroux[18]. They supplemented an X-ray fluorescence method with the quantitative GC method and found very good agreement between them. *Table 18.2* shows that there is some free palmitic and stearic acid present in the standard brominated oils, and by taking these values plus the dibromo and tetrabromo derivatives concentration, it was deduced that 12 out of 20 carbonated citrus drinks contained brominated sesame oil, six contained a mixture of brominated corn and olive oils and one each contained brominated olive plus brominated cotton seed oils. At this time, when the higher Canadian limit applied, these drinks were found to contain 10–50 mg/10 fl. oz.

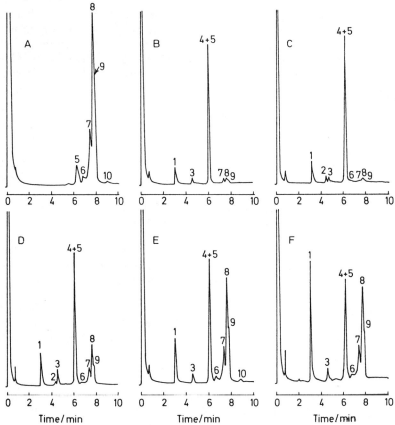

Figure 18.3 GLC chromatograms of brominated vegetable oils treated with sodium methoxide. A, methyl tetrabromostearate; B, brominated olive oil (laboratory prepared); C, brominated olive oil (commercial); D, brominated sesame oil (commercial); E, brominated corn oil (laboratory prepared); and F, brominated cotton seed oil (commercial). 1, C_{16} methyl ester; 2 and 3, C_{18} methyl esters; 4, methyl dibromostearate derivative; and 5–10, methyl tetrabromostearate derivatives. (After Conacher et al.[17]). For GC details see Table 18.1[19], except that the support material is 110/120 Gas Chrom Q

Two mon-carbonated beverages contained approximately 80 mg of brominated vegetable oil per 10 fl.oz.

When the Canadian limit was reduced, the existing GC method was not sensitive enough, and therefore Conacher[19] amended the method. There had been some interference in the previous method with co-eluted compounds on the chromatograph which interfered to a small degree with the C_{16} and C_{18} methyl esters, but it was not sufficient to affect the previous results. However, at low levels of brominated vegetable oils, this interference became relevant and Conacher found that by using standard methyl tetrabromostearate and methyl dibromostearate in place of the standard complete oil, the extent of the interference could be calculated and therefore more accurate results could be obtained.

Table 18.2 Fatty acid composition (%) of standard brominated vegetable oils (after Conacher et al.[18])

Brominated vegetable oil	Palmitic acid	Stearic acid plus C_{18} un-saturated acids	Dibromo derivatives	Tetrabromo derivatives
Olive	8.6	5.0	74.6	11.8
Sesame	5.9	5.0	42.7	46.4
Cotton seed	14.5	1.9	19.4	64.2
Corn	6.3	1.6	21.7	70.4

E. Sucrose diacetate hexaisobutyrate

Sucrose diacetate hexaisobutyrate has been used as a replacement for brominated vegetable oils in citrus soft drinks. It has similar properties to brominated vegetable oils in that it acts as a dispersing agent for the flavour citrus oils and stabilises the resulting cloud.

Conacher and Chadha[20] have produced a GC method for the estimation of sucrose diacetate hexaisobutyrate in soft drinks which is similar to that for the estimation of brominated vegetable oils. The drink was saturated with salt and extracted with diethyl ether, the solvent being washed successively with sodium hydroxide, hydrochloric acid and water before evaporation to dryness. The residue was dissolved in diethyl ether and chromatographed in order to make sure that the sample did not contain compounds which would elute together with the intended internal standard, trimyristin. The internal standard was then added, the mixture chromatographed on a 3% JXR stationary phase, and a quantitative estimate made of any sucrose diacetate hexaisobutyrate present. The emulsifier gave a complex array of peaks, all overlapping, but it was possible to get an area measurement under the composite peak. Out of ten drinks, five were found to contain between 12 and 14 mg per 10 fl. oz, which is up to the maximum allowed of 50 p.p.m. in soft drinks under Canadian legislation. At less then 1 mg per 10 fl. oz the method was only qualitative and as little as 0.01 mg per 10 fl. oz could be detected. There was a danger that components of a soft drink could co-elute with sucrose diacetate hexaisobutyrate, unbeknown to the analyst. Because of this possibility, the method was only treated as a guide to the concentration of the emulsifier, and if it was thought to be present in excess of the 50 p.p.m. maximum, another more precise method would have to be used.

If the described method is treated as one for screening, then Conacher, Chadha and Iyengar[21] have described a more specific procedure based on the estimation of the acetic and isobutyric acids present in sucrose diacetate hexaisobutyrate after a transesterification step. Sulphuric acid in decanol was used to form decyl acetate and decyl isobutyrate, which were chromatographed on a 3% SE-30 stationary phase using hexanoic acid as internal standard. The emulsifier was extracted from soft drinks in a similar way to that already described[20], and figures obtained by both the screening method and the more specific one were in good agreement.

F. Triethyl citrate and triacetin

Triethyl citrate and triacetin are additives used in egg whites to facilitate emulsion and therefore reduce the time required to produce whipped egg. These emulsifiers can be added to either liquid or dried egg whites and they also act as foam stabilisers.

Kogan and Strezleck[22] extracted triethyl citrate from egg whites at a pH of 3 with diethyl ether and triacetin from egg whites at a neutral pH, also with diethyl ether. The separate extracts were evaporated to low bulk and chromatographed on an SE-30 stationary phase. Chemical tests were conducted prior to GLC to verify the presence of these emulsifiers, and the IR spectra of the GLC eluted compounds were obtained for additional confirmation. No recovery analyses were performed and it is suggested that the method therefore be treated as a qualitative one.

References

1. *The Emulsifiers and Stabilisers in Food Regulations 1962,* Statutory Instrument No. 721, HMSO (1962)
2. Huebner, V. R., *J. Amer. Oil Chem. Soc.,* **36,** 262 (1959)
3. McInnes, A. G., Tattrie, N. H. and Kates, M., *J. Amer. Oil Chem. Soc.,* **37,** 7 (1960)
4. Wood, R. D., Raju, P. K. and Reiser, R., *J. Amer. Oil Chem. Soc.,* **42,** 161 (1965)
5. Sahasrabudhe, M. R. and Legari, J. J., *J. Amer. Oil Chem. Soc.,* **44,** 379 (1967)
6. Watts, R. and Dils, R., *J. Lipid Res.,* **10,** 33 (1969)
7. Blum, J. and Koehler, W. R., *Lipids,* **5,** 601 (1970)
8. Pentiti, A., *Sci. Aliment.,* **16,** 389 (1970)
9. Ma, R. M. and Morris, M. P., *Food Additives Analytical Manual,* 1968 (U.S. Dept of Health, Education and Welfare, Food and Drug Admin., Washington)
10. Neckermann, E. F. and Noznick, P.P., *J. Amer. Oil Chem. Soc.,* **45,** 845 (1968)
11. Sahasrabudhe, M. R., *J. Amer. Oil Chem. Soc.,* **44,** 376 (1967)
12. Sahasrabudhe, M. R. and Legari, J. J., *J. Amer. Oil Chem. Soc.,* **45,** 148 (1968)
13. Wetterau, F. P., Olsanski, V. L. and Smullin, C. F., *J. Amer. Oil Chem. Soc.,* **41,** 791 (1964)
14. Murphy, J. M. and Grisley, L. H., *J. Amer. Oil Chem. Soc.,* **46,** 384 (1969)
15. Sahasrabudhe, M. R. and Chadha, R. K., *J. Amer. Oil Chem. Soc.,* **46,** 8 (1969)
16. Lundquist, G. and Meloan, C. E., *Anal. Chem.,* **43,** 1122 (1971)
17. Conacher, H. B. S., Chadha, R. K. and Sahasrabudhe, M. R., *J. Amer. Oil Chem. Soc.,* **46,** 558 (1969)
18. Conacher, H. B. S., Meranger, J. C. and Leroux, J., *J. Ass. Off. Anal. Chem.,* **53,** 571 (1970)
19. Conacher, H. B. S., *J. Ass. Off. Anal. Chem.,* **56,** 602 (1973)
20. Conacher, H. B. S. and Chadha, R. K., *J. Ass. Off. Anal. Chem.,* **55,** 511 (1972)
21. Conacher, H. B. S., Chadha, R. K. and Iyengar, J. R., *J. Ass. Off. Anal. Chem.,* **56,** 1264 (1973)
22. Kogan, L. and Strezleck, S., *Cereal Chem.,* **43,** 470 (1966)

19

Other Food Additives

A. Introduction

Applications of GC to the analysis of food additives other than those covered in Chapters 17 and 18 are collected together in this chapter. The three classes of additives dealt with here have not attracted sufficient attention to warrant a separate chapter for each one.

The GC conditions of some of the important references in this chapter are listed in *Table 19.1*.

B. Artificial sweeteners and cyclohexylamine

Saccharin and dulcin have been used as artificial sweeteners in such foods as soft drinks since the turn of the century. The reasons for their utilisation were both economic and medicinal, viz. their substitution for higher-priced sugar and glucose and because of their low calorific value when compared with carbohydrates. The last-named reason has become more relevant in recent times, particularly in the eyes of weight-conscious populations, and cyclamates (cyclohexylsulphamic acid salts) were introduced to replace or synergise saccharin in diabetic soft drinks and preserves.

A number of countries introduced legislation to control the use of cyclamates in soft drinks, as was already the case with saccharin, dulcin being a prohibited sweetener in many countries, including the UK. It was following reports that cyclamates, administered in large doses to rats over a prolonged period, could cause cancer and also that similar feeding of hamsters could cause coronary sclerosis and soft tissue calcification that a number of countries introduced legislation prohibiting the use of cyclamates in food. In the UK the *Artificial Sweeteners in Food Regulations 1969*[1] permit the use of only saccharin and its sodium or calcium salts.

With these changes in fashion for the use of artificial sweeteners, analytical methods were developed, primarily for the determination of cyclamate, but often including saccharin. These changes coincided with the upsurge of interest in the application of GC to food analysis, and was successfully used.

Table 19.1 GC conditions used in the analysis of artificial sweeteners, flavourings and flavour potentiators

Class of compound	Ref.	Column dimensions	Stationary phase	Support material	Carrier gas and conditions	Temperature or temperature programme/°C	Detector
Cyclohexene	2	4ft x 0.25in	10% Apiezon L	100/115 AW silanised Embacel	Ar, 40ml/min	50	AI
Cyclohexene	3	5ft x –	10% PEG400	100/120 Celite	Ar, 45ml/min	50–120 at 16/min	FI
Methyl esters of cyclamic acid and saccharin	7	5ft x 0.125in	3% SE-30	100/120 Varaport 30	He, 35ml/min	125–275 at 15/min	FI
Methyl esters of cyclamic acid and saccharin plus free dulcin	8	2m x 3mm I.D.	2.5% SE-52	60/80 DMCS Chromosorb G	He, 30–35 ml/min	150	FI
Cyclohexylamine	9	12ft x 0.125in	10% Carbowax 20M plus 2.5% NaOH	90/100 Anakrom SD	N₂, 72ml/min	100	FI
2,4-Dinitrophenyl-cyclohexylamine	11	1.4m x 1.5mm I.D.	1.0% GE-XE-60 plus 0.1% Epikote 1001	60/80 AW silanised Chromosorb G	N₂, 180ml/min	215	EC

No.	Compound	Column	Phase	Support	Carrier gas	Temp.	Detector
18	Sorbitol hexa-acetate	6ft x 4mm I.D.	10% DC-200	100/120 Gas Chrom Q	N_2, 120ml/min	200	FI
19	TMS derivative of sorbitol	6ft x 0.125in O.D.	4.2% SE-30	60/80 silanised Diatoport S	He, 50ml/min	160–280 at 4/min	FI
21	Benzaldehyde	3ft x 0.25in O.D.	25% Apiezon M	60/80 Chromoport XXX	He, 30 p.s.i. inlet pressure	100	FI
25	Methyl salicylate and safrole and derivatives	10ft x 4mm I.D.	15% Reoplex 400	60/80 Gas Chrom P	N_2, 90ml/min	130, or 90–150at 2/min	FI
27	β-Asarone	3–6ft x 4mm I.D.	5% Reoplex 400	60/80 Chromosorb G	N_2, 90ml/min	180	FI
29	N-TFA butyl glutamate	6ft x 0.125in O.D.	3% OV-17	100/120 Aeropak 30	N_2, 30ml/min	170	FI
30	TMS derivative of glycyrrhetic acid	4ft x 4mm I.D.	1.5% OV-1	60/80 Gas Chrom Q	N_2, 75ml/min	200–260 at 6/min	FI
31	TMS derivatives of maltol, ethyl maltol	6ft x 0.125in O.D.	10% UCW98	80/100 Diatoport S	He, 120ml/min	130	FI

There has been subsequent speculation that the tissue changes occurring in laboratory animals fed on cyclamate could be due to the action of cyclohexylamine, from which it is synthesised and which is present in trace amounts in cyclamate. Cyclohexylamine is also a metabolite of cyclamate.

Sorbitol has been included in this section because it is an added sweetener to diabetic foods, although the term 'artificial' is not really appropriate.

1. Cyclamate

Most methods of cyclamate analysis depend on its reaction with nitrous acid to form cyclohexene and sulphuric acid. The latter can be determined by precipitation as sulphate or by volumetric or colorimetric methods. Although these methods are generally satisfactory, they can be time-consuming, and GC methods were produced with speed as well as sensitivity in mind.

The GC determination of cyclamates in soft drinks was pioneered by Rees[2], using the reaction referred to above and measuring the cyclohexene produced in solution. Volumes of comminuted orange drink samples, containing preferably about 25 mg cyclamate, were acidified with hydrochloric acid, clarified and cleaned up by extraction of unwanted material with chloroform and also with petrol. The final aqueous solution was shaken with sodium nitrite solution and with petrol containing benzene as internal standard. When effervescence ceased, the petrol solution containing benzene and the reaction product cyclohexene was chromatographed on an Apiezon L stationary phase.

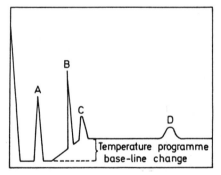

Figure 19.1 Gas chromatogram showing peaks of: A, cyclohexene; B, monochlorocyclohexane; C, cyclohexanone; D, cyclohexanol (after Richardson and Luton[3]). For GC details, see Table 19.1

Using essentially this same method, Richardson and Luton[3] proposed certain amendments to the determination of cyclamate in soft drinks. They considered that if the solution prior to extraction of the cyclamate gave a more acid reaction than a pH of 0.65, cyclamic acid was formed and this could be extracted into the organic solvents used in the clean-up step and thereby lost. They also preferred sulphuric acid to hydrochloric acid since this precluded the formation of monochlorocyclohexane. This method, which used PEG 400 as stationary phase, was operated successfully on soft drinks containing between 0.2 and 0.5% sodium cyclamate. *Figure 19.1* is a chromatogram showing the separation of cyclohexene, monochlorocyclohexane, cyclohexanone and cyclohexanol.

The same method was investigated by Dalziel, Johnson and Shenton[4] with a view to using strongly nitrosating conditions to form cyclohexyl nitrite in place of cyclohexene. This was not successful and, apart from minor adjustments, their method offers little extra to the original Rees method[2].

Instead of measuring cyclohexene in solution, Groebel and Wessels[5] measured it in the head space vapour and applied this method to the GC determination of cyclamate in fruit juices, soft drinks, wines and preserves. After some concentration of the sample, sulphuric acid and 10% aqueous methyl ethyl ketone were added and the mixture was diluted and centrifuged. A small volume of the clear supernatant was transferred to a special head space analysis flask and caused to react with sodium nitrite for 2 h at 24°C. The cyclohexene in the head space was subsequently determined on a PEG 1500 stationary phase. Cyclohexenyl nitrite, also formed in the reaction, did not interfere in the analysis.

Although cyclamate has been determined simultaneously with saccharin using TLC, by Dickes[6], this method was limited in some instances by interference from co-extracted impurities, and the GC method of Conacher and O'Brien[7] effected a clean and quantitative separation of these two artificial sweeteners. Soft drink samples were made alkaline and extracted with diethyl ether to remove interfering basic substances. The aqueous solution was acidified, stearic acid added as internal standard and the mixture extracted several times with ethyl acetate. After evaporation of the ethyl acetate solution, the cyclamate and saccharin were methylated using diazomethane, the resulting methyl esters

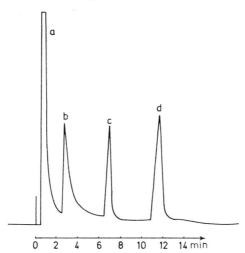

Figure 19.2 Gas chromatogram of cyclamate methyl ester (c), dulcin (b) and methyl saccharin (d) in methanol (a) on an unpolarised column (SE-52) at 150°C (after Koenig[8]). For GC details, see Table 19.1

being chromatographed on an SE-30 stationary phase. If diazomethane was added before the evaporation stage, recoveries of the sweeteners was over 90%, but they were 60–75% if there was no such addition. The method, although able to determine both cyclamate and saccharin simultaneously, was not as sensitive as the Rees method[2].

The methyl esters of cyclamate and saccharin, together with free dulcin, were separated using GC by Koenig[8]. He used diazomethane as the methylating agent and chromatographed the three sweeteners on a SE-52 stationary phase; *Figure 19.2* shows their typical separation.

2. Cyclohexylamine

Cyclohexylamine is the basis of the synthesis of cyclamate, and trace amounts are carried through into the final product. Cyclohexylamine has also been generated from cyclamate by the gut flora of certain laboratory test animals and is thought to be responsible for the tissue abnormalities found in feeding trials of cyclamate to laboratory animals. Much attention has therefore been paid to the cyclohexylamine levels in foods sweetened with cyclamates.

Cyclohexylamine may be determined by GC, either as the free amine or as its 2,4-dinitrophenyl derivative.

Howard *et al.*[9] examined a number of foods for the presence of cyclohexylamine, including fruit juices, carbonated beverages, canned fruits and various liquid and dry fruit bases for addition to food. Some of the liquid samples were made alkaline and steam-distilled into hydrochloric acid solution. The distillates were made alkaline and extracted into dichloromethane and the extracts concentrated for GC. Some of the liquid concentrates and aqueous solutions of cyclamates were made alkaline and extracted without recourse to

Table 19.2 Levels of cyclohexylamine (CHA) found in some artificially sweetened products and in sodium and calcium cyclamates (after Howard *et al.*[9])

Brand	Size	Calculated weight from declared cyclamate/g	CHA found/p.p.m.
Cola types:		Carbonated beverages	
A	10fl. oz	0.45	0.2
B	10fl. oz	0.60	1.1
C	10fl. oz	0.38	6.5
D	12fl. oz	0.72	0.0
E	10fl. oz	0.33	0.3
F	12fl. oz	0.26	0.1
Citrus types:			
G	10fl. oz	0.60	0.2
H	10fl. oz	0.38	0.3
I (lemon-lime)	10fl. oz	0.30	0.0
J (orange)	12fl. oz	1.16	0.2
K (orange)	12fl. oz	0.90	0.2
K (lemon)	12fl. oz	0.90	0.0
Miscellaneous types:			
L (grape)	12fl. oz	1.12	0.3
M (grape)	12fl. oz	0.90	0.2
N (grape)	32fl. oz	1.44	0.0
O (root beer)	12fl. oz	1.16	0.0
O (ginger ale)	12fl. oz	0.86	0.0
P (ginger ale)	16fl. oz	1.20	0.0
Q (almond-smash)	16fl. oz	1.20	0.1

Table 19.2 (*continued*)　　　　　　　　　　　　　*Other Food Additives*

Brand	Size	Calculated weight from declared cyclamate/g	CHA found/p.p.m.
		Dry beverage bases	
R (lemon-lime)	4.8 g	1.69	160* (0.4)†
R (grape)	4.8 g	1.60	480* (1.2)†
S (grape)	8.5 g	2.98	0.0
T (grape)	5.7 g	1.83	370* (1.1)†
S (strawberry)	8.5 g	2.98	0.0
U (lime)	4.7 g	1.32	204* (0.5)†
V (lemon)	7.1 g	0.93	162* (1.2)†
		Canned fruits	
W (peaches)	1 lb	0.60	0.3
W (cocktail)	1 lb	0.60	0.9
		Fruit juice drinks	
X (combination)	16 fl. oz	0.60	0.7
		Liquid and dry weight control formulations	
Y (liquid)	8 fl. oz	undeclared	0.0
Z (dry base)	0.72 fl. oz	1.75	2.8* (0.2)†
		Food sweetener preparations (liquid concentrates)	
AA	8 fl. oz	20.40	1.5
BB	6 fl. oz	15.12	5.5
CC	12 fl. oz	28.80	1.4
		Food sweetener preparations (dry base)	
CC	4.5 oz	8.10	9.5
DD	8 oz	38.40	2.6
		Artificial sweeteners	
EE	Sodium cyclamate		8.3
FF	Sodium cyclamate		7.3
EE	Calcium cyclamate		3.8
FF	Calcium cyclamate		54

*CHA content calculated on basis of dry base weight.
†CHA content calculated on basis of dilution according to directions.

steam distillation. The GC findings were confirmed by trapping out the cyclohexylamine, converting it to dithiocarbamate, and by analysis using IR spectroscopy. *Table 19.2* gives the cyclohexylamine content of various foods as found by this method. Dried base additives contained the highest levels of cyclohexylamine, a dry beverage base containing as much as 480 p.p.m. Directly consumable foods, such as carbonated beverages, contained as much as 6.5 p.p.m. cyclohexylamine. Fazio and Howard[10] conducted a collaborative study of this procedure and recommended it to be adopted as a first action method in the USA. Recoveries of 89–95% were obtained by the collaborators with beverages containing cyclohexylamine at the 0.5 and 1 p.p.m. level.

The 2,4-dinitrophenyl derivative of cyclohexylamine gives a product with strong electron-capturing properties, and this was exploited by Weston and Wheals[11] in the analysis of cyclohexylamine in soft drinks. It was collected in acid solution from distillation of an alkaline medium, and the distillate was extracted with toluene. The toluene was evaporated and the cyclohexylamine was dissolved in a sodium borate buffer solution and caused to react with 1-fluoro-2,4-dinitrobenzene in dioxan to yield the 2,4-dinitrophenyl derivative. Occasionally aniline was encountered in the analysis of soft drinks; it was believed to arise from the reduction of the orange azo dyes used to colour the samples. This aniline was removed by conversion to its bromo derivative and the co-extracted excess 1-fluoro-2,4-dinitrobenzene, which could cause background interference, was hydrolysed by alkali. This method gave a linear response for cyclohexylamine in the range $1-10$ μg per ml of soft drink, with a detection limit of 0.25 mg.

The same derivative and a similar extraction procedure were used by Solomon, Pereira and Duffield[12]. They chromatographed an extract using an EC detector system incorporating [63]Ni and found that the detection limit was 10 pg cyclohexylamine. This method was only applied to aqueous solutions and not to the analysis of cyclohexylamine in foodstuffs.

3. Saccharin

Saccharin has been used as an artifical sweetener for more than 50 years, and its safety in use has only recently been seriously questioned because of the backlash of the controversy regarding the safety in use of cyclamate. However, it is still a permitted additive in most countries, and the removal of cyclamate has left saccharin as the main low-calorie sweetener for dietetic and diabetic foods.

The chemical analysis of saccharin is relatively simple, depending on extraction from acid solution and subsequent estimation by colorimetry or titrimetry. Because interferences can be caused in chemical methods by co-extracted preservatives such as those of the benzoic acid family, analysts have produced GC methods using the *N*-methyl ester of saccharin.

Groebel[13] methylated saccharin using diazomethane and chromatographed the ester on a 20% SE-30 stationary phase, using ethyl *p*-methoxybenzoate as internal standard.

This method was expanded by Conacher and O'Brien[14], who used the same derivative of saccharin for its determination in soft drinks. Stearic acid was added as internal standard and, together with the saccharin, was extracted from acid solution with ethyl acetate prior to methylation and GLC on a 3% JXR stationary phase. Two peaks were observed with a ratio of areas of 17:3, the major one being used for quantitative purposes. This peak was *N*-methyl saccharin and the minor one was thought to be *O*-methyl saccharin. Preservatives such as sorbic acid, benzoic acid and 4-hydroxybenzoate esters did not interfere with the method, which was capable of determining down to 5 mg saccharin per 10 fl. oz soft drink. These workers attempted to chromatograph the TMS and the TFA derivatives, but without success.

Success was, however, achieved by Gerstl and Ranfft[15] in preparing and chromatographing the TMS derivative of saccharin. They used a mixed stationary phase of 3% OV-7 and 1.5% OV-22 and applied the method to the determination of saccharin in foods and animal feeding stuffs.

Saccharin was determined in wine using the *N*-methyl derivative by Piorr and Novakovic[16]. The sample was mixed with an equal volume of saturated salt solution, acidified and extracted with diethyl ether. After methylation, the solution was chromatographed on a 3% Carbowax 20M stationary phase: 0.05 mg saccharin per litre of wine could be detected.

4. Dulcin

Dulcin has been used in many countries as an artificial sweetener, although it has not maintained the same popularity as saccharin. It is not a permitted artificial sweetener in the UK.

Koenig[8] chromatographed dulcin with cyclamate and saccharin on an SE-52 stationary phase. Dulcin was chromatographed unchanged, together with the methyl esters of the other sweeteners, all three compounds being well separated (see *Figure 19.2*).

Groebel[13] also included dulcin in his application of GC to the analysis of saccharin.

5. Sorbitol

Where large amounts of sweetener are needed in a dietetic or diabetic food, saccharin is unsuitable because of its bitter after-taste at high concentrations. The large increase in the consumption of dietetic foods during the last few years has meant that the food technologist has resorted to using sorbitol and other sugar alcohols as low-calorie sweetening agents.

Sorbitol has been determined by colorimetric and gravimetric methods, but GC offers greater specificity in its capacity to separate similar sugar alcohols which might interfere in the analysis.

Jones, Smith and Sahasrabudhe[17] quantitatively determined the sorbitol in dietetic biscuits via GC of the acetate. Biscuits were crumbled and extracted with hexane to remove fat, and the residue was further extracted with 80% aqueous ethanol. The ethanolic solution was evaporated and the hexa-acetate derivative made for GC. Six samples of dietetic biscuits were found to contain between 13 and 30% sorbitol using this method.

From a previous study of the determination of sorbitol in bakery products, wines and vinegars, Hundley[18] conducted a collaborative exercise to ascertain whether his method was suitable for the determination of sorbitol in devil's food cake, yellow cake and sugar cookies. These foods were fortified with sorbitol at levels between 1.5 and 7% and the sugar alcohol was extracted with methanol. The sorbitol hexa-acetate was made, using acetic acid in pyridine as the acetylating agent, and GLC was performed on a 10% DC-200 stationary phase. The method was satisfactory but did not separate sorbitol and mannitol hexa-acetates.

A similar extraction procedure was used by Fernandez-Flores and Blomquist in the determination of sorbitol in dietetic and diabetic foods[19]. They prepared the TMS derivative and used an SE-30 stationary phase for its chromatography. The method is particularly suitable for the analysis of sorbitol where there is little or no other sugar present, such as in dietetic candies and cookies.

C. Flavourings and flavour potentiators

Flavourings are often made up of a very large number of organic compounds and, apart from a few major components, most of these are present in trace amounts. The difficulties inherent in the separation and qualitative establishment of flavouring components are such that classical chemical methods have little chance of being fully effective.

GC has been greatly utilised in studies of the natural flavours of many foods, and reference may be made to Chapters 9–16, where analysis of the flavours of the relevant foods is discussed. Even so, GC only allows a limited separation of complex flavour components and MS is coupled with GC in order to achieve a greater measure of success.

Many flavours have been extracted from natural foods or substances, refined, and added to other insipid foods to give them flavour. Essential oils (Chapter 12) and coffee and vanilla (Chapter 13) are examples of substances capable of yielding flavour for use in other foods.

This section of the chapter contains GC methods applicable to a miscellany of flavourings which have not been covered in the other chapters of this book, and also includes the application of GC to the analysis of some flavour potentiators.

Bidmead and Kratz[20] applied GC methods to find out whether or not natural flavouring matters had been sophisticated with synthetic flavourings or other organic substances. They found that natural pineapple juice contained ethyl acetate, ethyl propionate, isoamyl acetate and furfural, whereas a pineapple flavouring they examined contained a significant level of allyl hexanoate, showing it to be part synthetic. Another pineapple flavouring examined contained limonene and eugenol, these compounds, respectively, indicating the presence of lemon oil and clove oil.

A grape flavouring was analysed by GC and contained linalool, benzaldehyde, α-terpineol, citronellol, methyl-*N*-methylanthranilate, methyl-*N*-dimethylanthranilate and β-naphthyl methyl ester. It is known that lemon oil not only contains linalool, α-terpineol and citronellol, but also associated monoterpenes, and since the latter were not found in the flavouring, it was assumed that the other three lemon oil constituents had been added synthetically. The dimethylanthranilate derivative is not found naturally and occurs with the monomethylanthranilate in commercial preparations. It was therefore concluded that these compounds also had been added synthetically, together with the β-naphthyl methyl ether, which has not been found in natural flavourings.

The overall conclusion was that the grape flavouring was probably entirely synthetic. Two stationary phases were used in these analyses, viz. 20% Carbowax 20M and 20% SE-30, and identifications of peaks were confirmed by IR spectroscopy and MS.

The remainder of the applications of GC to analytical problems concerning flavourings are subdivided into the eight sub-sections that follow.

1. Benzaldehyde

Benzaldehyde is a major flavour ingredient of cherry and almond flavourings and UV spectrophotometric methods for its determination can sometimes prove to be difficult in the presence of benzoic acid. Brunelle and Martin[21] used

GLC of the free aldehyde on a 25% Apiezon M stationary phase and found that it was not necessary to use steam distillation to separate benzaldehyde from benzoic acid since separation was achieved during the chromatography. The concentration of benzaldehyde in natural cherry flavourings is approximately 0.2% but this was exceeded in the imitation flavourings examined. *Table 19.3* gives the benzaldehyde contents of various cherry and almond flavourings and essences as found by this method.

Table 19.3 (after Brunelle and Martin[21])

Flavouring or essence	Benzaldehyde/(g/100ml)
Cherry flavour	0.17
Cherry flavour	0.24
Cherry flavour	0.26
Pure almond extract	1.35
Imitation cherry flavour	0.51
Almond essence	2.72
Sweet cherry flavour	2.30
Cherry pit cordial flavour	0.30
Imitation wild cherry flavour	0.31
Imitation black cherry flavour	0.56
Kirsch cherry liqueur flavour	1.75
Cherry flavour base	0.68

2. Coumarin

The Food Standards Committee Report on Flavouring Agents, 1965[22], listed coumarin among the 16 substances it recommended to be prohibited in food in the UK. Bucci, Boniforti and Cesari[23] have reviewed the public health problems involved in the use of certain flavourings in food and produced a GLC method for coumarin. Using an SE-30 stationary phase and EC detection, it was possible to detect as little as 0.2 μg coumarin.

Pellerin *et al.*[24], in using a GLC method for the detection of preservatives and flavours in food, separated coumarin on an SF-96 stationary phase at 250°C.

3. Safrole, isosafrole, dihydrosafrole, dihydroanethole and methyl salicylate

These five flavouring agents were removed from non-alcoholic, commercial, carbonated beverages and detected using a GC method, by Larry[25]. Commercial beverages or syrup bases were decarbonated and steam distilled and the distillates extracted with chloroform. The chloroform extracts were concentrated and subjected to GLC on a Reoplex 400 stationary phase. The method was used to quantitatively determine methyl salicylate and this was found in ten test samples within the range 43–82 p.p.m. With some modifications to this method, including a change of internal standard and of chromatographic conditions to suit the quantitative determination of safrole, Larry[26] conducted another collaborative study. The revised method was recommended as a first action one in the USA for the quantitative determination of safrole and the semiquantitative determination of isosafrole, dihydrosafrole and dihydroanethole. The Food Standards Committee Report on Flavouring Agents, 1965[22], recommended that safrole, isosafrole and dihydrosafrole should not be used in food.

4. β-Asarone

β-Asarone is a component of oil of calamus, and the oil has been used as part of the flavouring added to vermouth. The FDA of the USA does not permit the presence of β-asarone in food because feeding trials on rats showed evidence of cancer.

Larry[27] steam-distilled samples of sweet and dry vermouth and extracted the distillates with 1:1 diethyl ether:hexane and, after concentration of the extract, subjected it to GLC on a 5% Reoplex stationary phase. The method was found to be quantitative for β-asarone in the range 5–100 p.p.m., but there was no record of the presence of this flavouring in any of the vermouth samples.

5. Allyl hexanoate

Allyl hexanoate has a strong pineapple-like flavour and has been used as an ingredient in synthetic pineapple flavourings which can be added to foods, such as yoghurt and sugar confectionery.

van den Dool, Hansen and van der Puijl[28] kept food samples in a closed flask at 65°C for 1 h, sampled the head space vapour for GLC analysis on an SE-30 stationary phase and determined the allyl hexanoate concentration down to the 5 p.p.m. level. Pineapple-flavoured sweets were ground and mixed with an equal weight of water prior to the head space sampling. As an alternative, samples were steam-distilled and the distillates were extracted into 2:1 diethyl ether:pentane. The extracts were concentrated and used for GC, the limit of detection being less than 1 p.p.m. allyl hexanoate.

6. Monosodium glutamate

Monosodium glutamate is a flavouring agent, but its main use is as a flavour potentiator in 'meaty' food products.

Analytical methods for the determination of monosodium glutamate usually depend on a measurement of glutamic acid. Gal and Schilling[29] devised a GC method for the determination of glutamic acid via its N-TFA derivative of the butyl ester and applied it to the analysis of monosodium glutamate in soups and seasonings. The sample was shaken with water, active charcoal was added and the mixture was filtered after 30 min. The filtrate was passed through a column of the H$^+$ form of Dowex 50W-X8 and the glutamic acid on the column was eluted with normal hydrochloric acid. The N-TFA butyl glutamate was prepared and chromatographed on a 3% OV-17 stationary phase. The method was suitable for monosodium glutamate concentrations of 2–17%, as would be found in vegetable and clear soups and also in seasonings.

7. Ammonium glycyrrhizinate

Ammonium glycyrrhizinate, besides being capable of increasing sugar sweetness, can be used as a flavour potentiator in chocolate- and caramel-flavoured beverages.

Larry, Fuller and Harrill[30] carried out recovery experiments with this ammonium salt added to caramel-containing beverages and used a GC method for analysis. After hydrolysis of the sample, the TMS derivative of glycyrrhetic acid was made and chromatographed on a 1.5% OV-1 stationary phase. The average recovery from beverages was 87%, but no commercial samples actually containing this flavour potentiator were examined.

8. Maltol and ethyl maltol

Both maltol and ethyl maltol have received attention because of their flavour-potentiating properties. Maltol is more a flavouring than a potentiator, whereas ethyl maltol is known to be a more powerful enhancer of flavour. Gunner, Hand and Sahasrabudhe[31] conducted recovery experiments after adding these compounds to apple juice. They used ethyl acetate as extractant and used the GC of the TMS derivatives for the analysis. No evidence was presented of the addition of these compounds to apple juice.

D. Mineral hydrocarbons

Mineral oil is used as a cold dipping emulsion for the treatment of raisins and sultanas in order to increase the drying rate of the grapes. This aids the prevention of dried fruits sticking together, raisins which have been mechanically de-seeded causing the largest problem. In the UK the *Mineral Hydrocarbons in Food Regulations 1966*[32] include a limit of 0.5% on those hydrocarbons which are allowed in dried fruit.

The natural composition of the cuticular waxes of dried fruits includes acids, alcohols, esters, aldehydes, hydrocarbons and the triterpene called oleanolic acid. The natural pattern of these compounds is changed by the infiltration of added mineral oil into the cuticle. Advantage of this difference was exploited by Radler[33], who produced a method for the detection of mineral oil in dried fruits. Surface lipids were extracted into chloroform and the hard wax was separated from the soft wax by fractionation with petrol. Alumina column chromatography was used to separate the soft wax part into hydrocarbon, ester plus aldehyde, alcohol and acid fractions. GC was used to determine the composition of the hydrocarbon and alcohol fractions and also the unsaponifiable portion of the lipids.

The cold dipping of sultanas in mineral oil led to an appreciably higher concentration of C_{16} and C_{18} fatty acids in the saponifiable matter when compared with that of unprocessed fruits, and a GC analysis showing a proportional increase in these two acids offers a method of detecting the presence of mineral oil.

References

1. *The Artificial Sweeteners in Food Regulations 1969*, Statutory Instrument No. 1817, HMSO (1969)
2. Rees, D. I., *Analyst*, **90**, 568 (1965)
3. Richardson, M. L. and Luton, P. E., *Analyst*, **91**, 520 (1966)
4. Dalziel, J. A. W., Johnson, R. M. and Shenton, A. J., *Analyst*, **97**, 719 (1972)
5. Groebel, W. and Wessels, A., *Dt. LebensmittRdsch.*, **68**, 393 (1972)
6. Dickes, G. J., *J. Ass. Publ. Anal.*, **3**, 119 (1965)
7. Conacher, H. B. S. and O'Brien, R. C., *J. Ass. Off. Anal. Chem.*, **54**, 1135 (1971)
8. Koenig, H., *Z. Anal. Chem.*, **255**, 123 (1971)
9. Howard, J. W., Fazio, T., Klimeck, B. A. and White, R. H., *J. Ass. Off. Anal. Chem.*, **52**, 492 (1969)
10. Fazio, T. and Howard, J. W., *J. Ass. Off. Anal. Chem.*, **53**, 701 (1970)
11. Weston, R. E. and Wheals, B. B., *Analyst*, **95**, 680 (1970)
12. Solomon, M. D., Pereira, W. E. and Duffied, A. M., *Anal. Lett.*, **4**, 301 (1971)
13. Groebel, W., *Z. Lebensmitt. Forsch.*, **129**, 153 (1966)
14. Conacher, H. B. S. and O'Brien, R. C., *J. Ass. Off. Anal. Chem.*, **53**, 1117 (1970)
15. Gerstl, R. and Ranfft, K., *Z. Anal. Chem.*, **258**, 110 (1972)
16. Piorr, W. and Novakovic, N., *Dt. LebensmittRdsch.*, **66**, 223 (1970)
17. Jones, H. G., Smith, D. M. and Sahasrabudhe, M., *J. Ass. Off. Anal. Chem.*, **49**, 1183 (1966)
18. Hundley, H. K., *J. Ass. Off. Anal. Chem.*, **56**, 66 (1973)
19. Fernandez-Flores, E. and Blomquist, V. H., *J. Ass. Off. Anal. Chem.*, **56**, 1267 (1973)
20. Bidmead, D. S. and Kratz, P. deC., *Cereal Sci. Today*, **11**, 486 (1966)
21. Brunelle, R. L. and Martin, G. E., *J. Ass. Off. Agric. Chem.*, **46**, 950 (1963)
22. *Food Standards Committee Report on Flavouring Agents*, Ministry of Agriculture, Fisheries and Food (1965)
23. Bucci, F., Boniforti, L. and Cesari, A., *Boll. Lab. Chim. Prov.*, **16**, 195 (1965)
24. Pellerin, F., Gautier, J. A., Castillo-Penna, M. and Blanc-Guenee, J., *J. Pharm. Belg.*, **20**, 181 (1965)
25. Larry, D., *J. Ass. Off. Anal. Chem.*, **52**, 481 (1969)
26. Larry, D., *J. Ass. Off. Anal. Chem.*, **54**, 900 (1971)
27. Larry, D., *J. Ass. Off. Anal. Chem.*, **56**, 1281 (1973)
28. van den Dool, H., Hansen, A. and van der Puijl, I., *Z. Lebensmitt. Forsch.*, **138**, 272 (1968)
29. Gal, S. and Schilling, P., *Z. Lebensmitt. Forsch.*, **148**, 18 (1972)
30. Larry, D., Fuller, M. J. and Harrill, P. G., *J. Ass. Off. Anal. Chem.*, **53**, 698 (1970)
31. Gunner, S. W., Hand, B. and Sahasrabudhe, M., *J. Ass. Off. Anal. Chem.*, **51**, 959 (1968)
32. *The Mineral Hydrocarbons in Food Regulations 1966*, Statutory Instrument No. 1073, HMSO (1966)
33. Radler, F., *J. Sci. Food Agric.*, **16**, 638 (1965)

Part 5

Food Contaminants

20

Pesticides

A. Introduction

A considerable amount of the general technology and understanding of GC has arisen from the analysis of pesticide residues by this technique. This is due not only to the sensitivity and specificity of GC, but also to its power of separating compounds of similar molecular structure. Consequently, a very large number of pesticides have been subjected to GC and, for convenience, are referred to by their trivial names.

Probably the biggest advertisement afforded GC as a microanalytical technique came its way in the early 1960s, when it was realised that it was capable of solving many of the problems in pesticide residue analysis. Numerous analytical institutions purchased GC equipment with residue analysis as the primary reason, and many gas chromatographers served their apprenticeships in the detection and determination of organochlorine (OC) pesticide residues. The interest in OC pesticide residue analysis remains, having reached its zenith in the mid- to late-1960s.

Because it was necessary to find an analytical technique capable of detecting nanogram quantities of pesticides, and since the conventional detectors of TC, AI and FI types were insensitive at this level, it was vital that GC be provided with more sensitive detector systems. In 1960 Lovelock and Lipsky[1] described the EC detector, which was found to be particularly sensitive to halogens and therefore suitable for the analysis of OC compounds. In the same period Coulson et al.[2] produced their MC detector, which can be made specific for halogens and sulphur and therefore can be used to advantage in OC residue analysis and also in the detection of thiophosphate insecticides.

GC was an established technique by the time residue analysts turned their attention to the determination of organophosphorus (OP) pesticide residues. Low levels of parathion could be determined by using EC detection since this OP insecticide contains a nitro group which is sensitive in that system, but the remainder of this class of insecticides was too insensitive to the standard detector systems. In 1964 Giuffrida[3] produced the AFID for use in the determination of OP pesticide residues in food, which placed the analysis of these compounds on a similar sensitivity footing to the OC compounds with their EC and MC detection.

Pesticides

No subject has done more to stimulate detector technology than pesticide residue analysis. Following the development of EC, MC and AFID, use has been made of FPD and also detectors specific for the determination of nitrogen-containing compounds. The latter system has been utilised in the multiresidue analysis of pesticides such as the carbamates and triazines.

Because of the persistence and, therefore, accumulation of OC pesticide residues and their possible hazards to health, many of them have been phased out. Alternative insecticides, such as those of the carbamate family, have assumed a new importance for this reason, and consequently much more attention has been paid to their detection and determination, and GC methods designed to cope with these multiresidues have been to the fore.

Multiresidue analysis is the most important facet of the determination of pesticides in food, which means that several compounds are frequently sought in a single analysis. It is therefore imperative that the extract used for GC should be as clean as possible, and for this reason emphasis has been placed on extraction and clean-up procedures in this chapter.

Pesticides have been divided on the basis of their functional elements, i.e., halogen, phosphorus and nitrogen, and some of these have been subdivided according to their pesticidal action, i.e. insecticides, fungicides and herbicides.

Table 20.1 includes the GC conditions pertaining to some of the more important references in this chapter.

B. Organohalogen pesticides

1. Organochlorine insecticides and acaricides

The concern shown internationally for the persistence of OC insecticide residues and their possible harmful effects on animal life has led to numerous surveys of their residue levels in all types of food. There has been a mass of published methods and other data relating to the application of GC to the determination of OC insecticide residues in food, and consequently the bibliography cited here has been selected to give examples of the different approaches in extraction, clean-up and detection.

The literature pertaining to the initial extraction of these residues mainly concerns the solvents acetone[4-8], hexane or petrol[9-11], acetonitrile[12-15], dimethylformamide[16-18], dimethyl sulphoxide[19-22] and propylene carbonate[23]. Clean-up of samples is usually carried out by solvent partition followed by column chromatography on Florisil, alumina or activated carbon. Numerous stationary phases have been utilised for the separation of OC hydrocarbons, general-purpose phases being SE-30, Apiezon L and QF-I. Detection systems have almost exclusively relied on EC, MCD and FPD.

In 1960 Goodwin et al.[4] applied GC to the detection of aldrin, dieldrin, *pp'*-DDT and γ-BHC in crop extracts at the C.05 p.p.m. level, using a 10% E301 stationary phase and an AI detector. This pioneer work was enlarged upon by Goodwin, Goulden and Reynolds[5], who reduced the loading of the stationary phase to 2.5% and incorporated 0.25% of Epikote resin in order to obtain inactivation of the kieselguhr support material, thereby reducing peak tailing. EC detection replaced the AI detector.

The procedure of acetone extraction, followed by addition of sodium sulphate solution and extraction of OC compounds into petrol, was used by Hamence, Hall and Caverly[6] in the examination of food samples. They introduced the clean-up step of further extraction into acetonitrile, addition of sodium sulphate solution to the acetonitrile extract and re-extraction into petrol, a method first used by Jones and Riddick[24]. The petrol extract was further cleaned up on Brockmann V activity alumina, using petrol to elute fractions containing different groups of OC compounds. As an alternative for the verification of the results by further GC on other colums, these workers used chemical reactions with re-chromatography on the same column.

The Goodwin *et al.*[5] method was the basis of three surveys of the OC residue levels of fruits, vegetables and other foods by the authors[25-27]. Four stationary phases were used, viz. 5% SE-30, 3% Apiezon L, 5% QF-1 and 2% XE-60, for the GC of the OC residues in order to confirm their identities. The cyanosilicone phase, XE-60, was found to be particularly valuable in the separation of OC fungicide residues, which were co-extracted with the OC insecticides.

The direct extraction of food and crop samples with acetonitrile, thus omitting acetone, and the partitioning into petrol or hexane, is the most popular method for extracting OC pesticide residues in the USA, and it is the official method of the Association of Official Analytical Chemists (AOAC)[28].

The method is used for the determination of multiple residues of both OC and OP residues. The sample, having been blended with acetonitrile, is extracted into petrol after the addition of water, and the petrol extract is cleaned up on Florisil, using mixtures of petrol and diethyl ether for elution. Analysis is carried out by GC with TLC and paper chromatography as supporting techniques. The stationary phase used is 10% DC-200 and detection is by EC for OC compounds. This official method was based on the original work of Burke and Giuffrida[29], who applied it to the determination of OC residues in vegetables such as maize, spinach, carrots, green beans, potatoes, broccoli, peas, lettuce, radish, green onions, kale and cabbage. The detection limit was generally 0.01 p.p.m. and recoveries of between 73 and 98% were obtained.

Although little work had been carried out until the 1970s on combining GC with MS in residue analysis, the fact that polychlorinated biphenyls (PCBs) can be co-extracted with OC pesticides and that the two classes of compound are difficult to separate by GC alone has created much greater interest in the combination of the techniques (see under PCBs, Chapter 21).

Food extracts were separated on conventionally packed non-polar columns by Bellman and Barry[30], and the effluents were led, via a glass–frit interface connection, to a MS for detection. It would not normally be necessary to use MS for GC residue analysis of most foods, but the combination offers an advantage in the examination of meat, fish and eggs, where PCBs are more liable to occur.

A useful identification method, which is often utilised before GC of an OC pesticide, is one depending on its distribution between the two immiscible solvents, hexane and acetonitrile. Beroza and Bowman[31] proposed the use of this method and measured the total content of an OC pesticide and also of that portion of it in the hexane phase after separation between the two solvents. These two concentrations were expressed as a ratio, and these workers listed ratios for numerous OC pesticides. Co-extracted food or crop material should not interfere with these ratios.

Table 20.1 GC conditions used in the analysis of pesticides

Class of compound	Ref.	Column dimensions	Stationary phase	Support material	Carrier gas and conditions	Temperature or temperature programme/°C	Detector
OC pesticides	5	2ft x 0.16in I.D.	2.5% E301 plus 0.25% Epikote 1001	100/120 Kieselguhr	N_2, 200ml/min	163	EC
OC and OP pesticides	28	6ft x 4mm I.D.	10% DC-200	80/100 Chromosorb W-HP	N_2, 120ml/min	200	EC (for OCs) AFI (for OPs)
OC fumigants and CS_2	37	6ft x 0.25in O.D.	10% DC-710	80/100 HMDS Chromosorb W	N_2, 20ml/min	60	EC
Methyl bromide and ethylene oxide	41	2m x 4.6mm I.D.	15% UCON LB-550-X	60/80 Chromosorb W	He, 80ml/min	85	FI
Hexachlorobenzene and isomers of BHC	44	2m x 3mm I.D.	2 parts 3%OV-61 2 parts 7.5% QF-1 1 part 3% XE-60	80/100 silanised Chromosorb P 80/100 Chromosorb W-HP 80/100 Anakrom AS	90:10 Ar:CH_4, 40ml/min	190	EC
TMS derivatives of OC herbicides	53	6ft x 1mm I.D.	5% DC-200	80/100 Chromosorb W-HP	He, 40ml/min	190	TC
OP pesticides	28	6ft x 4mm I.D.	10% DC-200	60/100 Chromosorb W-HP	N_2, 60ml/min	205	AFI
OP pesticides	77	2.4m x 4mm I.D.	5% OV-101 or OV-210	80/100 Gas Chrom Q	N_2, 160ml/min	150–300 at 10/min	FP
Phosphorus- and sulphur-containing pesticides	104	2.4m x 4mm I.D.	5% Dexsil 300	80/100 AW Chromosorb W	N_2, 160ml/min	150–300 at 10/min	FP

109	2,2,4-Trichloro-acetophenone	4ft x 6mm O.D.	5% SE-96	80/100 Chromosorb W	N₂, 6?/ml/min	200	EC
114	Organonitrogen pesticides	6ft x 4mm I.D.	10% DC-200	80/100 Chromosorb W-HP	He, 80ml/min	180	CC
131	Carbamates	6ft x 4mm I.D.	7.5% QF-1 plus 5% DC-200	80/100 Chromosorb W-HP	N₂, 60ml/min	160	AFI
131	Carbamates	6ft x 3.5mm I.D.	5% OV-101	80/100 Chromosorb W-HP	N₂, 60ml/min	155	FP
129	Methyl derivatives of methyl carbamates	5ft x 0.125in	80/100 Porapak P	80/100 Porapak P	He, 27ml/min	180	AFI
121	TCA derivative of the hydrolysis product of carbaryl	1.8m x 4mm I.D.	10% DC-200	60/80 Gas Chrom Q	N₂, 75–100 ml/min	190	EC
118	2,4-Dinitroaniline derivatives of carbamates	4ft x 0.25in O.D.	2% XE-60	50/60 Anakrom ABS	N₂, 180ml/min	190	EC
137	Triazine herbicides	1m x 2mm I.D.	8% Reoplex 400	80/100 HMDS Chromosorb W	He, 37ml/min	180	FI
148	Captan, captafol, folpet	6ft x 4mm I.D.	3% XE-60	80/100 Chromosorb W-HP	N₂, 100ml/min	178	EC
152	Methyl derivatives of dinitrophenolic pesticides	6ft x 4mm I.D.	10% DC-200	80/90 Anakrom ABS	N₂, 120ml/min	185	EC
154	Binapacryl and its methylated derivative following hydrolysis	1.8m x 4mm I.D.	3% GE-XE-60	80/100 silanised Chromosorb W	N₂, 300ml/min (measured at room temp.)	200	EC
155	Thiabendazole	1.2m x 6mm	10% DC-200	80/100 Varaport 30	–	210	FP
161	Pyrethrins, piperonyl butoxide and N-octyl-bicycloheptene dicarboximide	4ft x 0.25in I.D.	5% SE-30	60/80 AW, DMCS Chromosorb W	N₂, 40ml/min	190	FI

In 1965 Storherr and Watts[32] introduced the clean-up method termed sweep co-distillation as part of the determination of OP or OC pesticides and culminating in GC analysis. These workers noticed that crop extractives which had not been removed by standard clean-up procedures were deposited on the glass wool packing at the injection end of the GC column. Taking advantage of this observation, they produced a heated precolumn of glass wool, complete with injection port and carrier gas inlet. Four times 250 μl of ethyl acetate extracts (equivalent to 2 g crop sample) were introduced into the heated tube and nitrogen, at the rate of 600 ml/min, swept the pesticides into a cooling coil, to be collected for analytical GC. The co-extracted materials remained on the glass wool, the first few inches of which could be replaced, as necessary. No other clean-up procedure was needed, and good recoveries of added pesticides were obtained. Watts and Storherr[33] used this method for the determination of pesticide residues in milk, and Storherr *et al.*[34] modified the method for the determination of these residues in edible oils, by replacing the glass wool with glass beads in a larger pre-column. Once the sweep co-distillation apparatus has been constructed, it offers a clean-up procedure which is no more time-consuming than other procedures, and it has the advantage of requiring only small volumes of solvent for elution and column washing.

2. Halogenated and other fumigants

Fumigants are used extensively for the control of insects in stored grain and flour, and residues of parent compounds or metabolites have been known to persist for several weeks in these food products. There is also a possibility that nematocidal fumigants, used mainly in soil, could be transferred to root crops, e.g. potatoes.

By implication, fumigants are gases or vapours of low-boiling liquids, and, apart from ethylene oxide and carbon disulphide, most are simple, halogenated aliphatic hydrocarbons, e.g. chloroform, carbon tetrachloride, methyl bromide, ethylene dichloride and ethylene dibromide. Prior to GC, the methods used for quantitatively evaluating halogenated fumigants depended on hydrolysis with subsequent determination of the halogen by titrimetry and other wet chemical methods.

The nematocide 1,2-dibromo-3-chloropropane was satisfactorily chromatographed by Kanazawa and Sato[35] as early as 1961, but no application of the method to food analysis was indicated.

Bielorai and Alumot[36] described a GC method where ethylene dibromide was quantitatively determined in fumigated samples of cereals and lentils. The sample plus water was steam-distilled after the addition of a small quantity of benzene, and the distillate, which contained ethylene dibromide in benzene, was used for analysis on a DC-50 stationary phase. The lower limit of detection by their method was 4 p.p.m. ethylene dibromide, and it was found that this fumigant could still be detected 6 weeks after its application to cereals.

A similar method was used by Bielorai and Alumot[37] for the extraction of chloroform, carbon tetrachloride, trichloroethylene and carbon disulphide from cereal grain, toluene being preferred to benzene as the co-distilling solvent. *Figure 20.1* is a chromatogram showing the separation, after only 8 min, of this

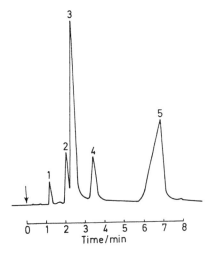

Figure 20.1 Chromatogram of fumigant mixture on silicone oil DC-710 column. 10% wt./wt. on 80–100 mesh Chromosorb W HMDS. Arrow indicates starting point. 1, Carbon disulphide; 2, chloroform; 3, carbon tetrachloride; 4, trichloroethylene; 5, toluene. (After Bielorai and Alumot[37]). For full GC details, see Table 20.1

mixture of fumigants plus toluene on a 10% DC-710 stationary phase. Chloroform and carbon tetrachloride were not completely separated, but this can be achieved on a SE-30 stationary phase.

Three extraction methods for multiple residues of organic fumigants in cereal grain were tried by Malone[38]. He compared a sweep co-distillation extraction procedure incorporating toluene as the co-distilling solvent, a steam distillation procedure[37], and a method involving refluxing the sample with acid and sweeping the fumigants with nitrogen through 30/60 Chromosorb and collecting them in toluene, cooled to $-86°C$. The last-named method was preferred and, using GC on a 30% DC-200 stationary phase, it was possible to detect 0.1 p.p.m. of chloroform, carbon tetrachloride, carbon disulphide, methyl bromide and ethylene dibromide. The limit of detection for ethylene dichloride was 4 p.p.m.

The treatment of food such as wheat or flour with the fumigants ethylene and propylene oxides can lead to the formation of chlorohydrins, owing to reaction between these oxides and the inorganic chlorides present in the food. The chlorohydrins produced from ethylene and propylene oxides, namely 2-chloroethanol and 1-chloro-2-propanol, respectively, can be analysed by GC using a 20% Carbowax 20M stationary phase. Ragelis, Fisher and Klimeck[39] extracted wheat flour and ground pepper samples by percolating diethyl ether through a column of the food. The ether extract of the flour was used directly for injection, but the pepper solution was cleaned up on Florisil before GC. A pepper sample previously treated with ethylene oxide showed a residue of 110 p.p.m. of 2-chloroethanol, whereas a propylene oxide-treated pepper gave a residue of 10 p.p.m. of 1-chloro-2-propanol.

2-Chloroethanol, which is produced from ethylene oxide, plus ethylene dibromide and also methyl bromide, were determined in flour and wheat grain by Heuser and Scudamore[40]. Samples were shaken with 5:1 acetone:water, and, after settling out, the supernatant was taken for GC analysis. In early work a special injection system was used, but this has been superseded by direct on-column injection. Excellent recoveries were obtained for added 2-chloroethanol and ethylene dibromide to both flour and wheat, the limit of detection being approximately 5ng for both compounds. Heuser and Scudamore[41] used the same extraction method for residues of methyl bromide in wheat grain and for residues of 2-chloroethanol in ethylene oxide-treated flour. As in their previous publication, 15% UCON LB-550-X stationary phase was used and operated at 85°C, under which conditions the fumigants eluted prior to the solvent peak.

The same authors[42] examined such foods as cereals, pulses, nuts and dried fruit for fumigant residues arising from treatment of these foods with methyl bromide and ethylene dibromide. The food extracts were caused to react with a solution of 0.66% v/v ethylene oxide in 4:1:1 acetonitrile:iso-propyl ether: 0.6N sulphuric acid. Ionic bromide residues were converted to 2-bromoethanol, which was determined by GC, whereas methyl bromide and ethylene dibromide did not react and were determined concurrently.

D-D, which is a mixture of *cis-* and *trans-*1,3-dichloro-1-propene and 1,2-dichloropropane, is a nematocidal fumigant used in soils, and Karasz and Gantenbein[43] examined potatoes which were grown in a D-D-treated soil. The potatoes were blended with acetone, water and salt solution were added and the aqueous mixture was extracted with hexane. After clean-up on silica gel, the hexane extract was analysed on a stationary phase comprising 15% QF-1 and 10% DC-200, using MCD. No residues of D-D were found in any potato sample.

3. Organo-halogen fungicides and molluscicides

The most widely applied halogen-containing fungicides are the chlorinated benzenes. Hexachlorobenzene is used to control bunt in cereals, and residues of hexachlorobenzene have been found in wheat and also in meat fat, dairy products and eggs, by various workers.

Di Muccio, Boniforti and Monacelli[44] were concerned about the chromatographic separation of hexachlorobenzene and the α, β, γ and δ isomers of BHC, which is important since these fungicidal and insecticidal residues can occur together. Many single-phase column packings were inadequate for this separation, and after a study of several mixed column materials, 2:2:1 3% OV-61 on 80/100 Gas Chrom P:7.5% QF-1 on 80/100 Chromosorb W-HP: 3% XE-60 on 80/100 silanised Anakrom ABS was the one chosen to give the desired separation of these chlorinated hydrocarbons. *Figure 20.2* is a chromatogram indicating the type of separation obtained.

The most-used chlorinated nitrobenzene is pentachloronitrobenzene (PCNB), also called quintozene. It is particularly useful in the control of botrytis in hothouse-grown lettuce, and since this commodity is marketed internationally, the analysis of residues of PCNB is very important. Since PCNB has a relatively short retention time on non-polar phases, such as SE-30, there is a possibility that it can be confused with other OC pesticides, e.g. γ-BHC.

In the authors' survey of pesticides in foods, carried out in 1968[27], 11 of the samples examined contained PCNB in excess of 1 p.p.m., and were either English or Dutch. In 1972 Baker and Flaherty[45] determined residues of this fungicide in lettuce, tomato and banana, using a hexane extraction, dimethylformamide

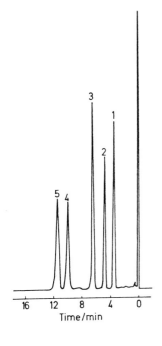

Figure 20.2 1, HCB; 2, α-BHC; 3, γ-BHC; 4, β-BHC; 5, δ-BHC (after Di Muccio et al.[44]). For GC details, see Table 20.1

partitioning and further clean-up on alumina. Confirmation of GC results, which were obtained on three different stationary phases, was made by reduction of PCNB to the corresponding aniline using lithium aluminium hydride. Retail lettuces of Dutch origin contained 0.78–2.6 p.p.m., and four English samples contained 0.70–1.2 p.p.m. This indicates that lettuce is generally no more free of PCNB residues than when the problem was highlighted in 1968, despite a recommendation that such residues should not exceed 0.3 p.p.m.[46]

Botrytis on strawberries has been effectively controlled by dichlofluanid, and Eades and Gardiner[47] examined frozen strawberries for dichlofluanid and its metabolite dimethylaminosulphanilide, by a GC method. The minced fruit was mechanically rolled with benzene for 15 min and the extract was analysed on a 5% DC-11 stationary phase. Residues of dichlofluanid were found in frozen and canned strawberries, the fruit having been treated with the fungicide in field trials. No dimethylaminosulphanilide was found in any sample.

4. Organohalogen herbicides

2,4-D (2,4-dichlorophenoxyacetic acid) and 2,4,5-T (2,4,5-trichlorophenoxy-acetic acid) have been used for many years as systemic herbicides and defoliants. With the realisation of the sensitivity advantages of GC incorporating EC detection and MCD, for the determination of halogen-containing insecticides, residue analysts attempted to use these techniques for the analysis of these herbicides. In order to obtain volatile compounds for GC, it was necessary to make the esters of these compounds.

329

Yip[48] extracted 2,4-D from green crops, e.g. frozen spinach, with a mixture of diethyl ether and petrol, after the addition of acid. 2,4-D was separated by a sodium bicarbonate extraction, and, after acidification, it was re-extracted into chloroform. The methyl ester was prepared by addition of diazomethane, and GC was performed on a 20% DC silicone grease stationary phase with MCD. Wheat samples were analysed similarly, except that the original extractant was ethanol. The limit of detection of the herbicide was 0.05 p.p.m.

MCPA (4-chloro-2-methylphenoxyacetic acid) and MCPB (4-(4-chloro-2-methylphenoxy)butyric acid) have similar properties to 2,4-D and 2,4,5-T, in that they are hormone-type herbicides. Gutenmann and Lisk[49] suggested an analysis of MCPA and MCPB, incorporating bromination of these acids, using a carbon tetrachloride solution saturated with iodine and containing 5% bromine, and esterification with BF_3 —methanol. The methyl esters were chromatographed on a 5% silicone stationary phase and detected by EC.

2,4-D, MCPA, MCPB and herbicides of the same group, e.g. 2,4-DB (4-(2,4-dichlorophenoxy)butyric acid), dichlorprop (2-(2,4-dichlorophenoxy)propionic acid), were separated and determined by Dubosq and Dedde[50]. Apiezon L on silica gel was the column material used to separate these herbicides from each other, except a mixture of MCPA and mecoprop, which required a DEGS stationary phase for their separation.

2,4-D and also 2,4,5-TP (2-(2,4,5-trichlorophenoxy)propionic acid) have been used as dilute sprays on citrus to prevent preharvest fruit drop. Meagher[51] described a method where citrus peel was blended with acetone and solvent extracted to yield water-soluble bound acids, free acids and esters. The free acid fraction was esterified with 2-butoxyethanol saturated with hydrochloric acid, and the esters were cleaned up on Florisil and subjected to GC on a 5% QF-1 stationary phase. Recoveries of 89—93% of these added herbicides were obtained at the very low levels.

Acetone was also used by Munro[52] to extract 2,4-D and 2,4,5-T from tomatoes and other fruits and vegetables. After extensive preliminary clean-up, the free acids were methylated and the esters were further cleaned using a sweep co-distillation procedure. GC was carried out using MCD and a 2.5% NPGS stationary phase.

The preparation of the methyl esters of 2,4-D, 2,4,5-T and other chlorinated herbicides can be a time-consuming process, particularly when a large number of analyses is required. The TMS derivatives of these herbicides were successfully made by Garbrecht[53], and the optimum conditions were finalised by Baur, Baker and Davis[54]. Because these derivatives were prepared rapidly, there was a large saving in analytical time in comparison with the preparation of alkyl derivatives. Garbrecht studied the separation of 2,4-D, 2,4,5-T, 2,4,5-TP, 2,4-DB, MCPA, MCPP and dicamba (2-methoxy-3,6-dichlorobenzoic acid) via their TMS derivatives and *Figure 20.3* is a chromatogram of that separation on a 5% DC-200 stationary phase at 190°C with TC detection.

2,3,6-TBA (2,3,6-trichlorobenzoic acid), a non-selective herbicide used to control certain deep-rooted, broad-leaved weeds, and pentachlorophenol, a general herbicide, plus 2,4-D and other chlorinated phenoxyaliphatic acids, were determined by Yip[55] in vegetable oils. The samples were extracted with sodium bicarbonate solution, which was acidified, and the liberated acids were extracted into chloroform. The methyl esters were prepared using diazomethane, and these

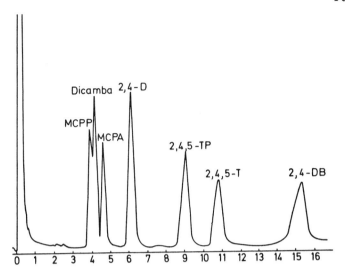

Figure 20.3 Gas chromatogram of six common chlorophenoxy acids and dicamba as trimethylsilyl esters (after Garbrecht[53]). For GC details see Table 20.1

derivatives were separated on a 5% DC-200 stationary phase using MCD. Recoveries in excess of 87% were obtained for these herbicides in the range 0.02–0.08 p.p.m.

TCA (sodium trichloroacetate) is a pre-emergence herbicide used particularly for the control of couch grass and wild oats. Chiba and Morley[56] determined TCA residues in wheat grain by a distillation procedure from dilute sulphuric acid solution, the TCA being converted to chloroform, which was collected in xylene. A 20% DC-710 stationary phase, operated at 86°C and incorporating MCD, was used for the GC of the chloroform. The acaricide dicofol also yields chloroform in the distillation procedure, but can be separately determined by its hydrolysis product, *pp'*-dichlorobenzophenone, which is measured spectrophotometrically.

Pentachlorophenol is a pesticide with properties of a herbicide, an insecticide and a fungicide. Stark[57] was also concerned about the industrial usage of pentachlorophenol and the inevitability of its passage into rivers and seas and, thereby, into fish. 0.1M potassium hydroxide was used by Stark to extract fish samples, and the solution was acidified and the pentachlorophenol extracted into toluene. After methylation, using diazomethane, the ester was determined by GC on a mixed stationary phase consisting of 4% SF-96 and 8% QF-1. The limit of detection in fish was 0.5 ng/g. Procedures were also described for the GC determination of the TMS derivative of pentachlorophenol, and also for its identification by MS.

Dalapon-sodium (sodium 2,2-dichloropropionate) is a selective contact herbicide which is translocated from foliage and roots and is used to control annual and perennial grasses.

In 1963 Getzendaner[58] determined dalapon residues at the 0.1 p.p.m. level in cranberries and bananas. Samples were extracted with water containing phosphoric acid and phosphotungstic acid, and the extracts were saturated with salt before extraction with diethyl ether. The diethyl ether extract was directly chromatographed on a PEGA stationary phase using a ^{90}Sr ionisation detector system.

Dichlobenil (2,6-dichlorobenzonitrile) is a herbicide used for the control of annual and perennial broad-leaved and grassy weeds. Meulemans and Upton[59] examined crop and fish samples for this herbicide, together with its metabolite, 2,6-dichlorobenzoic acid. Acidified samples were blended with a mixture of benzene and isopropanol, and ultimate clean-up was performed on attapulgite clay. The dichlobenil and methyl 2,6-dichlorobenzoate were chromatographed on a 5% DC-200 stationary phase, the limit of detection being 0.05 p.p.m.

A similar limit of detection in the GC determination of picloram (4-amino-3, 5,6-trichloropicolinic acid), a herbicide used for weed control in cereals, was achieved by Bjerke, Kutschinski and Ramsey[60]. Cereal grain samples were treated with aqueous potassium hydroxide solution and the extract was cleaned up on alumina, prior to esterification with diazomethane. The methyl ester was chromatographed on a phosphoric acid-impregnated LAC-446 stationary phase.

Bromacil (5-bromo-6-methyl-3-s.butyl uracil) is a herbicide employed for general weed control. Residues of bromacil have been determined in crops by Pease[61]. He solvent-extracted samples and, after intermediate clean-up steps, obtained a concentrated nitromethane extract. This extract was analysed on a 20% SE-30 plus 0.2% Epikote 1001 stationary phase using MCD, the limit of detection being 0.04 p.p.m.

5. Other chlorinated pesticides

The acaricide Aramite (2-(4-butylphenoxy(-1-methyl ethyl-2-chloroethyl sulphite) was determined in combination with other pesticides in crops by Archer[62]. Samples were extracted with benzene and cleaned up on a Florisil column. Having eluted the other pesticides, e.g. *p,p'*-DDT, present in the extract, with 1:9 diethyl ether:pentane, Aramite was eluted with 1:49 isopropanol:pentane. GC was performed on a 5% SE-30 stationary phase using EC detection.

Dichlozoline (3-(3,5-dichlorophenyl)-5,5-dimethyloxazolidine-2,4-dione) is a fungicide used in the control of botrytis on grapes, and there is a possibility that this fungicide might be present in the juice, must or wine made from the treated fruit. Lemperle and Kerner[63] extracted the fruit or wine and cleaned up the ultimate benzene extract on Florisil. In the GC analysis QF-1 was the stationary phase, with EC detection.

Pack *et al.*[64] extracted grapes, grape juice, must or wine with hexane and cleaned up the extract on silica gel. GC of dichlozoline was performed on a 5% QF-1 stationary phase with EC detection, the limit of detection being 0.01 p.p.m.

C. Organophosphorus pesticides

Whereas OC pesticides encompass insecticides, acaricides, fungicides and herbicides, OP pesticides are almost exclusively insecticides and acaricides. Following the application of GC methods to the analysis of OC pesticides, the analysis of OP insecticide residues in crops and other foods was approached in a similar manner. The development of extraction and clean-up methods for OP insecticides have, naturally, drawn on the experience gained in OC residue analysis, and it has been a relatively straightforward task to modify them for extraction and clean-up of OP compounds. The biggest problem was detection, and new systems were developed with extra sensitivity to phosphorus- and sulphur-containing compounds, e.g. AFID and FPD (see Chapter 3).

Coulson, Cavanagh and Stuart[65], as early as 1959, separated parathion, malathion and demeton in mixtures of OC and OP pesticides, using GC with IR spectroscopy as the detecting system. The EC detector, so suitable in OC residue analysis, has limited application in OP residue analysis, because of lack of sensitivity, and chromatographers turned their attention to other detecting systems.

Following the utilisation of the modified AFID, Giuffrida and Ives[66] compared it with the conventional FID in a study of the recovery of added OP insecticides to fruit and vegetables. Since that time, the AFID, incorporating sodium, potassium, rubidium or caesium as the alkali element, has been used extensively in OP analysis.

The MCD, developed by Coulson *et al.*[2], has become a popular system for detecting sulphur-containing OP insecticides, and Nelson[67-69] examined thiophosphate residues on fruits and vegetables by using a clean-up procedure which was applied to OC pesticide residues[70], followed by GC analysis incorporating this detector.

The microwave-powered helium plasma detector of Bache and Lisk[71,72] was originally applied to phosphorus-containing insecticides. After GC separation, the effluents were analysed according to their atomic emission at 2535.65 Å for phosphorus.

The more specific the detection system the less need there is for a clean-up of the extract, and the FPD was developed by Brody and Chaney[73] with this in mind.

Getz[74] found, by using FPD, that nanogram quantities of OP compounds could be detected in the presence of microgram quantities of co-eluted halogen compounds. Low fat or non-fat samples, e.g. lettuce and cotton leaf, were analysed by GC, without clean-up, although raw extracts of oily and waxy samples deposited fatty substances at the injector end of the column, ultimately causing peak broadening and tailing.

Pesticide residues in onions have presented the gas chromatographer with problems. The electron-capturing properties of di- and trisulphides naturally present in onions give large peaks on many of the phases used during OC pesticide analysis incorporating EC detection, and the MCD, set to detect sulphur in the detection of thiosulphate residues in onions, is equally undesirable. Getz asserted that FPD showed less interference than other detectors used in the determination of OP residues in onion, particularly when using the 526 nm filter.

As with OC pesticide residue analysis, so there have been different approaches to the extraction and clean-up of OP insecticide residues. Most analysts desire a method which encompasses the detection and determination of all OP insecticides and their metabolites.

In his analysis of OP insecticide residues in crops, Getz[74] used an extraction procedure based on his earlier work, which incorporated paper chromatographic identification[75]. Acetonitrile was used as the extracting solvent, and after filtration and evaporation, the residue was dissolved in ethyl acetate for column chromatographic clean-up on activated carbon.

In 1961 Laws and Webley[76] used alumina and activated carbon for the clean-up of phosphorus-containing residues. The dichloromethane extract was partitioned between petrol and 15% aqueous methanol, thus dividing the OP compounds into non-polar and polar sections, respectively. The petrol-soluble non-polar compounds were cleaned up on alumina and the water-soluble polar compounds, having been extracted into chloroform, were cleaned up on activated carbon.

This method was used by the authors in their scheme for the analysis of OP insecticide residues in fruits and vegetables[26]. Acetone was used as extraction solvent in order to remove OC as well as OP pesticide residues.

Thirty-nine foods were examined for multiresidues of pesticides containing phosphorus and/or sulphur by Bowman, Beroza and Hill[77]. Four extraction procedures were used, depending on the nature of the food: Soxhlet extraction with 9:1 chloroform:methanol and removal of solvent under vacuum at 60°C; the same Soxhlet extraction procedure but evaporating the solvent at 100°C; acetone extraction and partition into dichloromethane; and the solution of fats in hexane. All extracts were partitioned into acetonitrile for GC analysis on OV-101 and OV-210 stationary phases, using FPD in both phosphorus- and sulphur-sensing modes. *Figure 20.4* shows chromatograms of OP standards on OV-101 and OV-210 stationary phases and indicates the response differences of FPD set in the phosphorus and sulphur modes using the OV-101 phase.

The sweep co-distillation clean-up method of Storherr and Watts[32], described under OC Insecticides and Acaricides was devised, primarily, for the analysis of OP residues in crops. Watts and Storherr[33] and Storherr *et al.*[34] used the method for the clean-up of milk and oils, respectively, prior to the examination for OP insecticides. Storherr and Watts[78] conducted a successful collaborative study on the method and, for a laboratory which has to monitor a range of foods for OP pesticides, sweep co-distillation is a low-cost, efficient clean-up procedure. With initial ethyl acetate extraction, this procedure has been incorporated in the Official Methods of Analysis of the AOAC[28].

In OP insecticide residue analysis by GC the number of operating permutations is considerable, and includes the extraction solvent, the clean-up procedure, the stationary phase, the detecting system and, in some cases, derivative formation.

Acetonitrile as the initial extraction solvent finds particular favour when OC, as well as OP, pesticides are sought in food. Nevertheless several workers have used acetonitrile for extracting OP compounds alone. Storherr *et al.*[79] used acetonitrile to extract crop samples in the determination of malathion, parathion, parathion-methyl, diazinon and carbophenothion, and Storherr, Ott and Watts [80] blended non-fatty foods with acetonitrile, followed by clean-up on activated carbon. McCaulley[81] extracted fruits and vegetables with the same solvent and

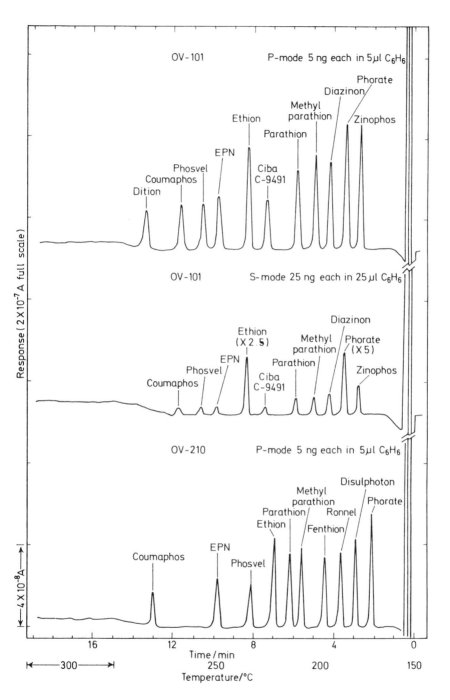

Figure 20.4 Chromatograms of pesticide standards (after Bowman et al.[77]). For GC details, see Table 20.1

335

vacuum-sublimed the OP compounds from the plant material extractives, prior to GC and IR spectroscopic identification.

Other solvents used in the extraction and separation of OP pesticide residues in crops or food items have been chloroform[82-85], dichloromethane[86,87], ethyl acetate[87-89], methanol[90-93], benzene[94-96], hexane[97-99] and propylene carbonate[23].

Although Florisil is used as a column clean-up material for both OC and OP pesticides, Versino, van der Venne and Vissers[100], in an assessment of eight clean-up materials, considered that the most satisfactory one for OP clean-up was Nuchar C-190N charcoal plus Solka Floc BW-40 cellulose, as cited by McLeod *et al.*[101]

A method for eliminating clean-up procedure for the determination of malathion in cabbage was put forward by Crossley[102]. A chloroform extract was made of the sample and injected on to a precolumn containing 10% silicone oil. After 7.5 min, the carrier gas flow was reversed in the precolumn only, thus preventing extraneous material from entering the analytical column, which contained an 8% E-301 stationary phase.

The choices of clean-up procedure and detector in the GC of OP pesticides residues have overshadowed other parameters of the analysis. Many stationary phases are suitable for analysis of single OP compounds, but the choice of a phase to separate and determine large numbers of OP compounds becomes limited. Bowman and Beroza[103,104] tabulated the retention data of 138 pesticides on OV-101, OV-17, OV-210 and OV-225 stationary phases and 146 pesticides on Dexsil 300. Dexsil 300 is a polycarboranesiloxane of exceptionally high thermal stability which enables a wide range of temperature programming to be performed.

Alkaline hydrolysis of most OP pesticides yields either dialkyl phosphate or *O,O* dialkyl phosphorothionate. The hydrolysis product can be methylated with diazomethane to give methyl esters. A comparison of the chromatogram before and after these reactions shows a shift in peaks which will confirm, or otherwise, the identity of the OP compound. This approach was the basis of work conducted by St. John and Lisk[105], Askew, Ruzicka and Wheals[106] and Shafik, Bradway and Enos[107].

Trichloronate (2,4,5-trichlorophenyl ethyl phosphonothionate) is hydrolysed in alkaline solution at 60°C to yield 2,4,5-trichlorophenol. The phenol can be acetylated and determined by GC, incorporating EC detection. Katague and Anderson[108] used this method for the analysis of trichloronate in crops.

Claborn and Ivey[109] acid-hydrolysed chlorfenvinphos and chromatographed the resulting 2,2′,4′-trichloroacetophenone on a 5% SF-92 stationary phase using an EC detector. The method was applied to the analysis of chlorfenvinphos in animal tissues and milk.

Animal tissues and milk were the materials examined for naphthalophos by Thornton and Schumann[110]. This OP compound was hydrolysed to naphthostyril, which was brominated, the product being determined by GC analysis.

D. Organonitrogen pesticides

1. Carbamates and substituted ureas

Carbamates, e.g. carbaryl, carbofuran, propoxur, aminocarb and zectran, are used as insecticides. Carbaryl has been suggested as an alternate insecticide to

p,p'-DDT, the use of the latter having been restricted or banned in many countries. Some substituted ureas and carbamates, e.g. monuron, diuron, pyrazon, barban and propham, have herbicidal properties varying from total weed control (monuron) to selective weed control of cereals (barban).

The detection and determination of sub-p.p.m. amounts of multiresidues of carbamates and substituted ureas in crops and foods has lagged behind that of OC and OP multiresidue analysis. One reason is that the concern regarding the persistence of OC residues, in particular, which is reflected in the mass of literature on the determination of OC pesticide residues, has overshadowed the development of similar methods for these pesticides which are relatively non-persistent. The lack of a detector sensitive enough to determine nanogram amounts of carbamates and substituted ureas, which is frequently necessary, was also a discouragement to gas chromatographers engaged in this field. The detector problem was overcome, in some instances, by the preparation of halogen-containing derivatives for EC detection, although this often incorporated other chemical reactions in order to obtain a suitable product for derivative formation, and this meant lengthy methods. Successful attempts were made to develop detectors which were specific to nitrogen-containing compounds by Martin[111] and by Coulson[112,113].

Martin's detection system incorporates a furnace where the nitrogen released from the compound is reduced to ammonia, which is titrated microcoulometrically. Coulson used his electrolytic conductivity detector (CCD), after reductive pyrolysis of the compound using a nickel wire catalyst. GC methods incorporating CCD, applied to nitrogen-containing compounds, including carbamates and substituted ureas, have been collaboratively studied by both the Environmental Protection Agency and the FDA of the USA, and the retention data have been collected by Laski and Watts[114].

Of all the carbamates, most attention has been paid to the analysis of carbaryl. In 1964 Ralls and Cortes[115] extracted carbaryl from green beans using dichloromethane, cleaned up the extract on Florisil and brominated the carbamate with a solution of bromine in carbon tetrachloride to give a mixture of derivatives which could be conveniently detected by GC, incorporating EC detection. The method was not quantitative, but was considered as a screen at the 1 p.p.m. level. Similarly, carbaryl was extracted by dichloromethane from snap beans and then hydrolysed in alkaline solution to give the corresponding naphthol, by van Middlelem, Norwood and Waites[116]. The naphthol was brominated to yield a compound sensitive to EC detection. This method was suggested as a general one for many carbamates. The same derivative, followed by esterification, was made by Gutenmann and Lisk[117], after acetone extraction of samples, e.g. apple, broccoli, beans, maize, chicken and trout. Some interfering substances were precipitated by using a solution of ammonium chloride in phosphoric acid. The limit of detection for the GC determination of carbaryl in the foods examined was 0.1 p.p.m.

A similar procedure was used by Holden, Jones and Beroza[118] prior to the determination of methyl and dimethyl carbamate residues in crops. Dichloromethane was the extractant, and some interfering materials were removed by precipitation with ammonium chloride—phosphoric acid solution. Further clean-up was effected on activated charcoal and, after hydrolysis, 1-fluoro-2,4-dinitrobenzene in benzene was added to yield 2,4-dinitroaniline derivatives.

These derivatives were chromatographed on a 2% XE-60 stationary phase and with EC detection.

Hydrolysis of certain carbamates followed by derivatisation can yield products more sensitive to EC detection, and this principle has been exploited by several workers in carbamate residue analysis[119-124]. For the same reason, trifluoroacetylation of carbamates has been used[125-128].

Moye[129] used on-column transesterification of *N*-methyl carbamates to form methyl *N*-methyl carbamates in order to determine these pesticides. The crop sample, e.g. lettuce, was blended with methanol and the extract was made 5mM to sodium hydroxide before injection on to a column, the first 6 in of which contained 80/120 glass beads and the remainder 80/100 Porapak P. Detection was by AFID incorporating a rubidium tip. Reproducible high yields were obtained using on-column transesterification of eight *N*-methyl carbamates, namely carbaryl, mobam, UC 10854, carbofuran, zectran, methiocarb, tranid, aldicarb and aminocarb. The same method of derivatisation was used by van Middlelem, Moye and Janes[130] for the determination of carbofuran and 3-hydroxycarbofuran in lettuce. Dichloromethane extraction was used followed by clean-up on a column of Nuchar Attaclay. Before the transesterification it was necessary to acid-hydrolyse the water-soluble conjugated carbamate compounds to yield aglycones, which were extractable in organic solvents.

Fruit and vegetable samples were examined for herbicidal carbamates using a GC method, without recourse to derivative formation by Onley and Yip[131]. Samples were extracted with ethanol or ethanol—water for samples of low moisture content, and the carbamates were extracted into petrol and concentrated. The concentrated petrol extract was cleaned up on a column of magnesium oxide and cellulose, and the eluate was used for GC. EC detection was used for dichlormate, and the analysis of sulphur-containing carbamates was performed using FPD. Other carbamates were determined using AFID. *Figure 20.5* shows chromatograms of potato samples fortified with the carbamates EPTC, vernolate, molinate and cycloate, using either AFID or FPD.

After hexane extraction and alumina clean-up, Cerny and Blumenthal[132] determined the sprout inhibitors propham and chlorpropham by GC. 3% OV-17 was used as the stationary phase and detection was by AFID incorporating a rubidium tip.

2. Thiocarbamates

The common thiocarbamates, namely zineb, maneb, nabam, ziram and thiram, are fungicides used to control foliar disease of many crops. Residues have been analysed almost exclusively by a colorimetric method which depends on the breakdown of the thiocarbamate to yield carbon disulphide, which is caused to react with copper to form copper diethyldithiocarbamate. The problems associated with this method include its lack of specificity and occasional misleading positive results with green-leaved samples.

In 1971 Onley and Yip[133] applied a GC method to the analysis of ethylene thiourea residues. Ethylene thiourea is a degradation product of thiocarbamates, and although its determination is neither a specific nor a quantitative indication of the former presence of thiocarbamates, it is the best guide available at the present time. In this method the sample of fruit, vegetable or milk was extracted with a mixture of ethanol and chloroform. The extract was partially cleaned up on a cellulose column and the eluate was caused to react with a methanolic solution of 1-bromobutane. The resulting derivative gave a sharp peak on the chromatogram when using a silicone stationary phase and AFID. The analysis was supported by TLC findings, and the limit of detection was 0.02 p.p.m. ethylene thiourea.

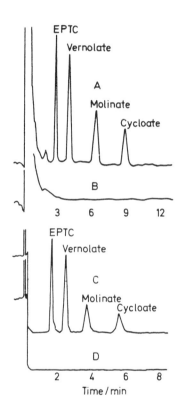

Figure 20.5 Chromatograms of potatoes, unfortified and fortified with 0.1 p.p.m. EPTC, vernolate, molinate and cycloate. A (fortified) and B (unfortified) were obtained with the thermionic detector on 301 mg sample. C (fortified) and D (unfortified) were obtained with the flame photometric detector on 20 mg sample. (After Onley and Yip[131]). For GC details, see Table 20.1

Using a GC method, Newsome[134] determined ethylene thiourea in apples by converting it to the S-benzyl compound, after extracting the fruit with methanol. The trifluoroacetate was made and chromatographed on a 2% BDS stationary phase and using EC detection. Recoveries were 89% or better in the range 0.01–1 p.p.m. of added ethylene thiourea to apples.

Methanol was also the extracting solvent used by Haines and Adler[135] to remove ethylene thiourea from crop samples, e.g. apple, banana, tomato and potato. Alumina was used to clean up the extract and 2-butylthio-2-imidazoline was formed by reaction with 1-bromobutane and sodium borohydride in dimethylformamide. GC was carried out on a 20% SE-30 stationary phase, using FPD, and the limit of detection was 0.01 p.p.m.

3. Triazines

The triazines are herbicides, almost exclusively. Some triazines, e.g. simazine, are pre-emergence herbicides for the control of broad-leaved and grassy weeds, whereas others are both pre- and post-emergence herbicides. Anilazine, being an exception, is a triazine fungicide, used to control certain fungi on foliage. Most of the GC applications to triazine residue analysis have concerned maize and other cereal crops where weed control is particularly important.

As early as 1962, Chilwell and Hughes[136] analysed six triazines by GC. Simazine, atrazine, trietazine, atraton, prometon and prometryne were separated on 120/150 glass beads coated with either 0.5% Apiezon L or 0.1% PEGA.

The same six triazines, plus ametryne, propazine and desmetryne, were determined in crops by Delley *et al.*[137] The sample was extracted with methanol and the extract evaporated, and water was added. After making the solution alkaline with sodium carbonate, it was extracted with diethyl ether and eventually cleaned up on Brockmann activity III alumina. The eluate was evaporated and dissolved in carbon tetrachloride and then cooled to $-20°C$. At this temperature the solution was passed through a column of sodium bisulphate and the triazines were eluted with cooled chloroform. GC was carried out on a 8% Reoplex 400 stationary phase using FID.

Many workers have used methanol to extract triazines from crops, particularly maize. Westlake, Westlake and Gunther[138] extracted ACD 15M (2-chloro-4-isopropylamino-6-hydroxy-methylamino-1,3,5-triazine) from maize ears with methanol. The extract was evaporated and the residue was dissolved in dichloromethane for GC on a 2% Reoplex 400 stationary phase, using CCD, with a detection limit of 0.1 p.p.m.

Methanol was also used as extractant by Schroeder *et al.*[139] They determined hydroxycyprazine residues in maize, after chlorination, followed by GC incorporating a nitrogen-selective detector. After extraction of the samples, water was added to the methanol extract, which was washed with toluene, acidified and partially cleaned up on a cation exchange column. The eluate was evaporated to dryness and the residue was dissolved in chloroform before further clean-up on silica gel. Phosphorus pentachloride in dimethylformamide was used as the chlorinating medium, and the product was cleaned up on an alumina column prior to GC. The method was used to detect down to 0.03 p.p.m. of this chlorinated derivative.

Beynon[140] developed a method for determining cyanazine and its metabolites in crops, using aqueous methanolic extraction. Clean-up was achieved by solvent partition using either ethyl acetate or diethyl ether, and by column chromatography on either alumina or Florisil. GC was performed on a CHDMS stationary phase using EC detection.

The fungicide anilazine was determined in crop extracts, such as strawberry, tomato, cucumber and potato, by Wales and Mendoza[141]. The sample was blended with acetonitrile and the extract was partitioned with hexane before reaction with methanolic sodium hydroxide. The product was analysed by GC on a stationary phase consisting of a mixture of 4% SE-30 and 6% QF-1. This phase, although the choice of many, was found to be not ideal because of the partial adsorption of the anilazine, and Mendoza and Shields[142] suggested that 1.5% OV-17 plus 1.95% QF-1 was a better mixture.

In their description and use of the CCD, Laski and Watts[114] included some triazines, e.g. atrazine, propazine, ametryne, Igran and anilazine.

The microwave emission detector was described by Bache and Lisk[143] for the analysis of triazine residues. The technique was applicable to those triazines which contain chlorine or sulphur, the emission of these elements being measured after excitation in a low-pressure helium plasma system.

4. Substituted phthalimides

The substituted phthalimides captan, captafol and folpet are fungicides used mainly on foliage.

In 1967 Kilgore, Winterlin and White[144] and Kilgore and White[145] used GC methods to analyse captan in fruits and vegetables and captafol in fruits, respectively, Benzene was used to blend samples, extracts containing captafol being cleaned up on a column of charcoal-Attaclay before GC analysis on a 5% DC-11 stationary phase using EC detection. Extracts of fruits containing captan residues were chromatographed without clean-up, but vegetables of high oil content were blended with acetonitrile and subjected to clean-up on Florisil, 1% acetone in benzene being the eluant. Captan was subsequently determined on a 10% DC-200 stationary phase using EC detection. The limit of detection for both fungicides was 0.01 p.p.m., recoveries of the added captafol and captan to crop materials averaging 93 and 92%, respectively.

The GC separation of captan and folpet was studied by Bevenue and Ogata[146], who chose 3% XE-60 as the best stationary phase for this purpose. Papaya fruit samples were fortified with both fungicides at the 1 p.p.m. level and blended with benzene, and activated carbon was added to the extracts to effect clean-up, prior to filtration and GC. Recoveries of 85–95% were obtained.

Pomerantz and Ross[147] used two stationary phases, viz. 10% DC-200 and a mixture of 10% DC-200 and 15% QF-1, for the separation of captan, captafol and folpet, plus the metabolites tetrahydrophthalimide and phthalimide.

Pomerantz, Miller and Kava[148] blended carrot, cabbage and soyabean samples with acetonitrile before a clean-up by solvent partition between dichloromethane and petrol. Florisil was used as a column chromatographic clean-up material, elution being effected with 20% dichloromethane in petrol, followed by 50% dichloromethane in petrol containing 1.5% acetonitrile. Folpet was eluted with both of these solvent mixtures, whereas captan and captafol were eluted only in the second solvent mixture. Co-extracted pesticides, e.g. p,p'-DDT, were eluted in the first solvent mixture. GC was performed on either a 5% QF-1 or a 3% XE-60 stationary phase, using EC detection, and recoveries of added fungicides were in excess of 80%.

A similar extraction and solvent partition clean-up procedure was used by Baker and Flaherty[149] for the GC determination of folpet, captan and captafol residues in certain fruits. A column of silica gel was used to further clean up the extract, 1:1 hexane: diethyl ether being used as eluant. Treatment with acidic permanganate oxidised captan and captafol, leaving folpet unchanged, and examination of the chromatograms before and after the oxidation distinguished folpet from the other two pesticides.

Pesticides

A description of the CCD for specific determination of nitrogen-containing compounds was given by Laski and Watts[114]. The tabulated retention data on a DC-200 stationary phase included captan and folpet.

5. Cyano- and nitro-phenols and binapacryl

The cyano- and nitro-phenols are mainly used as contact herbicides on broad-leaved weeds in cereals and other crops. DNOC (2-methyl-4,6-dinitrophenol) is also used as an ovicide.

In 1965 Gutenmann and Lisk[150] determined DNOC, dinoseb, ioxynil and bromoxynil residues in crop samples, using a GC method. Samples were extracted with acetone and, after filtration and evaporation of the extract, residues were caused to react with diazomethane to form methyl derivatives for GC. EC detection was used after chromatographic separation on a 5% silicone oil stationary phase. The limit of detection was 0.1 p.p.m. by this method.

The methyl esters of DNOC and dinoseb were also used by Boggs[151] for the GC analysis of these dinitrophenolic herbicides in fruits and vegetables. Dimethylformamide was used as extracting solvent.

Yip and Howard[152] used chloroform to extract DNOC, dinoseb, dinosam and dinex from beans, peas, apples and oranges. They treated the evaporated extract with diazomethane to form methyl esters before clean-up on columns of acidic Celite and Florisil. GC was performed on a 10% DC-200 stationary phase and with EC detection. Recoveries of the added herbicides in the range 0.1–0.2 p.p.m. were in excess of 82%.

The methyl esters of the dinitrophenolic or cyanophenolic herbicides were formed because they are derivatives which chromatograph without the severe peak tailing encountered when chromatographing the parent compounds. Hrivnak and Stota[153] separated by GC, without derivative formation, DNOC, dinoseb, dinex, bromoxynil, ioxynil and 2,4-dinitro-6-phenylphenol, using a stationary phase of 3% 2,2-dimethylpropane-1,3-diol succinate incorporating 1% phosphoric acid to prevent peak tailing.

Binapacryl (2-(1-methyl-n-propyl)-4,6-dinitrophenyl 2-methylcrotonate) is an acaricide used in the control of red spider and other mites. It also has fungicidal properties, being used to control powdery mildew of top fruits. Baker and Hoodless[154] used a GC method, plus chemical confirmation, for the analysis of binapacryl in fruits, such as apple, cherry, peach, pear and plum. Samples were blended with a mixture of hexane, diethyl ether and dimethylformamide and, after washing with water, the organic layer was cleaned up on a column of silica gel. GC was carried out on one of two stationary phases, viz. 3% XE-60 or 15% DC-200, using EC detection. For confirmation, binapacryl was hydrolysed with methanolic potassium hydroxide and methylated with diazomethane, and the resulting 2-(1-methyl-n-propyl)-4,6-dinitroanisole was subjected to GC on the XE-60 stationary phase.

6. Other nitrogen-containing pesticides

GC methods have been used to determine residues of the fungicides thiabenda-zole[155,156] (see also Chapter 17), and diphenylamine[157], and also to

342

determine residual concentrations of the insecticide nicotine[158] and of fenaza-flor[159] (phenyl 5,6-dichloro-2-trifluoromethyl benzimidazole-1-carboxylate), a non-systemic acaricide.

E. Other pesticides

Other important pesticides which do not fall into any other category in this chapter and which have been determined by GC methods are the pyrethrin insecticides with their synergists piperonyl butoxide and *N*-octylbicycloheptene-dicarboximide[160-162], the acaricide Omite[163] (the 2-(*p*-tert.-butylphenoxy) cyclohexyl 2-propyne ester of sulphurous acid), acaricidal polybutenes[164] and the slug killer metaldehyde[165].

References

1. Lovelock, J. E. and Lipsky, S. R., *J. Amer. Chem. Soc.*, **82**, 431 (1960)
2. Coulson, D. M., Cavanagh, L. A., de Vries, J. E. and Walther, B., *J. Agric. Food Chem.*, **8**, 399 (1960)
3. Giuffrida, L., *J. Ass. Off. Agric. Chem.*, **47**, 293 (1964)
4. Goodwin, E. S., Goulden, R., Richardson, A. and Reynolds, J. G., *Chem. & Ind.*, 1220 (1960)
5. Goodwin, E. S., Goulden, R. and Reynolds, J. G., *Analyst*, **86**, 697 (1961)
6. Hamence, J. H., Hall, P. S. and Caverly, D. J., *Analyst*, **90**, 649 (1965)
7. Giuffrida, L., Bostwick, D. C. and Ives, N. F., *J. Ass. Off. Anal. Chem.*, **49**, 634 (1966)
8. Beckman, H., Bevenue, A., Carroll, K. and Erro, F., *J. Ass. Off. Anal. Chem.*, **49**, 996 (1966)
9. Klein, A. K., Watts, J. O. and Damico, J. N., *J. Ass. Off. Agric. Chem.*, **46**, 165 (1963)
10. Porter, M. L., Young, S. J. V. and Burke, J. A., *J. Ass. Off. Anal. Chem.*, **53**, 1300 (1970)
11. Carr, R. L., *J. Ass. Off. Anal. Chem.*, **54**, 525 (1971)
12. Johnson, L., *J. Ass. Off. Agric. Chem.*, **45**, 363 (1962)
13. Sawyer, A. D., *J. Ass. Off. Anal. Chem.*, **49**, 643 (1966)
14. Saha, J. G., *J. Ass. Off. Anal. Chem.*, **49**, 768 (1966)
15. Onley, J. H. and Yip, G., *J. Ass. Off. Anal. Chem.*, **52**, 526 (1969)
16. de Faubert Maunder, M. J., Egan, H., Godly, E. W., Hammond, E. W., Roburn, J. and Thomson, J., *Analyst*, **89**, 168 (1964)
17. Noren, K. and Westoo, G., *Acta. Chem. Scand.*, **22**, 2289 (1968)
18. Tunistra, L. G. M. T. and Roos, J. B., *Neth. Milk Dairy J.*, **24**, 65 (1970)
19. Haenni, E. O., Howard, J. W. and Joe, F. L., *J. Ass. Off. Agric. Chem.*, **45**, 67 (1962)
20. Eidelman, M., *J. Ass. Off. Agric. Chem.*, **45**, 672 (1962)
21. Wood, N. F., *Analyst*, **94**, 399 (1969)
22. Iwata, Y., Westlake, W. E. and Gunther, F. A., *J. Ass. Off. Anal. Chem.*, **54**, 739 (1971)
23. Schnorbus, R. R. and Phillips, W. F., *J. Agric. Food Chem.*, **15**, 661 (1967)
24. Jones, L. R. and Riddick, J. A., *Anal. Chem.*, **24**, 569 (1952)
25. Dickes, G. J. and Nicholas, P. V., *J. Ass. Publ. Anal.*, **5**, 52 (1967)
26. Dickes, G. J. and Nicholas, P. V., *J. Ass. Publ. Anal.*, **6**, 60 (1968)
27. Dickes, G. J. and Nicholas, P. V., *J. Ass. Publ. Anal.*, **7**, 14 (1969)
28. *Official Methods of Analysis of the Association of Official Analytical Chemists*, 11th edn, AOAC, Washington, 475 (1970)
29. Burke, J. and Giuffrida, L., *J. Ass. Off. Agric. Chem.*, **47**, 326 (1964)
30. Bellman, S. W. and Barry, T. L., *J. Ass. Off. Anal. Chem.*, **54**, 499 (1971)

31. Beroza, M. and Bowman, M. C., *Anal. Chem.*, **37**, 291 (1965)
32. Storherr, R. W. and Watts, R. R., *J. Ass. Off. Agric. Chem.*, **48**, 1154 (1965)
33. Watts, R. R. and Storherr, R. W., *J. Ass. Off. Anal. Chem.*, **50**, 581 (1967)
34. Storherr, R. W., Murray, E. J., Klein, I. and Rosenberg, L. A., *J. Ass. Off. Anal. Chem.*, **50**, 605 (1967)
35. Kanazawa, J. and Sato, R., *Bunseki Kagaku*, **10**, 1350 (1961)
36. Bielorai, R. and Alumot, E., *J. Sci. Food Agric.*, **16**, 594 (1965)
37. Bielorai, R. and Alumot, E., *J. Agric. Food Chem.*, **14**, 622 (1966)
38. Malone, B., *J. Ass. Off. Anal. Chem.*, **52**, 800 (1969)
39. Ragelis, E. P., Fisher, B. S. and Klimeck, B. A., *J. Ass. Off. Anal. Chem.*, **49**, 963 (1966)
40. Heuser, S. G. and Scudamore, K. A., *Chem. & Ind.*, 1557 (1967)
41. Heuser, S. G. and Scudamore, K. A., *Analyst*, **93**, 252 (1968)
42. Heuser, S. G. and Scudamore, K. A., *Pesticide Sci.*, **1**, 244 (1970)
43. Karasz, A. B. and Gantenbein, W. M., *J. Agric. Food Chem.*, **19**, 1270 (1971)
44. Di Muccio, A., Boniforti, L. and Monacelli, R., *J. Chromatog.*, **71**, 340 (1972)
45. Baker, P. B. and Flaherty, B., *Analyst*, **97**, 378 (1972)
46. *Tech. Rep. Ser.*, World Health Organisation, No. 458 (1970)
47. Eades, J. F. and Gardiner, K. D., *Chem. & Ind.*, 1359 (1967)
48. Yip, G., *J. Ass. Off. Agric. Chem.*, **45**, 367 (1962)
49. Gutenmann, W. H. and Lisk, D. J., *J. Ass. Off. Agric. Chem.*, **46**, 859 (1963)
50. Dubosq, F. and Dedde, M., *Chim. Anal.*, **50**, 40 (1968)
51. Meagher, W. R., *J. Agric. Food Chem.*, **14**, 374 (1966)
52. Munro, H. E., *Pesticide Sci.*, **3**, 371 (1972)
53. Garbrecht, T. P., *J. Ass. Off. Anal. Chem.*, **53**, 70 (1970)
54. Baur, J. R., Baker, R. D. and Davis, F. S., *J. Ass. Off. Anal. Chem.*, **54**, 713 (1971)
55. Yip, G., *J. Ass. Off. Agric. Chem.*, **47**, 1116 (1964)
56. Chiba, M. and Morley, H. V., *J. Ass. Off. Anal. Chem.*, **49**, 341 (1966)
57. Stark, A., *J. Agric. Food Chem.*, **17**, 871 (1969)
58. Getzendaner, M. E., *J. Ass. Off. Agric. Chem.*, **46**, 269 (1963)
59. Meulemans, K. J. and Upton, E. T., *J. Ass. Off. Anal. Chem.*, **49**, 976 (1966)
60. Bjerke, E. L., Kutschinski, A. H. and Ramsey, J. C., *J. Agric. Food Chem.*, **15**, 469 (1967)
61. Pease, H. L., *J. Agric. Food Chem.*, **14**, 94 (1966)
62. Archer, T. E., *Bull. Environ. Contam. Toxicol.*, **3**, 71 (1968)
63. Lemperle, E. and Kerner, E., *Z. Anal. Chem.*, **256**, 353 (1971)
64. Pack, D. E., Lee, H., Bouco, L., Sumida, S., Hisada, Y. and Miyamoto, J., *J. Ass. Off. Anal. Chem.*, **56**, 53 (1973)
65. Coulson, D. M., Cavanagh, L. A. and Stuart, J., *J. Agric. Food Chem.*, **7**, 250 (1959)
66. Giuffrida, L. and Ives, F., *J. Ass. Off. Agric. Chem.*, **47**, 1112 (1964)
67. Nelson, R. C., *J. Ass. Off. Agric. Chem.*, **47**, 289 (1964)
68. Nelson, R. C., *J. Ass. Off. Anal. Chem.*, **49**, 763 (1966)
69. Nelson, R. C., *J. Ass. Off. Anal. Chem.*, **50**, 922 (1967)
70. Mills, P. A., Onley, J. H. and Gaither, R. A., *J. Ass. Off. Agric. Chem.*, **46**, 186 (1963)
71. Bache, C. A. and Lisk, D. J., *J. Ass. Off. Anal. Chem.*, **50**, 1246 (1967)
72. Bache, C. A. and Lisk, D. J., *Anal. Chem.*, **37**, 1477 (1965)
73. Brody, S. S. and Chaney, J. E., *J. Gas Chromat.*, **4**, 42 (1966)
74. Getz, M. E., *J. Gas Chromat.*, **5**, 377 (1967)
75. Getz, M. E., *J. Ass. Off. Agric. Chem.*, **45**, 393 (1962)
76. Laws, E. Q. and Webley, D. J., *Analyst*, **86**, 249 (1961)
77. Bowman, M. C., Beroza, M. and Hill, K. R., *J. Ass. Off. Anal. Chem.*, **54**, 346 (1971)
78. Storherr, R. W. and Watts, R. R., *J. Ass. Off. Anal. Chem.*, **51**, 662 (1968)
79. Storherr, R. W., Getz, M. E., Watts, R. R., Friedman, S. J., Erwin, F., Giuffrida, L. and Ives, F., *J. Ass. Off. Agric. Chem.*, **47**, 1087 (1964)
80. Storherr, R. W., Ott, P. and Watts, R. R., *J. Ass. Off. Anal. Chem.*, **54**, 513 (1971)
81. McCaulley, D. F., *J. Ass. Off. Agric. Chem.*, **48**, 659 (1965)
82. Bowman, M. C. and Beroza, M., *J. Agric. Food Chem.*, **15**, 465 (1967)
83. Winterlin, W., Mourer, C. and Beckman, H., *J. Agric. Food Chem.*, **18**, 401 (1970)

84. Crossley, J., *J. Ass. Off. Anal. Chem.*, **53**, 1036 (1970)
85. Gauer, W. O. and Seiber, J. N., *Bull. Environ. Contam. Toxicol.*, **6**, 183 (1971)
86. Steller, W. A. and Pasarela, N. R., *J. Ass. Off. Anal. Chem.*, **55**, 1280 (1972)
87. Elgar, K. E., Marlow, R. G. and Mathews, B. L., *Analyst*, **95**, 875 (1970)
88. Watts, R. R., Storherr, R. W., Pardue, J. R. and Osgood, T., *J. Ass. Off. Anal. Chem.*, **52**, 522 (1969)
89. Williams, I. H., Kore, R. and Finlayson, D. G., *J. Agric. Food Chem.*, **19**, 456 (1971)
90. Eberle, D. O. and Novak, D., *J. Ass. Off. Anal. Chem.*, **52**, 1067 (1969)
91. Eberle, D. O. and Hoermann, W. D., *J. Ass. Off. Anal. Chem.*, **54**, 150 (1971)
92. Crisp, S. and Tarrant, K. R., *Analyst*, **96**, 310 (1971)
93. Committee for Analytical Methods (MAFF, London), *Analyst*, **98**, 19 (1973)
94. Dawson, J. A., Donegan, L. and Thain, E. M., *Analyst*, **89**, 495 (1964)
95. Boone, G. H., *J. Ass. Off. Agric. Chem.*, **48**, 748 (1965)
96. Bowman, M. C. and Beroza, M., *J. Ass. Off. Anal. Chem.*, **49**, 1154 (1966)
97. Horler, D. F., *J. Stored Prod. Res.*, **1**, 287 (1966)
98. Elms, K. D., *J. Stored Prod. Res.*, **3**, 393 (1967)
99. Sherman, M. and Herrick, R. B., *J. Agric. Food Chem.*, **20**, 985 (1972)
100. Versino, B., van der Venne, M-T. and Vissers, H., *J. Ass. Off. Anal. Chem.*, **54**, 147 (1971)
101. McLeod, H. A., Mendoza, C., Wales, P. and McKinley, W. P., *J. Ass. Off. Anal. Chem.*, **50**, 1216 (1967)
102. Crossley, J., *Chem. & Ind.*, 1969 (1966)
103. Bowman, M. C. and Beroza, M., *J. Ass. Off. Anal. Chem.*, **53**, 499 (1970)
104. Bowman, M. C. and Beroza, M., *J. Ass. Off. Anal. Chem.*, **54**, 1086 (1971)
105. St. John, L. E. and Lisk, D. J., *J. Agric. Food Chem.*, **16**, 408 (1968)
106. Askew, J., Ruzicka, J. H. and Wheals, B. B., *J. Chromatog.*, **41**, 180 (1969)
107. Shafik, M. T., Bradway, D. and Enos, H. F., *Bull. Environ. Contam. Toxicol.*, **6**, 55 (1971)
108. Katague, D. B. and Anderson, C. A., *J. Agric. Food Chem.*, **14**, 505 (1966)
109. Claborn, H. V. and Ivey, M. C., *J. Agric. Food Chem.*, **13**, 354 (1965)
110. Thornton, J. S. and Schumann, S. A., *J. Agric. Food Chem.*, **20**, 635 (1972)
111. Martin, R. L., *Anal. Chem.*, **38**, 1209 (1966)
112. Coulson, D. M., *J. Gas Chromat.*, **3**, 134 (1965)
113. Coulson, D. M., *J. Gas Chromat.*, **4**, 285 (1966)
114. Laski, R. R. and Watts, R. R., *J. Ass. Off. Anal. Chem.*, **56**, 328 (1973)
115. Ralls, J. W. and Cortes, A., *J. Gas Chromat.*, **2**, 132 (1964)
116. van Middlelem, C. H., Norwood, T. L. and Waites, R. E., *J. Gas Chromat.*, **3**, 310 (1965)
117. Gutenmann, W. H. and Lisk, D. J., *J. Agric. Food Chem.*, **13**, 48 (1965)
118. Holden, E. R., Jones, W. M. and Beroza, M., *J. Agric. Food Chem.*, **17**, 56 (1969)
119. Baunok, I. and Geissbuehler, H., *Bull. Environ. Contam. Toxicol.*, **3**, 7 (1968)
120. Argauer, R. J., *J. Agric. Food Chem.*, **17**, 888 (1969)
121. Butler, L. I. and McDonough, L. M., *J. Agric. Food Chem.*, **16**, 403 (1968)
122. Butler, L. I. and McDonough, L. M., *J. Ass. Off. Anal. Chem.*, **53**, 495 (1970)
123. Tilden, R. L. and van Middlelem, C. H., *J. Agric. Food Chem.*, **18**, 154 (1970)
124. Adler, I. L., Gordon, C. F., Haines, L. D. and Wargo, J. P., *J. Ass. Off. Anal. Chem.*, **55**, 802 (1972)
125. Lau, S. C. and Marxmiller, R. L., *J. Agric. Food Chem.*, **18**, 413 (1970)
126. Stanley, C. W., Thornton, J. S. and Katague, D. B., *J. Agric. Food Chem.*, **20**, 1265 (1972)
127. Stanley, C. W. and Thornton, J. S., *J. Agric. Food Chem.*, **20**, 1269 (1972)
128. Seiber, J. N., *J. Agric. Food Chem.*, **20**, 443 (1972)
129. Moye, H. A., *J. Agric. Food Chem.*, **19**, 452 (1971)
130. van Middlelem, C. H., Moye, H. A. and Janes, M. J., *J. Agric. Food Chem.*, **19**, 459 (1971)
131. Onley, J. H. and Yip, G., *J. Ass. Off. Anal. Chem.*, **54**, 1366 (1971)
132. Cerny, M. and Blumenthal, A., *Mitt. Geb. Lebensmitt. Hyg.*, **63**, 289 (1972)
133. Onley, J. H. and Yip, G., *J. Ass. Off. Anal. Chem.*, **54**, 165 (1971)
134. Newsome, W. H., *J. Agric. Food Chem.*, **20**, 967 (1972)

135. Haines, L. D. and Adler, I. L., *J. Ass. Off. Anal. Chem.,* **56**, 333 (1973)
136. Chilwell, E. D. and Hughes, D., *J. Sci. Food Agric.,* **8**, 425 (1962)
137. Delley, R., Friedrich, K., Karlhuber, B., Szekely, G. and Stammbach, K., *Z. Anal. Chem.,* **228**, 23 (1967)
138. Westlake, W. E., Westlake, A. and Gunther, F. A., *J. Agric. Food Chem.,* **18**, 685 (1970)
139. .Schroeder, R. S., Patel, N. R., Hedrich, L. W., Doyle, W. C., Riden, J. R. and Phillips, L. V., *J. Agric. Food Chem.,* **20**, 1286 (1972)
140. Beynon, K. I., *Pesticide Sci.,* **3**, 389 (1972)
141. Wales, P. J. and Mendoza, C. E., *J. Ass. Off. Anal. Chem.,* **53**, 509 (1970)
142. Mendoza, C. E. and Shields, J. B., *J. Ass. Off. Anal. Chem.,* **54**, 986 (1971)
143. Bache, C. A. and Lisk, D. J., *J. Gas Chromat.,* **6**, 301 (1968)
144. Kilgore, W. W., Winterlin, W. and White, R., *J. Agric. Food Chem.,* **15**, 1035 (1967)
145. Kilgore, W. W. and White, E. R., *J. Agric. Food Chem.,* **15**, 1118 (1967)
146. Bevenue, A. and Ogata, J. N., *J. Chromatog.,* **36**, 529 (1968)
147. Pomerantz, I. H. and Ross, R., *J. Ass. Off. Anal. Chem.,* **51**, 1058 (1968)
148. Pomerantz, I. H., Miller, L. J. and Kava, G., *J. Ass. Off. Anal. Chem.,* **53**, 154 (1970)
149. Baker, P. B. and Flaherty, B., *Analyst,* **97**, 713 (1972)
150. Gutenmann, W. H. and Lisk, D. J., *J. Ass. Off. Agric. Chem.,* **48**, 1173 (1965)
151. Boggs, H. M., *J. Ass. Off. Anal. Chem.,* **49**, 772 (1966)
152. Yip, G. and Howard, S. F., *J. Ass. Off. Anal. Chem.,* **51**, 24 (1968)
153. Hrivnak, J. and Stota, Z., *J. Gas Chromat.,* **6**, 9 (1968)
154. Baker, P. B. and Hoodless, R. A., *Analyst,* **98**, 172 (1973)
155. Mestres, R., Campo, M. and Tourte, J., *Ann. Falsif. Exp. Chim.,* **63**, 160 (1970)
156. Hey, H., *Z. Lebensmitt. Forsch.,* **149**, 79 (1972)
157. Gutenmann, W. H. and Lisk, D. J., *J. Agric. Food Chem.,* **11**, 468 (1963)
158. Martin, R. J., *J. Ass. Off. Anal. Chem.,* **50**, 939 (1967)
159. Crofts, M., Harris, R. J. and Whiteoak, R. J., *Pesticide Sci.,* **3**, 29 (1972)
160. Bevenue, A., Kawano, Y. and De Lano, F., *J. Chromatog.,* **50**, 49 (1970)
161. Kawano, Y. and Bevenue, A., *J. Chromatog.,* **72**, 51 (1972)
162. Miller, W. K. and Tweet, O., *J. Agric. Food Chem.,* **15**, 931 (1967)
163. Westlake, W. E., Gunther, F. A. and Jeppson, L. R., *J. Agric. Food Chem.,* **19**, 894 (1971)
164. Gaston, L. K. and Gunther, F. A., *Bull. Environ. Contam. Toxicol.,* **3**, 1 (1968)
165. Selim, S. and Seiber, J. N., *J. Agric. Food Chem.,* **21**, 430 (1973)

21
Other Food Contaminants

A. Introduction

Besides pesticides (Chapter 20), there are many other diverse categories of food contaminants which require methods for the analysis of their residues. There has been considerable international interest in the toxicology, and, hence, the analysis, of such contaminants as polychlorinated biphenyls (PCBs), methyl mercury, *N*-nitrosamines, solvents, mycotoxins, hormones and disinfectants, and GC has become the technique of choice in many instances for their determination.

However, there is insufficient data to warrant a separate chapter for each of these contaminants, and therefore they have been assembled in this chapter.

Table 21.1 includes GC details of some of the more important references in this chapter.

B. Polychlorinated biphenyls (PCBs)

PCBs show plasticising and dielectric properties and are used in the manufacture of paints, lacquers, rubbers, waxes, resins, plasticisers, lubricants, capacitor dielectric fluids and fireproofing preparations. PCBs, like OC pesticides, have found their way into animals and animal products which may be used for human consumption, viz. fish, eggs and game birds. Fish accumulate most of their PCBs from river water and concentrate them in the fatty tissues and the liver.

PCBs are, in general, less toxic than DDT, but they are more persistent and because of their similar properties and structure, plus their relatively high electron-capturing capacity, they cause interference in the standard GC methods for the determination of OC pesticides (see Chapter 20). Much of the current literature on the GC determination of PCBs is therefore concerned with their separation from OC pesticides.

PCBs are often referred to by one of their trade names, Aroclor, and each Aroclor is a mixture of several PCBs, making their individual analyses, besides their differentiation from OC pesticides, a difficult problem. In the standard GC methods for OC pesticides PCBs most often clash with members of the DDT family.

Table 21.1 GC conditions used in the analysis of various food contaminants

Class of compound	Ref.	Column dimensions	Stationary phase	Support material	Carrier gas and conditions	Temperature or temperature programme/°C	Detector
PCBs	12	2ft x 0.125 in	10% DC-200	80/100 DMCS Gas Chrom Q	N_2, 85ml/min	180	EC
PCBs and their degradation products	8	1.22m x 4mm I.D.	4% SE-30 plus 4% QF-1	70/80 Anakrom ABS	95:5 Ar:CH_4, 60ml/min	180	EC
PCB degradation products	12	3ft x 0.125in	—Carbowax 400	80/100 Porasil S	H_2, 20ml/min	178	FI
Methyl mercuric chloride	14, 15	5ft x 0.125in	10% Carbowax 20M	Chromosorb W	N_2, 65ml/min	between 130 and 145	EC
Methyl mercuric bromide	18	0.4:m x 4mm	2% BDS	100/120 AW, DMCS Chromosorb W	N_2, 80–100 ml/min	120	EC
Methyl mercuric iodide	23	4ft x 0.25in	7% Carbowax 20M	AW, DMCS Chromosorb W	N_2, 60ml/min	170	EC
N-nitrosamines	29	5.5m x 3mm I.D.	15% FFAP	80/100 Chromosorb W	He, 25ml/min	140	CC
N-nitrosamines	29	5.5m x 3mm I.D.	15% Carbowax 20M	80/100 Chromosorb W	He, 25ml/min	120	AFI
Nitramines	35	5ft x –	10% PEG 20M	AW, HMDS Celite	Ar, 36ml/min	150	EC
NN-di-n-alkyl-hepta-fluorobutanamides	31	2.75m x 2mm I.D.	15% FFAP	80/100 Chromosorb W	N_2, 50ml/min	60	EC
NN-di-n-alkyl-hepta-fluorobutanamides	31	5.8m x 2mm I.D.	20% FFAP	80/100 Chromosorb W	N_2, 50ml/min	110	EC
Hexane, acetone and isopropanol	50, 58,	2ft x 0.25in O.D.	80/100 Porapak P	80/100 Porapak P	He, 60–70 ml/min	70–180, various	FI

348

Compound	Ref.	Column	Liquid phase	Support	Carrier gas	Temp. (°C)	Detector
Hexane, acetone and isopropanol	50, 58, 60	1ft x 0.25in O.D.	80/100 Porapak Q	80/100 Porapak Q	He, 60–70 ml/min	70–180, various programmes	FI
Aliphatic chloro-hydrocarbons	55	6ft x 6mm O.D.	150/200 Porapak Q	150/200 Porapak Q	N_2, 90 ml/min	160	MC
Methyl nitrite	59	6ft x 4mm O.D.	1ft. of glass beads 5ft. of 120/150 Porapak R	1 ft. of glass beads 5ft. of 120/150 Porapak R	N_2, 35ml/min	150	FI
Isopropanol	58	6ft x 0.25in O.D.	40% Castorwax	60/80 AW Chromosorb W	He, 15ml/min	80	FI
Dibutyl phthalate	66	2m x 3mm	4% silicone oil	80/100 AW, DMCS Chromosorb G	N_2, 28ml/min	140–250 at 22.5/min	FI
Patulin chloroacetate	68	6ft x 4mm I.D.	3% JXR	—	—	130	FI
Ipomeamarone	69	6ft x 0.25in O.D.	10% UC-W98	100/120 Gas Chrom Q	He, 60/min	180	TC
Diethyl stilboestrol trifluoroacetate	71	5ft x 0.125in	3% OV-17	100/120 Varaport 30	N_2, 40ml/min	190	EC
Methyl 6-chloropicolinate	75	6ft x 3mm I.D.	4% LAC 446 plus 0.5% H_3PO_4	80/100 AW Chromosorb W	N_2, 60ml/min	148	EC
Chlorobutanol	76	6ft x 4mm I.D.	5% Igepal 880	80/100 silanised Gas Chrom Q	N_2, 35ml/min	135	EC
TFA derivative of 2-aminoquinoxaline	77	1.25m x 4mm I.D.	5% NPGS	100/120 AW silanised Chromosorb W	N_2, 240ml/min measured at room temp.	190	EC
Pyribenzamine	78	8ft x 4mm I.D.	2% Carbowax 20M	100/120 Gas Chrom S	N_2, 60ml/min	218	FI
Indene	79	1.8m x —	20% SE-30	60/80 Chromosorb W	N_2, 60ml/min	100	FI

349

In 1967 Holmes, Simmons and Tatton[1], in an article concerned with the presence of chlorinated hydrocarbons in British wildlife, pointed out that analysts had long been aware that some samples, particularly those of bird liver, bird fat, eggs and fresh-water fish, produced an additional ten or so peaks on the chromatograms of standard OC pesticide analyses. When it was discovered that these extra peaks were due to PCBs, it was also found that some of the peaks thought to be entirely due to OC pesticides, on both silicone- and Apiezon-type columns, could be either partially or even completely due to PCBs. These workers suggested that a preliminary clean-up using TLC was helpful to the subsequent GC analysis of both PCBs and OC pesticides.

It was Jensen[2], in 1966, who was first to discover the presence of PCBs in what were apparently solely OC pesticide residues, and Jensen *et al.*[3] carried out a general survey of the PCB levels in marine animals from Swedish waters, using GC methods.

Risebrough, Reiche and Olcott[4] found that the chromatograms resulting from the examination of bird and fish samples contained mutual interference by DDT-type insecticides and PCBs. It was possible to hydrolyse DDT and its metabolites with ethanolic potassium hydroxide and record the reduction in certain peak heights due to these compounds. The presence of PCBs was confirmed by MS. These workers asserted that it was obvious that some previously reported figures for the levels of the DDT family in foods and other substrates must have been erroneously high, having been confused with PCBs.

After the GC of mixtures of PCBs and OC pesticides, Shaw[5] trapped the effluents on 80/100 glass beads which were wetted with hexane, and cooled them in solid carbon dioxide. The hexane and condensed water were vacuum distilled at room temperature and the tube containing the glass beads coated with the chlorinated compounds was connected to an MS which was capable of distinguishing the DDT family from PCBs.

Hutzinger, Jamieson and Zitko[6] also described the MS differentiation of PCBs and OC pesticides, and overlapping GC peaks containing 10–100ng of mixtures of PCBs, OC pesticides and chlorinated naphthalenes have been similarly analysed by Bonelli[7].

In an examination of the PCB content of salmon and herring oils Hannan, Bills and Herring[8] used Aroclor 1254 as a standard, which gave 13 peaks on a stationary phase comprising 4% SE-30 and 4% QF-1. Several of the peaks were found to coincide with those of DDT, DDE, TDE and their isomers. The samples of oils were extracted in petrol and partitioned with acetonitrile. The extract was cleaned up by Florisil column chromatography, the PCBs plus DDE being separated from other members of the DDT family. During the subsequent GC determination, the effluents were led from the column and irradiated in a UV source and the products were rechromatographed. This second chromatography gave fingerprint degradation patterns for PCBs, which distinguished them from DDE and other interfering compounds.

Florisil column chromatographic clean-up was also used by Monod *et al.*[9] to separate PCBs from DDT metabolites in an examination of sardines from the Bay of Marseilles. Prior to the clean-up, the samples were Soxhlet-extracted with 2:1 hexane:acetone, and subsequent GLC was carried out on both DC-200 and QF-1 stationary phases.

A survey of the PCB contamination of fish and birds found in parts of the

River Rhine and off the Netherlands coast was conducted by Koeman, ten Noever de Brauw and de Vos[10]. They Soxhlet-extracted the samples in petrol, partitioned the PCBs into dimethylformamide and cleaned up the extracts on Florisil before GC and MS identification.

The composition of a technical Aroclor was determined by Rote and Murphy[11] in order to judge the different GC responses of the individual PCBs present. The total chlorine content is often different for each PCB, and this is reflected in the different sensitivity given by the EC detector for each compound, and therefore workers who assume that all PCBs are equally sensitive produce erroneous quantitative results.

PCBs can be differentiated from OC pesticides after their degradation by reaction with a palladium catalyst. Asai *et al.*[12] applied this technique to eight Aroclors and to members of the DDT family. The degradation, which can produce reactions of hydrogenation, dehydrogenation and hydrogenolysis, was achieved by attaching a column, containing a neutral palladium catalyst on Gas Chrom Q, to the GC injection port. PCBs degrade to biphenyl and cyclohexyl-benzene, whereas the DDT family yields different compounds. *Figure 21.1* shows the carbon skeleton chromatograms produced at 260 and 300°C from

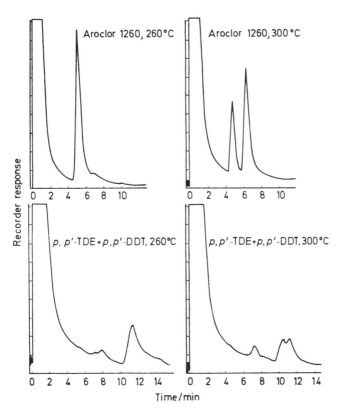

Figure 21.1 Carbon skeleton chromatograms at two catalyst temperatures of Aroclor 1260, and a mixture of p,p'-TDE and p,p'-DDT (after Asai et al.[12]). For GC details, see Table 21.1

Aroclor 1260 and also by a mixture of *p,p'*-DDT and *p,p'*-TDE, indicating the marked difference between the two patterns. This is a simple way of distinguishing these two contaminant classes.

Specht[13] has recently developed a GC method for the determination of PCBs migrating from food packaging materials (See Section F of this chapter).

C. Methyl mercury

Mercury compounds have found their way into rivers and seas as a result of the depositing of mercury-containing industrial effluents and by the leaching of organomercurial fungicides from soil. Fish and other aquatic life can accumulate mercury residues, most of which is converted into methyl mercury, and this has a relatively high mammalian toxicity, the symptoms being neurosis and paralysis.

The determination of total mercury in food and biological materials has been carried out by such methods as colorimetry, flameless atomic absorptiometry and neutron activation analysis, and these methods have also been applied to samples containing methyl mercury. The weakness of these methods is evidently one of non-specificity, and GC methods were established for the specific analysis of methyl mercury in fish, egg and meat by Westoo[14, 15].

The isolation procedure, as applied to fish, was based on the extraction of methyl mercury with benzene from an aqueous acidified extract of the muscle tissue. The benzene solution was concentrated by distillation and the methyl mercuric chloride was extracted into ammoniacal solution. This solution was acidified and re-extracted with benzene, and this final extract used for GLC on a 10% Carbowax 20M stationary phase, using an EC detector. This procedure for the extraction of methyl mercury was modified so that it could be applied to a wider range of foods, including meat and eggs, and incorporated a step to break down the link between methyl mercury and sulphur which occurred in the non-volatiles of these foods. In the method either mercuric thiols were formed by addition of mercuric chloride and then removed or the methyl mercury was extracted via its water-soluble cysteine complex after the addition of cysteine acetate, leaving the thiols behind. An alumina column clean-up step was also incorporated into the original method. *Table 21.2* gives the methyl mercury and total mercury contents of meats, egg and fish, and shows that the fresh-water fish (Swedish) contain most of the mercury as methyl mercury. The lowest percentage of methyl mercury shown in the table is that of an egg yolk sample at 50% of the total mercury.

Johansson, Ryhage and Westoo[16] used essentially this modified Westoo method for the analysis of methyl mercury within the range 0.15–3.2 p.p.m. total mercury in fish, and found that the calculated total mercury content from the GC method agreed well with a GC–MS method and also with activation analysis figures.

The Westoo method was also effectively implemented by Kamps and McMahon[17] in the analysis of methyl mercury in tuna and swordfish. They used the cysteine extraction modification and GLC on a PDEAS stationary phase. The normal limit of detection of the method was found to be 0.08 p.p.m. methyl mercuric chloride, and results on nine tuna samples were in the range 0.22–0.55 p.p.m. as total mercury. The one swordfish sample contained 2.37 p.p.m. as total mercury.

Certain fish samples of high lipid content can cause gelatinous emulsions between aqueous homogenates and benzene, and Newsome[18] used hydrobromic acid plus potassium bromide in place of hydrochloric acid to overcome this when using the Westoo procedure. He used the cysteine extraction modification and a 2% BDS stationary phase for GLC analysis.

Table 21.2 Some methyl mercury contents in foods (after Westoo[15])

Foods	µg of total mercury/g of food	Methyl mercury in foods mg Hg/kg	% of total Hg
Meat (ox)	0.074	0.068	92
Meat (hen)	0.051	0.037	73
Meat (hen)	0.023	0.017	74
Liver (pig)	0.130	0.095	73
Liver (pig)	0.140	0.095	68
Liver (pig)	0.096	0.075	78
Egg yolk	0.010	0.005	50
Egg yolk	0.010	0.009	90
Egg white	0.023	0.020	87
Egg white	0.025	0.019	76
Egg white	0.012	0.011	92
Egg white	0.025	0.024	96
Egg white	0.012	0.011	92
Muscle tissue of perch	0.75	0.70	93
Muscle tissue of perch	0.22	0.20	91
Muscle tissue of perch	0.30	0.25	83
Muscle tissue of perch	0.42	0.38	90
Muscle tissue of perch	0.29	0.25	86
Muscle tissue of perch	3.25	2.99	92
Muscle tissue of perch	3.10	2.81	91
Muscle tissue of pike	3.35	3.11	93
Muscle tissue of pike	2.67	2.57	96
Muscle tissue of pike	0.68	0.60	88
Muscle tissue of pike	1.90	1.81	95
Muscle tissue of pike	0.56	0.55	98
Muscle tissue of pike	0.79	0.72	91
Muscle tissue of haddock	0.052	0.043	83
Muscle tissue of haddock	0.033	0.025	76
Muscle tissue of cod	0.036	0.028	78
Muscle tissue of cod	0.026	0.022	85

The methyl mercury contents of canned tuna, sardine and mackerel were determined by Fabbrini *et al.*[19], using a GC method after its extraction and conversion into a mercaptide complex. Twenty per cent of the tuna samples contained more than 0.5 p.p.m. organic mercury, but lesser concentrations were found in the sardine and mackerel samples.

In an academic study of the identification of methyl mercury, ethyl mercury and phenyl mercury, Nishi and Horimoto[20] caused the organic mercury compounds to react with a solution of an inorganic sulphide plus a metallic suspension. The original GC peaks disappeared after such treatment, and this subtractive technique was used to identify the organic mercury compounds. The disappearance of the original peaks was also noted when a column containing the metallic powder was placed between the GC column and the detector.

An atomic absorption spectrometric detection technique was used by Gonzales and Ross[21] in determining the concentration of effluent alkyl mercury compounds during GC analysis of fish samples. The methyl mercury was extracted by benzene from an aqueous acidic slurry of the fish sample, and was subjected to GC without recourse to a clean-up procedure. The methyl mercuric halide was separated on the column and either burnt in a flame photometric apparatus, the effluent gas being analysed in an atomic absorptiometer, or burnt in oxygen in a furnace at 780°C, the effluent gas being analysed by flameless atomic absorptiometry. Although the sensitivity afforded by this means of detection is not so great as that of the EC detector, it is adequate and relatively simple for the screening of methyl mercury in fish.

Having used the Westoo extraction procedure for methyl mercury in fish, Bache and Lisk[22] performed GC using a detection system comprising a microwave-powered inert gas plasma which analyses mercury by emission spectrometry using the 2537 Å line. In comparing this detection system with the EC detector, two Coho samples gave the same figures for both methods, viz. 0.20 and 0.18 p.p.m. methyl mercuric chloride.

A method was devised by Uthe, Solomon and Grift[23] in order to minimise the number of extraction steps and the volume of solvent as used in the Westoo procedure. The fish tissue was homogenised in an aqueous medium containing an acidic solution of sodium bromide, and the resulting methyl mercuric bromide was extracted with toluene. The methyl mercuric bromide was extracted from the toluene via its thiosulphate complex with aqueous ethanol, and after the addition of potassium iodide the aqueous solution was extracted with benzene, which removed methyl mercuric iodide. GLC was performed on a 7% Carbowax 20M stationary phase and using a specially designed EC detector. Down to 0.01 p.p.m. methyl mercury could be detected by this method.

D. N-Nitrosamines

N-nitrosamines may be formed by reaction between secondary amines and nitrous acid. It is well established that some of the N-nitrosamines, particularly N-nitrosodimethylamine, have hepatotoxic and carcinogenic properties, and there has been some concern that these compounds might be present in foods. Such concern is not misplaced when it is realised that nitrite and secondary amines can occur together in certain foods, nitrite and nitrate often being permitted preservatives or colour fixatives in such foods as smoked or cured meats and fish. Nitrate is often permitted at higher levels than nitrite and it has been shown that some nitrate is reduced to nitrite in the food matrix. Some foods contain amines, e.g. fish, whereas some processed foods could produce amines when undergoing heat treatment.

It is no surprise, therefore, to find that many food analysts have examined foods such as fish, meat, bacon and cheese for N-nitrosamines.

Walters[24] has reviewed the various methods used in the detection and estimation of N-nitrosamines in food. These methods include polarography, spectrophotometry, ion exchange chromatography, TLC, GC and MS. Polarography has the required sensitivity but does not have the necessary specificity. The same applies to some of the other techniques, with the marked exception of the combination of GC with MS. The most experienced analysts in the field of *N*-nitrosamine analysis agree that this combination is the only foolproof method for the positive identification of nitrosamines. It is significant that GC–MS methods for the analysis of these compounds have been developed from 1970 onwards, and it is suggested that figures in the literature prior to this decade should be treated with reserve.

It was also realised that a method was required which could detect sub-p.p.m. levels of nitrosamines in food. In 1970 Foreman, Palframan and Walker[25] performed successful recovery experiments with dimethyl- and diethylnitrosamines added to corned beef at the 1 p.p.m. level, having used an isolation procedure based on distillation from the alkaline homogenate of the sample. The microporous polymer beads Chromosorb 101 were used as the column packing which meant that aqueous samples could be injected, the water being eluted in a narrow band at the front of the chromatogram. These workers felt that with a suitable clean-up procedure, and with a concentration technique, a detection limit of the order of 0.01 p.p.m. could be achieved.

Gough and Webb[26, 27] have described a membrane separator which is capable of transferring nanogram quantities of nitrosamines from the GC column to the MS. They gave data for the determination of eight nitrosamines, and by incorporating a peak cutting system into the chromatograph a detection limit corresponding to 1μg per kg of original sample could be achieved. The same authors[28] used this technique for the analysis of non-volatile nitrosamines. This combined GC–MS method was used by Crosby *et al.*[29] in the confirmatory analysis of steam-volatile *N*-nitrosamines at that level of detection in bacon, fish, meat and cheese samples.

The minced food was put through a steam distillation procedure, and the distillate was made alkaline and extracted with dichloromethane and the extract concentrated at 50°C for GC. Each extract was analysed with two different sets of GLC conditions, viz. a stationary phase of 15% FFAP and a CCD set in the reductive mode for nitrogen, and a stationary phase of 15% Carbowax 20M and an AFID with rubidium tip. When a sample was suspected of containing a nitrosamine, it was rechromatographed on the FFAP stationary phase with MS detection. Experiments showed that *N*-nitrosodimethylamine, *N*-nitrosodiethylamine, and *N*-nitrosopiperidine were 70–90% recovered after their addition to foods, and that *N*-nitrosopyrrolidine was only 30% recovered at best. These workers concluded that these foods can contain minute concentrations of those nitrosamines which are known to be toxic to animals at higher levels.

In an evaluation of the two detectors used in this work Palframan, Macnab and Crosby[30] concluded that the CCD set in the reductive mode for nitrogen was more selective and less disturbed by small changes in operating parameters than the AFID. They described a flow-diverting system in conjunction with the AFID to enable it to function when volumes greater than 2μl of chlorinated solvents would otherwise extinguish the flame.

At this time the same laboratory produced a GC method for the determination of steam-volatile *N*-nitrosamines via the heptafluorobutanoyl derivatives of their secondary amines. Alliston, Cox and Kirk[31] successfully explored the possibility of using these derivatives for GC with EC detection. Nitrosamines in samples of cheese, bacon, meat and fish were isolated by steam distillation in neutral solution followed by two separate distillations from alkaline and acid solutions in order to remove volatile acidic and basic interfering compounds. The final distillate was evenly divided and one-half was reduced in an electrolysis cell which was described together with photograph. This electrolytic reduction produced the secondary amines corresponding to the original nitrosamines. Both halves of the distillate were then caused to react with heptafluorobutanoyl chloride in cyclohexane, and the resulting heptafluorobutanamides of the secondary amines were chromatographed. The unreduced portion of the distillate gave peaks due to adventitious amines, and the difference between the peak heights of reduced and unreduced portions gave an accurate evaluation of the nitrosamines.

Two sets of GLC conditions were used, both incorporating a FFAP stationary phase but operated with different loadings at 60°C and 110°C. *NN*-di-n-butyl-heptafluorobutanamide had a retention time close to a reagent impurity when chromatographed at 60°C, but was resolved at 110°C. The 110°C column could not be generally used, however, since *NN*-dimethyl- and *NN*-diethylheptafluoro-butanamides were not resolved. These authors[31] have tabulated the results of the analysis of various foods for six *N*-nitrosamines, indicating trace amounts as being 0.5–1.0 μg per kg food. The two highest results were 11.0 μg *N*-nitrosopyrrolidine per kg of fried pig's liver and 6.0 μg *N*-nitrosopyrrolidine per kg of stale cod, neither sample having been cured with nitrite. Good agreement was obtained between this method and that of Crosby *et al.*[29] on those samples which were analysed by both methods.

Brooks, Alley and Jones[32] suggested the utilisation of the same derivatives of nitrosamines because of the high proportion of fluorine therein, conferring good electron-capturing properties. The nitrosamines were caused to react with heptafluorobutyric anhydride in the presence of pyridine to yield derivatives which were approximately 6000 times as sensitive to EC detection as compared with FI detection.

Telling, Bryce and Althorpe[33] determined five volatile *N*-nitrosamines in meat products by a GS–MS method. Meat product samples were homogenised and vacuum-distilled, and the distillate was extracted with dichloromethane. The dichloromethane extract was evaporated in hexane and the solution was used for GLC on a 10% Carbowax 20M stationary phase, splitting the effluents 9:1 FI detector: MS, the latter being set to monitor the NO^+ peak. *Figure 21.2* is a chromatogram, produced by this method, of a cleaned-up pork luncheon meat extract, showing peaks which could have been attributed to *N*-nitrosamines had MS not been used to check them out. 2-Hydroxy acetone was present and not *N*-nitrosodimethylamine; nonanol was present and not *N*-nitrosomethyl-ethylamine; benzaldehyde was present and not *N*-nitrosomethyliso-butylamine; α-terpineol was present and not *N*-nitrosopropylbutylamine. Depending on the nitrosamine, detection limits were between 25 and 65 μg per kg of sample. This sensitivity was subsequently improved upon by Bryce and Telling[34], who monitored molecular ions or other fragments with MS, these being chosen because they were more abundant than NO^+. This lowered the detection limit of nitrosamines to between 1 and 5μg per kg of sample.

A screening method for the presence of volatile *N*-nitrosamines was described by Telling[35] and was based on their oxidation to nitramines which were particularly sensitive to EC detection. After a vacuum distillation isolation procedure

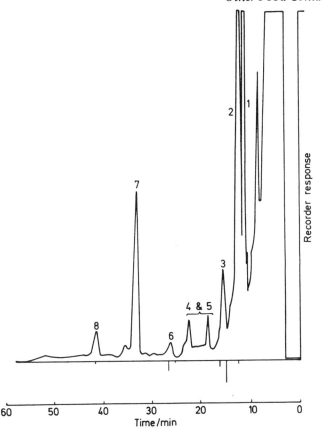

Figure 21.2 Gas chromatography of pork luncheon meat extract prepared by steam distillation and ion exchange (after Telling et al.[33]). Instrument, Pye 104 FID. Column, 5 ft 10% Carbowax 20M on acid-washed HMDS-treated Celite. Temperature, isothermal at 170°C. Carrier gas, nitrogen. Inlet pressure, 28 p.s.i. Flow rate, approximately 36 ml/min. Peak 1, acetoin; 2, hydroxy acetone; 3, nonanol; 4 and 5, terpene $C_{10}H_{18}O$; 6, benzaldehyde; 7, sesquiterpene and furfuryl alcohol; 8, α-terpineol. The corresponding retention times of nitrosamines are indicated

and extraction of the distillate with dichloromethane, extracts of vegetable or meat samples were cleaned up on alumina prior to the oxidation step carried out using trifluoroperoxyacetic acid. The excess reagent was removed and the sample was further cleaned up on alumina before GC with EC detection. Although the extraction and clean-up is somewhat lengthy, this method offers a useful screen for nitrosamines in foods at the 10 μg/kg level. A sample suspected of containing a nitrosamine in excess of this level would be quantitatively analysed by a GC–MS method.

Smoked fish, particularly smoked, nitrite-treated chub, was the subject of analysis for N-nitrosodimethylamine by Howard, Fazio and Watts[36]. Samples were digested with methanolic potassium hydroxide and the nitrosamine was

357

distilled from alkaline solution. This solution was extracted with dichloromethane and subjected to clean-up on a column comprising Celite impregnated with hydrochloric acid. GLC was carried out on an alkaline stationary phase of Carbowax 1540, with a modified AFID, and the limit of detection was 10μg per kg sample. Five experimental lots of smoked, nitrite-treated chub were examined for *N*-nitrosodimethylamine, but none was found at this level of sensitivity.

This method was used by Fazio *et al.*[37], with a modified clean-up procedure, in the examination of marine fish for *N*-nitrosodimethylamine. MS was used to confirm the identity of this nitrosamine, which was found in the range 4–26 μg/kg in sable, salmon and shad, whether raw or smoked or nitrite and/or nitrate treated.

The same method was used by Fazio, White and Howard[38] for the determination of *N*-nitrosodimethylamine in nitrite and/or nitrate-processed meats. These samples included sausage, ham, bacon, canned meat, fresh meat, corned meat and baby foods. One sample of smoked ham contained 5 μg *N*-nitrosodimethylamine per kg sample, the result being confirmed by MS.

The same method was also used by Fiddler *et al.*[39] for the determination of the same nitrosamine in ham. Examination of ten samples showed that *N*-nitrosodimethylamine was not present above the detection limit of 25 μg per kg sample.

There have been reports that spray-dried milk can contain nitrite and/or nitrate, and since the heating process in dried milk production might cause the formation of secondary amines, there is a remote chance that nitrosamines could be formed. Reineccius and Coulter[40] carried out recovery experiments of nitrosamines added to non-fat dried milk at the 10 μg/kg level in order to test their method, which depended on the oxidation of nitrosamines to nitramines after a steam distillation isolation procedure. No volatile nitrosamines were detected in milk that had been dried by indirect steam heat or by direct gas firing, although the latter process gave an unidentified peak which was shown to be not a nitrosamine.

A method which was sensitive down to 3 μg *N*-nitrosodimethylamine per kg of sample was applied to apple and to milk by Newell and Sisken[41]. The nitrosamine was isolated by vacuum distillation from the samples and adsorbed on to Porapak Q prior to GSC performed on a stationary phase of Porapak Q-S. The effluent of *N*-nitrosodimethylamine was removed from the column and catalytically reduced to NH_3 which was determined microcoulometrically. No figures were given for concentrations of *N*-nitrosodimethylamine in the two foods.

N-nitrosamines have recently been determined in alcoholic beverages[42] and in wheat flour and dairy products[43].

Pensabene *et al.*[44] have listed the GC, MS and IR spectroscopic data of 25 *N*-nitrosamines. They separated the aliphatic *N*-nitrosamines on a 5% Carbowax 20M-terephthalic acid stationary phase and aromatic *N*-nitrosamines on a 5% OV-1 stationary phase.

E. Solvents

Solvents are used for a variety of reasons in the preparation of foods. In some instances solvents are deliberately present to perform such functions as vehicles for the incorporation into foods of additives, e.g. flavourings, essential oils, colourings, vitamins and antioxidants. Solvents used for these purposes are permitted in many countries and include ethanol and isopropanol.

The concern of this section of the chapter is with the solvent residues which can be left behind in foods after processing, e.g. dichloromethane in decaffeinated coffee, hexane and trichloroethylene in refined vegetable oils and spice oleoresins and hexane and acetone in oilseed meals and flours.

In the UK the *Solvents in Food Regulations 1967*[45] permit the use of seven solvents at any concentration in any food, all other solvents being prohibited. In US legislation certain of those solvents not permitted in the UK, e.g. certain hydrocarbons and chlorinated hydrocarbons, can be present up to maximum limits in specified foods.

Because the residue analyst is looking for very low concentrations of solvents, gravimetric and volumetric methods are not applicable, and those methods dependent on vapour pressure measurement, gas detection or flash point data lack sensitivity and specificity. GC methods have been successfully applied to the following solvent residues in foods: hexane and other aliphatic hydrocarbons, chlorinated aliphatic hydrocarbons, acetone, methanol, isopropanol.

In an attempt to cover the analysis of a number of solvent residues encountered in oils and oleoresins, Dean *et al.*[46] described GC methods, incorporating a solid sampling device, which was designed to obviate the necessity of a clean-up procedure and, at the same time, preserve the cleanliness of the GC columns. The solid sampler is described in detail, with diagrams. The sample was introduced into a glass capillary which was sealed and then completely enclosed in the sampler. The sampler was connected to the GC column, brought to equilibrium temperature and the glass capillary broken by a plunger and the volatile solvents were swept by the carrier gas on to the column. GC was carried out either on a 8% Antarox CO-990 stationary phase or on Porapak S. The Antarox column was used for analysis of solvents with boiling points above 100°C and the Porapak column was especially suitable for the analysis of low boiling esters, hydrocarbons and alcohols and any other type of solvent with a boiling point less than 100°C.

Although these workers did not include chlorinated hydrocarbons in their lists of solvents, the analysis of hexane and trichloroethylene in olive and castor oils could be performed down to the 5 p.p.m. level on the Porapak S column at 200°C.

1. Hexane and other aliphatic hydrocarbons

Hexane and other closely related aliphatic hydrocarbons are extensively used as solvents for the extraction of edible oils in refining processes, and therefore residues might occur in both the oil and the remaining meal or flour. The residual hexane in hexane-extracted soyabean flakes was estimated by Black and Mustakas[47] using a GC method. Iso-octane was the extracting solvent for the hexane residues, the resulting solution being used for injection without recourse to a clean-up procedure. Three different stationary phases and GC conditions were used to confirm the hexane identity.

Soyabean and cottonseed meals and flours were examined for hexane and acetone residues by Fore and Dupuy[48]. A mixture of acetone, hexane and water is used in the manufacturing process to extract the oil from cottonseed flours, and it is therefore possible for these flours to contain residues of these solvents. Iso-octane was tried as extracting solvent but was not as effective as 95:5 dimethylformamide:water, although it was not certain whether or not all hexane and acetone had been extracted from the flour using this solvent mixture. GSC was performed on two columns of Porapak P and Porapak Q, respectively, the cottonseed meals or flours containing hexane in the range 0–1500 p.p.m. and the soyabean meals or flours containing that solvent in the range 6–80 p.p.m.

Dupuy and Fore[49] also used a volatilisation procedure in order to determine hexane in oilseed meals and flours and to support results obtained by the previous method. Ten per cent of water was added to the sample in a serum bottle, which was sealed and incubated at 110°C for 2 h. The head space was then sampled for GSC analysis on the same Porapak columns as used in their previous method. The quantitative recovery of hexane by this method was superior to that of extraction or distillation methods and it also had the advantage of using a small sample, viz. 2g.

In an improvement to this volatilisation method, Fore and Dupuy[50] placed 40mg of flour or meal into a glass injection port liner and inserted this into the injection port before injecting 80 μl of water into the area above the sample. Volatile components, including residual hexane, were swept on to the Porapak column and separated. Injection of 20 μl water between analyses was sufficient to keep the column clean. This is an improvement on their previous methods because it covers a wider range of hexane, viz. 1–2500 p.p.m., and is quicker, each analysis being completed in 20 min.

Many edible oils, e.g. cottonseed, corn, peanut, soyabean and safflower, are obtained from raw materials by solvent extraction processes using hexane or similar aliphatic hydrocarbons and, in some instances, aliphatic chlorinated hydrocarbons. Watts and Holswade[51] determined the residual hydrocarbon solvents, 2-methylpentane, 3-methylpentane, hexane and methyl cyclopentane in solvent-extracted edible oils using a GLC method. Direct injection of the oil samples was made on to a 10% didodecyl phthalate stationary phase, peaks due to oil components occurring early in the chromatogram and not interfering with the peaks due to the residual solvents.

The direct injection of the sample was bound to cause loss in sensitivity in subsequent analyses, as the oily substances built up at the injector end of the column, and a steam distillation clean-up procedure is recommended prior to GC to minimise this contamination and increase the sensitivity fivefold. This modification was obviously more time-consuming but the detection limit for aliphatic hydrocarbons in the edible oils was reduced to the order of 10 p.p.m.

The analysis of hexane in solvent-extracted oils was carried out by Nosti Vega, Gutierrez Rosales and Gutierrez-Quijano[52] using the head space sampling technique. Two grams of oil containing hexane up to 1.2% or 0.5 g of oil containing 1.2–2.2% hexane was incubated at 85°C before head space sampling and GLC analysis on a 5% SE-30 stationary phase.

Hop extract is used as a flavouring agent in the beer brewing industry and hexane is used to extract the hop to produce the extract. The US Code of Federal Regulations allow not more than 25 p.p.m. residual hexane in certain modified hop extracts, and Litchman, Turano and Upton[53] used a head space sampling technique prior to GSC for its analysis. The sample was weighed into a serum vial, water was added containing a small proportion of methanol, and the mixture was incubated at 70°C for 1 h. The head space was separated on a Porapak Q column, experiments showing that hexane was recovered quantitatively in the range 2 to 29 p.p.m.

2. Chlorinated aliphatic hydrocarbons

Chlorinated aliphatic hydrocarbons are used to extract edible vegetable oils and spice oleoresins, and the decaffeination of coffee may be carried out by dichloromethane extraction.

Trichloroethylene was determined in oils by Tous and Martel[54], who used direct injection of the samples on to a short precolumn of glass beads in order to minimise contamination of the main column, which was composed of 10% isodecylphthalate on silanised Chromosorb P. The limit of detection for trichloroethylene, and also hexane, was 1 p.p.m.

Dichloromethane, ethylene dichloride and trichloroethylene are commonly used to extract the oleoresins from spices. Under US Code of Federal Regulations a total of not more than 30 p.p.m. of these chlorinated hydrocarbons is permitted in the finished product. Roberts[55] conducted a collaborative study on the GSC determination of these solvents in the oleoresins of paprika, ginger and capsicum. The samples were extracted with ethanol and injected on to a 150/200 mesh Porapak Q column. As a result of the study, Roberts considered that a coarser Porapak Q and also the use of an internal standard would improve the method.

Caffeine is removed from raw coffee in the production of decaffeinated coffee and dichloromethane has been used as a solvent in this removal process. Gal and Schilling[56] examined decaffeinated coffee for residues of dichloromethane using a GSC method. The ground sample was steam distilled and the distillate was used for injection on to a column packed with 30/60 mesh silica gel. This medium was operated at 150°C and separated the dichloromethane, which was detected by FID. The limit of detection for this method was 1mg dichloromethane per kg decaffeinated coffee.

3. Acetone

Acetone is another solvent which has been used in some countries to extract oleoresins. Orsi[57] determined acetone residues in paprika oleoresins using GC, working to a maximum permitted allowable concentration of 30 p.p.m. The samples were mixed with water plus sodium sulphate and distilled, the distillate being extracted with benzene after the addition of potassium carbonate. GLC was carried out on a PEGA stationary phase.

Acetone is also used to detoxify mould damage of oilseed meals and therefore these are subsequently checked for the presence of acetone residue. As described in sub-section 1 (Hexane and Other Aliphatic Hydrocarbons), Fore and Dupuy's method[48] is applicable to acetone residues as well as those of hexane. Acetone levels of between nil and 80 p.p.m. were found in cottonseed flours, but no residues were present in three soyabean meals.

Dupuy, Rayner and Fore[58] produced a simple volatilisation method for the determination of acetone residues in oilseed meals and flours. This closely resembled the same laboratory's method[49] for hexane residues as described under sub-section 1 of this section. The incubation temperature prior to head space sampling of acetone vapour was kept down to 70°C because overheating of oilseed meals and flours might produce acetone by breakdown of some of the natural substances of which the food is made.

4. Methanol

In the production of hop extract additives for utilisation in the beer brewing industry, methanol is sometimes used, and Litchman and Upton[59] devised a GC method to check for methanol residues in such additives. Methanol was reacted with nitrite to form methyl nitrite which gave a 400-fold increase in response with FI detection as compared with the response of unreacted methanol. A standardised procedure was used to produce sufficient head space methyl nitrite for analysis, using a 4 min interval after the addition of nitrite to the sample.

5. Isopropanol

Isopropanol, like acetone, has also been used to detoxify mould damage in oilseed meals and flours, and therefore the level of residues of this solvent are checked in those samples. Fore, Rayner and Dupuy[60] extracted the samples with ethanol, which also served as an internal standard, and carried out GSC analyses on both Porapak P and Porapak Q columns. Peanut meal, cottonseed meal, cottonseed flour and fish flour were satisfactorily examined by this method.

In 1966 the United States FDA approved the manufacture and sale in the USA of fish protein concentrate as a food additive. Methods of manufacture of fish protein concentrate involve the use of isopropanol, and Smith and Brown[61] determined residues of this solvent in the protein concentrate by GLC on a Castorwax stationary phase. The samples were heated at 180°C for 20 min in a closed glass capillary to release the isopropanol for GLC. Quantitative recoveries were obtained for this method in the range 80–130 p.p.m. isopropanol, but no results obtained on fish protein concentrates were given.

F. Food packaging materials

The dearth of literature concerning the examination of food for packaging material residues can be attributable either to the fact that no residues exist in foods or the fact that they are present in minute quantities, which has presented too big a challenge to the analyst, including the gas chromatographer.

The literature concerning the applications of GC to the analysis of these materials is almost exclusively German and is concerned with the possibility of the migration of plasticisers into fatty foods.

In 1963 Wandel and Tengler[62] investigated the possible transference of diethyl phthalate from plastic food wrappings into various foods. They preferred GC for the analysis of this plasticiser to gravimetric or spectrophotometric methods, and showed that there was negligible transference even if the food samples underwent the most favourable storage conditions to prompt such a situation.

Trichlorethylene was used to extract diethyl phthalate and the evaporation of the extract was controlled according to the fat content of the sample, i.e. samples containing fat were evaporated less. Extracts were analysed on a column of 1:9 Reoplex:silica gel, and a limit of detection of the order of 1 or 2 p.p.m. was possible.

A year later Wandel and Tengler[63] extended their investigation to the determination of bis-(2-ethylhexyl)phthalate and tributyl O-acetyl citrate residues in foods. These are plasticisers which could be transferred from the food packaging to fatty foods. Dichloromethane was used to extract these esters from foods and the extracts were separated by GLC on a LAC 2-R-446 stationary phase. In the event of too many interfering peaks on the chromatogram, the esters were further extracted into methanol and hydrolysed with *p*-toluene sulphonic acid. The resulting 2-ethylhexanol and butanol were chromatographed on either LAC 2-R-446 or polyoxyethylene P 1500 stationary phase. The detection limits were of the order of 2 p.p.m. for these plasticisers.

Tengler[64] has more recently extended this approach to include pyrolysis of monomeric plasticisers.

Extreme conditions of storage of cheese and lard in lacquered aluminium foil caused transference of a proportion of the plasticisers from the lacquers to the foods, in experiments conducted by Pfab[65]. Aluminium foils were coated with nitrocellulose containing dibutyl phthalate and its vinyl chloride—vinyl acetate copolymer containing dicyclohexyl phthalate, and these were wrapped around cheese and lard samples which were stored, under pressure, at $25°C$ for 1 month. After this time, dichloromethane was used to extract the fat from the samples and the fat was saponified. The butanol and cyclohexanol were separated from the saponification mixture by steam distillation and chromatographed on a 15% Zitrophol O-11 stationary phase with isoamyl alcohol as internal standard.

Experiments were also carried out by Rohleder and von Bruchhausen[66] in order to assess an extraction method and the determination of dibutyl phthalate in fats. Butter fat containing this plasticiser was dissolved in acetone and the solution was cooled in an ice—salt bath to freeze out most of the fat. After centrifugation, the supernatant was taken for analysis by GLC on a DC silicone oil stationary phase. A precolumn consisting of a mixture of SE-30 and dicyclohexyl phthalate on Chromosorb was used to protect the analytical column from deposition of the fatty materials still present in the acetone extract. This method was capable of determining high boiling point plasticisers down to a level of 30 p.p.m.

The prime attention in the application of GC methods to food packaging materials has been paid to plasticisers, and part of the reason could be that these, if present, would be readily extracted by fats and oils and could impart a foreign flavour to the food.

The migration into foods of organic constituents of food packaging materials other than plasticisers is probably not so important, but Reith, van der Heide and Ligtenberg[67] investigated the transfer of the higher alcohols present in polyvinyl chloride foil into fats and oils. Experiments were described where dodecanol was transferred to samples of margarine and peanut oil, the alcohol being extracted and subsequently analysed by a GC method.

The possible migration of PCBs from some packaging materials into fatcontaining foods prompted Specht[13] to develop a GC method for the analysis of these contaminants.

G. Mycotoxins

GC has been applied to the determination of three different mycotoxins: patulin, ipomeamarone and trichothecenes.

1. Patulin

Patulin is a carcinogenic, mutagenic metabolite produced by certain strains of penicillium. Since storage rot in apples is caused by a similar penicillium to that known to produce patulin, there is a possibility that the mycotoxin could be transmitted to the juice made from unsound fruit. Pohland, Sanders and Thorpe[68] produced a GC method for patulin analysis after evaluating three derivatives, viz. TMS ether, acetate and chloroacetate. The latter was chosen as the most suitable derivative which was formed after extraction of the patulin from apple juice samples with ethyl acetate. GLC was performed on a 3% JXR stationary phase. None of the apple juice samples examined contained patulin.

2. Ipomeamarone

Ipomeamarone is a hepatotoxin produced by fungal, insect or chemical attack of the sweet potato. Boyd and Wilson[69] found that a GC method offered greater sensitivity than other methods. Blemished sweet potatoes were shredded and homogenised with 95:5 chloroform:methanol, and the homogenate was filtered. The filtrate was used for GLC on a 10% UC-W98 stationary phase. Five marketable sweet potato samples were examined and found to contain between 0.1 and 7.6mg ipomeamarone per g sample. The significance of these results in respect of toxic levels was not assessed.

3. Trichothecenes

Mouldy grain can give rise to trichothecenes which are tetracyclic sesquiterpenoid metabolites of certain fungi found in grain.

The normal analysis of these mycotoxins is by biological assay, which is both laborious and capable of yielding erratic results. Ikediobi *et al.*[70] obtained success with a GC method incorporating the formation of the TMS derivatives of those trichothecenes which contained free hydroxyl groups. The mouldy corn grain was extracted with ethyl acetate, the extract was concentrated, and further cleaned up by extraction in a series of solvents, before the derivative formation. Trimethylsilyl glucose was used as internal standard and GLC was performed on a 5% SE-30 or SE-52 stationary phase. The method was applied to the determination of the mycotoxins in grain infected with the organism *Fusarium tricinctum.*

H. Stilboestrol and derivatives

Diethyl stilboestrol is a synthetic oestrogen used as an animal feed additive to increase the feeding efficiency in cattle and lambs. This hormone has carcinogenic properties, and residues in animals killed for human consumption are consequently undesirable and are specifically prohibited in Canadian food legislation.

Coffin and Pilon[71] described a GC method for the determination of diethyl stilboestrol in beef, chicken and lamb tissues at the 2–10 µg/kg level. The sample was extracted with acetone and the diethyl stilboestrol β-glucoronide, in which form the oestrogen is often present, was hydrolysed with hydrochloric acid. After

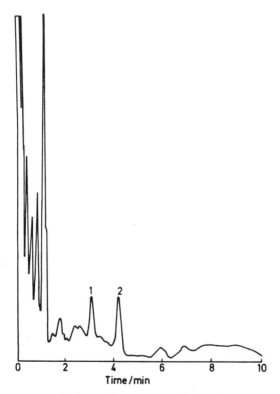

Figure 21.3 Gas chromatogram of the trifluoroacetate derivatives of 1 ng diethylstilboestrol: 1, cis-diethyl-stilboestrol trifluoroacetate; 2, trans-diethylstilboestrol trifluoroacetate (after Coffin and Pilon[71]). For GC details, see Table 21.1

chloroform extraction and column chromatographic clean-up, the GLC of the trifluoroacetate was carried out on a 3% OV-17 stationary phase. The trifluoro-acetates of both *cis* and *trans* isomers must be obtained on the chromatogram to confirm their respective identities, and *Figure 21.3* is a chromatogram of 1ng of the derivative, showing the isomeric peaks. The GLC method was preferred to the mouse uterine bioassay method, which was non-specific and difficult to apply on a routine basis. No figures were given for analyses on commercial meat or poultry samples.

The TMS derivative of stilboestrol was used by Guenther[72] in the detection and separation of this oestrogen in foods of animal origin. GLC was performed on a 2% OV-17 stationary phase, and identity of stilboestrol was confirmed by GC on other stationary phases and also by TLC, IR spectroscopy and MS techniques.

I. 6-Chloro-o-cresol

6-Chloro-*o*-cresol is an impurity often found in cresylic acid disinfectants. There have been incidents of this disinfectant impurity being found in foods, after the taint had been recognised by its odour and flavour characteristics. GC methods were used to confirm the organoleptic results in samples of chicken and biscuits.

Patterson[73] aspirated broiler chicken carcasses with nitrogen during thawing, and collected the volatiles in liquid nitrogen. The condensed volatiles were dissolved in benzene and chromatographed on a column containing three stationary phases of differing polarity in parallel. The retention time of the 6-chloro-*o*-cresol was confirmed by rechromatographing the extract and extinguishing the flame of the detector and sniffing the effluent. The findings were further confirmed by MS. The broiler chicken had obviously become contaminated with the phenol via a disinfectant used in their environment.

The same compound was found in biscuits which were claimed to have a disinfectant-type taint. Griffiths and Land[74] devised a method to extract and analyse 6-chloro-*o*-cresol in biscuits using GC coupled to MS. The biscuit sample slurry was steam-distilled, and the distillate was continuously extracted with pentane, the extract being evaporated to low bulk for chromatographic examination on three different stationary phases, viz. OV-1, Apiezon L and Carbowax 20M. The effluent was split between a FI detector and a sniffing port to enable the effluent identity to be confirmed by its odour. The odour threshold of 6-chloro-*o*-cresol is 60 000 times less than that of 4-chloro-*o*-cresol, the main component of the disinfectant in which it is an impurity. Although MS can be coupled to GC to further confirm the identity of this phenol, it was possible to taint a gram of biscuit with $1 \times 10^{-4} \mu g$ 6-chloro-*o*-cresol, a level at which even the MS was insensitive.

J. 6-chloropicolinic acid

The soil nitrification inhibitor 2-chloro-6-(trichloromethyl) pyridine could be used on soils in areas where the grass may be subsequently consumed by cattle, with the possibility of transference of its major metabolite, 6-chloropicolinic acid, into the milk.

Jensen[75] carried out feeding experiments on cattle, adding 6-chloropicolinic acid to their diet, and analysed milk and cream samples for the residue. Milk or cream was acidified and extracted with diethyl ether, cleaned up on a column of alumina and methyl 6-chloropicolinate was formed by reaction with diazomethane. GLC was performed on a 4% LAC 446 stationary phase and by using an EC detector. Added 6-chloropicolinic acid was recovered quantitatively from milk and cream and the limit of detection was 0.02 p.p.m. No residues of this compound were found in any ordinary milk or cream samples.

K. Chlorobutanol

Chlorobutanol is an antibacterial preservative used together with antibiotic substances in preparations for the treatment of bovine mastitis. Residues of this compound are not allowed in milk under US food legislation, and Wiskerchen and Weishaar[76] have devised a GC method for its determination. Samples of milk were steam-distilled in the presence of tungstosilicic acid and the distillate was extracted with diethyl ether. After careful concentration using iso-octane as a 'keeper' to prevent loss of chlorobutanol, this extract was chromatographed on a 5% Igepal 880 stationary phase, the method having a limit of detection of 0.01 p.p.m.

L. Sulphaquinoxaline

Sulphaquinoxaline is a coccidiostat which may be incorporated in poultry feeding stuffs at concentrations up to 100 mg/kg. A method was required which was more sensitive than the standard colorimetric, TLC or IR spectroscopic ones, preferably capable of determining the coccidiostat at a concentration of 0.1mg per kg feeding stuff.

Crisp[77] produced such a method, based on the GC of the TFA derivative of sulphaquinoxaline, and applied it to the determination of the coccidiostat in poultry meat and eggs. The minced meat or blended egg samples were homogenised in acetonitrile, which was dried and evaporated. The residue was dissolved in sodium hydroxide solution, extracted with chloroform and then with diethyl ether and was eventually hydrolysed into 2-aminoquinoxaline before preparation of the TFA derivative. GLC was performed on a 5% NPGS stationary phase and using EC detection. No meat or egg samples were reported to contain sulphaquinoxaline.

M. Pyribenzamine

Pyribenzamine is a drug of the antihistamine family and has been used to treat certain allergies in cattle and other animals. It is administered to cattle by intramuscular injection and there is a possibility that it could be transmitted to the milk. Luders et al.[78] described a method for pyribenzamine analysis in milk, sensitive to a level of 2 p.p.10^9. Pyribenzamine was extracted from an alkaline solution of milk with 4:1 diethyl ether:dichloromethane. It was extracted into acid, followed by re-extraction into the solvent, and was chromatographed in dimethylformamide solution using a 2% Carbowax 20M stationary phase. No samples of milk were reported to contain pyribenzamine.

N. Indene

The determination of monomeric indene in coumarone-indene resin, which is used as a chewing gum ingredient, was made by Groebel[79] using a GLC method. He separated indene on a 20% SE-30 stationary phase, using ethereal solutions of chewing gum samples. Indene was found in the range 0.002–0.059%. The suggested maximum limit was 0.01%.

O. Trace elements

So many other specific and sensitive techniques are available for the analysis of trace elements that there appears to be only limited scope for the use of GC in this field. Some research has been carried out on the GC of organometallic complexes, and methods have been applied to the determination of some transition elements in steel. β-diketones, such as acetylacetone and its trifluoro derivative, can be used to complex chromium, nickel and cobalt, and these complexes can be chromatographed. Few, if any, GC methods have been applied to the analysis of trace elements in foods.

References

1. Holmes, D. C., Simmons, J. H. and Tatton, J. O'G., *Nature,* **216,** 227 (1967)
2. Jensen, S., *New Scientist,* **32,** 612 (1966)
3. Jensen, S., Johnels, A. G., Olsson, M. and Otterlind, G., *Nature,* **224,** 247 (1969)
4. Risebrough, R. W., Reiche, P. and Olcott, H. S., *Bull. Environ. Contam. Toxicol.,* **4,** 192 (1969)
5. Shaw, D. G., *Bull. Environ. Contam. Toxicol.,* **8,** 208 (1972)
6. Hutzinger, O., Jamieson, W. D. and Zitko, V., *Nature,* **226,** 664 (1970)
7. Bonelli, E. J., *Anal. Chem.,* **44,** 603 (1972)
8. Hannan, E. J., Bills, D. D. and Herring, J. L., *J. Agric. Food Chem.,* **21,** 87 (1973)
9. Monod, J. L., Raybaud, H., Venot, C. and Carrara, G., *Bull. Soc. Pharm. Marseille,* **20,** 155 (1971)
10. Koeman, J. H., ten Noever de Brauw, M. C. and de Vos, R. H., *Nature,* **221,** 1126 (1969)
11. Rote, J. W. and Murphy, P. G., *Bull. Environ. Contam. Toxicol.,* **6,** 377 (1971)
12. Asai, R. I., Gunther, F. A., Westlake, W. E. and Iwata, Y., *J. Agric. Food Chem.,* **19,** 396 (1971)
13. Specht, W., *Dt. LebensmittRdsch.,* **70,** 136 (1974)
14. Westoo, G., *Acta Chem. Scand.,* **20,** 2131 (1966)
15. Westoo, G., *Acta Chem. Scand.,* **21,** 1790 (1967)
16. Johansson, B., Ryhage, R. and Westoo, G., *Acta Chem. Scand.,* **24,** 2349 (1970)
17. Kamps, L. R. and McMahon, B., *J. Ass. Off. Anal. Chem.,* **55,** 590 (1972)
18. Newsome, W. H., *J. Agric. Food Chem.,* **19,** 567 (1971)

19. Fabbrini, A., Modi, G., Signorelli, L. and Simiani, G., *Boll. Lab. Chim. Prov.*, **22**, 339 (1971)
20. Nishi, S. and Horimoto, Y., *Bunseki Kagaku*, **20**, 16 (1971)
21. Gonzalez, J. G. and Ross, R. T., *Anal. Lett.*, **5**, 683 (1972)
22. Bache, C. A. and Lisk, D. J., *Anal. Chem.*, **43**, 950 (1971)
23. Uthe, J. F., Solomon, J. and Grift, B., *J. Ass. Off. Anal. Chem.*, **55**, 583 (1972)
24. Walters, C. L., *Lab. Pract.*, **20**, 574 (1971)
25. Foreman, J. K., Palframan, J. F. and Walker, E. A., *Nature*, **225**, 554 (1970)
26. Gough, T. A. and Webb, K. S., *J. Chromatog.*, **64**, 201 (1972)
27. Gough, T. A. and Webb, K. S., *J. Chromatog.*, **79**, 57 (1973)
28. Gough, T. A. and Webb, K. S., *J. Chromatog.*, **95**, 59 (1974)
29. Crosby, N. T., Foreman, J. K., Palframan, J. F. and Sawyer, R., *Nature*, **238**, 342 (1972)
30. Palframan, J. F., Macnab, J. and Crosby, N. T., *J. Chromatog.*, **76**, 307 (1973)
31. Alliston, T. G., Cox, G. B. and Kirk, R. S., *Analyst*, **97**, 915 (1972)
32. Brooks, J. B., Alley, C. C. and Jones, R., *Anal. Chem.*, **44**, 1881 (1972)
33. Telling, G. M., Bryce, T. A. and Althorpe, J., *J. Agric. Food Chem.*, **19**, 937 (1971)
34. Bryce, T. A. and Telling, G. M., *J. Agric. Food Chem.*, **20**, 910 (1972)
35. Telling, G. M., *J. Chromatog.*, **73**, 79 (1972)
36. Howard, J. W., Fazio, T. and Watts, J. O., *J. Ass. Off. Anal. Chem.*, **53**, 269 (1970)
37. Fazio, T., Damico, J. N., Howard, J. W., White, R. H. and Watts, J. O., *J. Agric. Food Chem.*, **19**, 250 (1971)
38. Fazio, T., White R. H. and Howard, J. W., *J. Ass. Off. Anal. Chem.*, **54**, 1157 (1971)
39. Fiddler, W., Doerr, R. C., Ertel, J. R. and Wasserman, A. E., *J. Ass. Off. Anal. Chem.*, **54**, 1160 (1971)
40. Reineccius, G. A. and Coulter, S. T., *J. Dairy Sci.*, **55**, 1574 (1972)
41. Newell, J. E. and Sisken, H. R., *J. Agric. Food Chem.*, **20**, 711 (1972)
42. Castegnaro, M., Pignatelli, B. and Walker, E. A., *Analyst*, **99**, 156 (1974)
43. Riedmann, M., *J. Chromatog.*, **88**, 376 (1974)
44. Pensabene, J. W., Fiddler, W., Dooley, C. J., Doerr, R. C. and Wasserman, A. E., *J. Agric. Food Chem.*, **20**, 274 (1972)
45. *The Solvents in Food Regulations 1967*, Statutory Instruments Nos. 1582 and 1939, HMSO (1967)
46. Dean, A. C., Bradford, E., Hubbard, A. W., Pocklington, W. D. and Thomson, J., *J. Chromatog.*, **44**, 465 (1969)
47. Black, L. T. and Mustakas, G. C., *J. Amer. Oil Chem. Soc.*, **42**, 62 (1965)
48. Fore, S. P. and Dupuy, H. P., *J. Amer. Oil Chem. Soc.*, **47**, 17 (1970)
49. Dupuy, H. P. and Fore, S. P., *J. Amer. Oil Chem. Soc.*, **47**, 231 (1970)
50. Fore, S. P. and Dupuy, H. P., *J. Amer. Oil Chem. Soc.*, **49**, 129 (1972)
51. Watts, J. O. and Holswade, W., *J. Ass. Off. Anal. Chem.*, **50**, 717 (1967)
52. Nosti Vega, M., Gutierrez Rosales, F. and Gutierrez-Quijano, R., *Grasas Aceit.*, **21**, 276 (1970)
53. Litchman, M. A., Turano, L. A. and Upton, R. P., *J. Ass. Off. Anal. Chem.*, **55**, 1226 (1972)
54. Tous, J. G. and Martel, J., *Grasas Aceit.*, **23**, 1 (1972)
55. Roberts, L. A., *J. Ass. Off. Anal. Chem.*, **51**, 825 (1968)
56. Gal. S. and Schilling, P., *Z. Lebensmitt. Forsch.*, **145**, 30 (1971)
57. Orsi, F., *Konserv-Paprikaipar*, 112 (1970)
58. Dupuy, H. P., Rayner, E. T. and Fore, S. P., *J. Amer. Oil Chem. Soc.*, **48**, 155 (1971)
59. Litchman, M. A. and Upton, R. P., *Anal. Chem.*, **44**, 1495 (1972)
60. Fore, S. P., Rayner, E. T. and Dupuy, H. P., *J. Amer. Oil Chem. Soc.*, **48**, 140 (1971)
61. Smith, P. and Brown, N. L., *J. Agric. Food Chem.*, **17**, 34 (1969)
62. Wandel, M. and Tengler, H., *Dt. LebensmittRdsch.*, **59**, 326 (1963)
63. Wandel, M. and Tengler, H., *Dt. LebensmittRdsch.*, **60**, 335 (1964)
64. Tengler, H., *Farbe Lack*, **73**, 153 (1967)
65. Pfab, W., *Dt. LebensmittRdsch.*, **63**, 72 (1967)
66. Rohleder, K. and von Bruchhausen, B., *Dt. LebensmittRdsch.*, **68**, 180 (1972)
67. Reith, J. F., van der Heide, R. F. and Ligtenberg, H. J. H. M., *Z. Lebensmitt. Forsch.*, **126**, 49 (1964)

68. Pohland, A. E., Sanders, K. and Thorpe, C. W., *J. Ass. Off. Anal. Chem.*, **53**, 692 (1970)
69. Boyd, M. R. and Wilson, B. J., *J. Agric. Food Chem.*, **19**, 547 (1971)
70. Ikediobi, C. O., Hsu, I. C., Bamburg, J. R. and Strong, F. M., *Anal. Biochem.*, **43**, 327 (1971)
71. Coffin, D. E. and Pilon, J-C., *J. Ass. Off. Anal. Chem.*, **56**, 352 (1973)
72. Guenther, H. O., *Fleischwirtschaft*, **52**, 625 (1972)
73. Patterson, R. L. S., *Chem. & Ind.*, 609 (1972)
74. Griffiths, N. M. and Land, D. G., *Chem. & Ind.*, 904 (1973)
75. Jensen, D. J., *J. Agric. Food Chem.*, **19**, 897 (1971)
76. Wiskerchen, J. E. and Weishaar, J., *J. Ass. Off. Anal. Chem.*, **55**, 948 (1972)
77. Crisp, S., *Analyst*, **96**, 671 (1971)
78. Luders, R. C., Williams, J., Fried, K., Rehm, C. R. and Tishler, F., *J.Agric. Food Chem.*, **18**, 1153 (1970)
79. Groebel, W., *Z. Lebensmitt. Forsch.*, **129**, 292 (1966)

Appendices

Appendix 1

Some GC Liquid Stationary Phases and Equivalents

Hydrocarbons	Apiezon L, Apiezon M
Silicones, methyl	Silicone oil, SE-30, E-301, DC-200, JXR, OV-1, OV-101, SF-96
Silicones, methyl including 20% phenyl	DC-550, OV-7
Silicones, methyl including 35% phenyl	DC-710, OV-11
Silicones, methyl including 50% phenyl	SP-2250, OV-17
Silicones, chlorophenyl	SE-54, DC-560, SP-400, SE-52
Silicones, fluoro	QF-1, LSX-3-0295, OV-210
Silicones, cyano	XE-60, OV-225
Polyesters	Dinonyl phthalate (DNP), di-iso-decyl phthalate
Polyesters	Ethylene glycol adipate (EGA), butanediol succinate (BDS), LAC 6R-860, LAC 13R-741
Polyesters	Diethylene glycol succinate (DEGS), EGSS-X, LAC 3R-728
Polyesters	Ethylene glycol succinate (EGS), LAC 4R-886
Polyesters	Diethylene glycol adipate (DEGA), Reoplex 400, LAC 1R-296
Polyesters	Diethylene glycol adipate (DEGA) cross-linked with pentaerythritol, LAC 2R-446
Polyesters	Phenyldiethanolamine succinate (PDEAS), 2,2-(phenylimino) diethanol succinate polyester (PDEAS)
Polyesters	Neopentyl glycol succinate (NPGS), LAC 18R-767

Polyesters		Cyclohexane dimethanol succinate (CHDMS), polydimethylol cyclohexane succinate (CHDMS), LAC 12R-796
Polyethylene glycols		Carbowax 400-20,000. The numbers refer to the average molecular wt. Lower molecular wt. generally means higher polarity and lower maximum operating temperature.
Polypropylene glycols and esters		Ucon fluids. Products with the suffix LB are water-insoluble and with suffixes 50 HB and 75 H are water-soluble.
Polypropylene glycols and esters		Ucon 50 HB-660, Ucon 50 HB-3520, poly-propylene glycol sebacates
Others:	(i)	Triton X-100, Emulphor O, Dowfax 9N9, Tergitol, Ucon 50 HB-2000
	(ii)	Free fatty acid phase (FFAP), SP-1000
	(iii)	Sucrose diacetate hexaisobutyrate, tricresyl phosphate, neopentyl glycol sebacate, LAC 17R-770
	(iv)	1,2,3-tris (2-cyanoethoxy)propane (TCEP), Fractonitrile 111, β,β'-oxydipropionitrile

Appendix 2 Some GC Support Materials and Equivalents

	Johns-Manville	Applied Science	JJ's	Analabs	Others	Treatment
Pink supports	Chromosorb P	Gas-Chrom R	Diatomite S	Anakrom P Anakrom C22		May be obtained untreated or deactivated with acid, alkali or silylating agents
White supports	Chromosorb W	Gas-Chrom C	Diatomite C	Anakrom 545		May be obtained untreated or deactivated with acid, alkali or silylating agents
White supports	—	Gas-Chrom S	Diatomite CT	Anakrom U		Untreated
White supports	—	Gas-Chrom A	Diatomite CT-AW	Anakrom A		Acid washed
White supports	—	Gas-Chrom P	Diatomite CT-AAW	Anakrom AB		Acid and alkali washed
White supports	—	Gas-Chrom Z	—	Anakrom ABS		Acid washed and silanised (the Anakrom is also alkali washed)
Special quality supports	Chromosorb W (HP)	Gas-Chrom Q	Diatomite CQ Diatomite CLQ	Anakrom Q Anakrom SD	Aeropak 30 Diatoport S Supelcoport Varaport 30	Specially prepared and silanised
High density supports	Chromosorb G	—	Diatomite M	—		May be obtained untreated or deactivated with acid, alkali or silylating agents

Appendix 3

Common Names of Fatty Acids and their Carbon Numbers

Carbon number	Systematic name	Common name
4:0	butanoic	butyric
6:0	hexanoic	caproic
8:0	octanoic	caprylic
10:0	decanoic	capric
10:1	dec-9-enoic	–
12:0	dodecanoic	lauric
12:1	dodec-9-enoic	–
14:0	tetradecanoic	myristic
14:1	tetradec-5-enoic	–
14:1	tetradec-9-enoic	–
16:0	hexadecanoic	palmitic
16:1	hexadec-9-enoic	palmitoleic
16:3	hexadeca-6,10,14-trienoic	hiragonic
17:0	heptadecanoic	margaric
18:0	octadecanoic	stearic
18:1	octadec-6-enoic	petroselenic
18:1	octadec-9-enoic (*cis*)	oleic

Carbon number	Systematic name	Common name
18:1	octadec-9-enoic (*trans*)	elaidic
18:1	octadec-11-enoic	vaccenic
18:1	12-hydroxy-octadec-9-enoic	ricinoleic
18:2	octadeca-9,12-dienoic	linoleic
18:3	octadeca-9,11,13-trienoic	elaeostearic
18:3	octadeca-9,12,15-trienoic	linolenic
20:0	eicosanoic	arachidic
20:1	eicos-9-enoic	gadoleic
20:4	eicosa-5,8,11,14-tetraenoic	arachidonic
22:0	docosanoic	behenic
22:1	docos-11-enoic	cetoleic
22:1	docos-13-enoic	erucic
24:0	tetracosanoic	lignoceric
26:0	hexacosanoic	cerotic
28:0	octacosanoic	–

Index